13·10

J. KENNETH PEARCE, B.S.F., University of Washington, is Professor Emeritus of Logging Engineering at that institution, where he taught for many years. A Professional Engineer, he has studied logging operations in many countries, and has served as advisor to the World Bank on the improvement of logging operations. He has also acted as consultant to a number of forestry industry corporations.

GEORGE STENZEL, M.F., Yale University, is Professor of Logging Engineering at the University of Washington, where he has been teaching for many years. He has had extensive experience in both the public and private sectors of the forest products industry, and has published many articles relating to timber harvesting.

LOGGING AND
PULPWOOD PRODUCTION

J. KENNETH PEARCE
UNIVERSITY OF WASHINGTON

GEORGE STENZEL
UNIVERSITY OF WASHINGTON

THE RONALD PRESS COMPANY · NEW YORK

Library of Congress Catalog Card Number: 73–190210
PRINTED IN THE UNITED STATES OF AMERICA

Preface

In writing this book the authors' objectives were to provide both a textbook for forestry schools and a manual for the logging industry. We have endeavored to make the book suitable for use as a textbook in the course in timber harvesting offered by the four-year professional forestry schools and by some two-year community colleges, and also in technician training courses in logging or pulpwood production. The book is also directed toward the independent logger or pulpwood producer interested in improving his production methods and management, and to foresters and operators in the developing countries who are interested in the mechanization of their logging operations. For this reason some of the older methods which are now little used in North America, but are suitable for use in tropical countries where labor rates are lower, are included.

Logging, the production of round logs from standing timber and their transportation to the mill or plant is termed timber extraction in many countries. In managed forests it is sometimes referred to as harvesting the forest crop. The order of the chapters on logging in this book coincides with the consecutive steps in logging which are as follows:

1. Acquiring timber to log—Chapter 1
2. Planning the logging operation—Chapter 2
3. Engineering and constructing the haul roads—Chapters 3 and 4
4. Felling trees and bucking them into logs—Chapter 6
5. Skidding or yarding the logs to a landing or assembly point for loading—Chapters 7 and 8
6. Loading the logs on vehicles for major transportation—Chapter 9
7. Hauling the logs from the forest to their destination—Chapter 10

The colloquial names of the various operations involved in logging vary with the region.

Photographs for use as illustrations were requested from the manufacturers of well-known makes of logging machines and equipment. The photographs used as figures were those of the best quality which showed the machine in action in the forest. Special mention is due to the following individuals: Mr. A. Ross Bentley, Hiwassee Land Company, Calhoun, Tennessee, who took the photographs credited to that company especially for use as illustration in this book; Ken Schell, retired Director of Technical Publications, Skagit Corporation, Sedro-Woolley, Washington, who supplied the photographs credited to his company.

The illustrations of logging machinery and equipment, and the references to trade names in the text, represent only a sample of the many manufacturers in the industry. Due to the time lag between preparation of the manuscript and publication of the book, the pictures may not show the manufacturer's current models of such machines. Brochures of their current models may be obtained from the manufacturers or their dealers.

We are indebted to Dr. Barney Dowdle, Associate Professor of Forest Economics, University of Washington, for Chapter 13, Logging Cost Analysis.

J. KENNETH PEARCE
GEORGE STENZEL

Seattle, Washington
May, 1972

Contents

LOGGING AND
PULPWOOD PRODUCTION

1

Timber Acquisition

Forest Resource Reports No. 14 (U.S.D.A., F.S. 1958) and No. 17 (U.S.D.A., F.S. 1965) provide a very thorough and complete appraisal of the timber situation and outlook in the United States. The information presented here highlights certain salient items from these reports relative to our timber resources. The reader is referred to the Reports for more detailed information.

Ownership

In spite of more than three centuries of settlement and development forests still cover 759 million acres, or approximately one-third of the total land area of the United States. Two-thirds of the forested area of 759 million acres, or 509 million acres, is classed as commercial forest land (Fig. 1–1). Commercial, in this respect, refers to land which is suitable and available for the growing of continuous crops of trees which provide the forest products industry with raw material.

This commercial forest land classification includes areas which vary from highly productive timberlands to so-called poor sites which are marginal for the growing of timber. Furthermore, some of the areas classed as commercial are economically inaccessible for

1

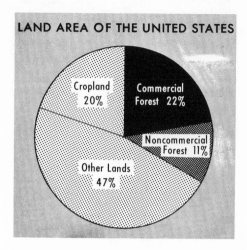

Fig. 1—1. Land area of the United States. (After U.S. Forest Service, Forest Resource Report No. 17.)

timber harvesting operations under present cost-price relationships. Still other areas are either non-stocked or support low-quality or sparse stands of timber. Many of the commercial forest lands are used for recreation or other purposes in addition to providing the forest industries with timber products. However, all of the forest land classed as commercial has the potential for producing timber

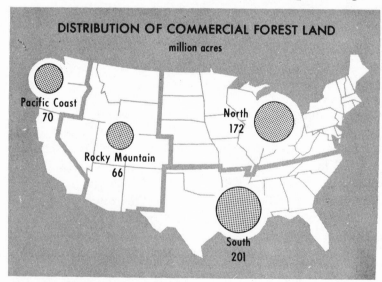

Fig. 1—2. Distribution of commercial forest land. (After U.S. Forest Service, Forest Resource Report No. 17.)

crops at some level of management. Figure 1-2 shows the distribution of commercial forest land in the United States.

The remaining one-third of the total forested area in the nation, 250 million acres, is classified as "non-commercial," and is so identified because of low productivity for timber growing, or in the case of certain public lands, because of legal reservation for recreation and other non-timber uses.

Area, volume of growing stock, and cut by ownership are shown in Fig. 1-3. Here the area of commercial forest land on a country-

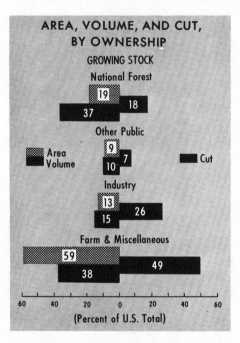

Fig. 1-3. Area, volume and cut, by ownership. (After U.S. Forest Service, Forest Resource Report No. 17.)

wide basis is shown in percent for the various types of ownership together with the corresponding expressions of growing stock and cut. The 59 percent in farm and miscellaneous ownership is somewhat equally divided between farm and other non-industrial private owners. These "other" owners include a great variety of business and professional people, housewives, wage earners, mining and landholding companies. Most of these owners are engaged in types of livelihood not directly connected with timber growing. Railroads and timber holding companies, particularly in the South and Pacific

Coast sections, own a considerable acreage of commercial forest land. Even though such organizations resemble forest industry holdings in many respects, they are classed as non-industrial private owners. However, most of the commercial forest land in farm and miscellaneous holdings are in the category of what might be called small ownerships. Ownership of commercial forest land on a sectional basis is shown in Fig. 1–4.

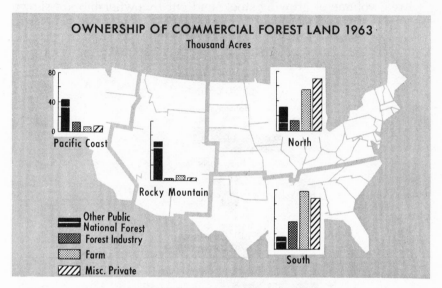

Fig. 1–4. Ownership of commercial forest land (1963). (After U.S. Forest Service, Forest Resource Report No. 17.)

Combining the 59 percent in farm and miscellaneous ownership with the 13 percent owned by industry comprises 72 percent of the total commercial forest area. This portion of the commercial forest area is distributed among some 4.5 million owners of which 3.4 million, or 75 percent, are classed as farm owners, with ownership concentrated in holdings of 500 acres or less. In contrast, the forest industry ownership pattern involves much larger holdings, usually above 5,000 acres in size and in many instances extending to much beyond 50,000 acres.

The ownership of commercial forest land varies in different parts of the country. Heavy concentrations of farm ownerships and other non-industrial private ownerships are located in the North and in the South. The South also has well over one-half of all forest industry ownership with the balance somewhat evenly divided between the North and West. In contrast, public ownership is least in the South and greatest in the West.

In summary the United States Forest Service (U.S. Department of Agriculture, 1958) has concluded the following:

1. Commercial forest land is largely in private ownership.
2. Small ownerships predominate (less than 5,000 acres).
3. Young growth stands predominate on private lands.
4. Sawtimber is about equally divided between public and private ownership.
5. Most of commercial forest land in forest industry and farm ownerships is in the South.
6. In many central and southeastern states more than half of commercial forest area is in farm ownership.
7. "Other" private holdings account for more than half of the commercial forest area in the Middle Atlantic States and southern New England.

Volume

On a nationwide basis the commercial forests contained 699 billion cubic feet of sound wood as of January 1, 1963 (U.S.D.A., 1965). Growing stock accounted for 90 percent of this total or 628 billion cubic feet. The remaining 10 percent consisted of sound wood volume in cull trees and salvageable dead trees. From the standpoint of timber class, nearly two-thirds of the total timber inventory was classed as sawtimber trees having a volume of 2,537 billion **board feet**° (**International** $\frac{1}{4}$ **in. Log Rule**). The distribution of sawtimber volume is shown in Fig. 1–5.

Ownership of sawtimber on a sectional basis is shown in Fig. 1–6. A comparison between Figs. 1–5 and 1–6 provides a means whereby certain relationships among ownership, sawtimber volume, and broad geographical location become apparent.

The great bulk of hardwood sawtimber volume is in private ownership and is somewhat evenly distributed among farm, forest industry, and other private ownerships. In contrast, well over half of the softwood timber volume is in public ownership with farm ownership of relative unimportance; see Figs. 1–5 and 1–6. The two principal groups which control the softwood sawtimber volume are non-farm private owners and the national forests.

° Words in heavy type are defined in the Glossary at the end of the book.

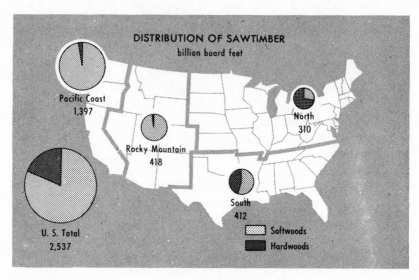

Fig. 1–5. Distribution of sawtimber volume. (After U.S. Forest Service, Forest Resource Report No. 17.)

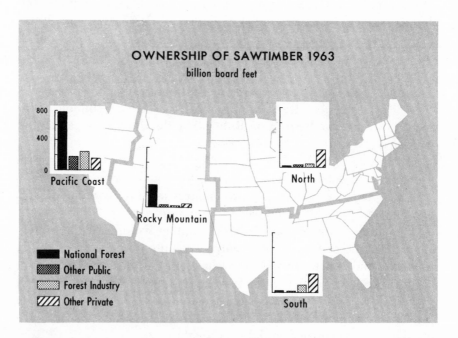

Fig. 1–6. Ownership of sawtimber (1963). (After U.S. Forest Service, Forest Resource Report No. 17.)

Species

Sawtimber volumes are concentrated in a relatively few primary species or groups. Table 1–1 presents the volume of **growing stock** and sawtimber on commercial forest land by species as of January 1, 1963. Douglas-fir ranks first, followed by western hemlock, western true firs, ponderosa and Jeffrey pines, southern pines, and western spruce. These species account for over two-thirds of the total sawtimber volume.

Considered from a growing stock standpoint, the ranking is somewhat different. While Douglas-fir still ranks first, the oaks are second, followed by southern pines, western hemlock, western true firs, ponderosa and Jeffrey pines, western spruce. These species account for over one-half of the total volume of growing stock.

Quality

While the national total volume of timber is impressive, it includes a large variety of species and tree qualities which exhibit a wide range of usefulness to the forest products industry. Volume figures without some reference to quality characteristics are not very meaningful since quality is a critical factor to the forest industries.

It is generally recognized that old-growth timber stands, which are composed of large slower-growing trees, possess quality characteristics that young-growth stands, which are composed of the smaller, faster-growing trees, do not have. Even though 10 percent of the total commercial forest area supports old-growth timber, presently these high-volume stands of timber are the major source of high-quality wood.

Young-growth timber stands, which occupy 90 percent of the commercial forest area, are becoming of increasing importance as the remaining old-growth stands are harvested; consequently, the quality factor of young-growth timber is an important consideration.

While tree size is not yet a major factor in certain subregions of the West and Coastal Alaska, other sections of the country have experienced a continuing decrease in the diameter of the average sawtimber tree. This situation results in a general lowering of log quality since log diameter is one of the important considerations in log grading.

TABLE 1–1. Volume of Growing Stock and Sawtimber on Commercial Forest Lands by Species. January 1, 1963. (Source: U.S.F.S., F.R.R. No. 17).

SPECIES	GROWING STOCK		SAWTIMBER	
	Volume	Proportion	Volume	Proportion
	Million cu.ft.	Percent	Million bd.ft.	Percent
Eastern softwoods:				
Southern pines	60,355	9.6	211,925	8.4
Spruce and fir	13,055	2.1	20,629	.8
White and red pines	6,245	1.0	21,255	.8
Cypress	3,961	.6	15,346	.6
Other	10,378	1.7	22,229	.9
Total..............	93,994	15.0	291,384	11.5
Eastern hardwoods:				
Select white and red oaks	28,563	4.5	77,867	3.1
Other oaks	32,407	5.2	85,387	3.4
Hickory	11,076	1.8	28,488	1.1
Hard maple	10,378	1.6	25,764	1.0
Ash, walnut, and black cherry	10,054	1.6	22,923	.9
Sweetgum	10,024	1.6	25,879	1.0
Yellow-poplar	6,753	1.1	21,202	.8
Yellow birch	4,854	.8	11,594	.5
Other	62,452	9.9	131,356	5.2
Total..............	176,561	28.1	430,460	17.0
Total, eastern..............	270,555	43.1	721,844	28.5
Western softwoods:				
Douglas-fir	106,073	16.9	602,622	23.8
Western hemlock	49,902	7.9	269,935	10.6
True firs	48,244	7.7	234,780	9.3
Ponderosa and Jeffrey pines	45,448	7.2	241,722	9.5
Spruce	28,883	4.6	155,404	6.1
White and sugar pines	9,052	1.5	53,083	2.1
Redwood	5,542	.9	31,257	1.2
Other	46,944	7.5	177,835	7.0
Total..............	340,088	54.2	1,766,638	69.6
Western hardwoods	17,239	2.7	48,317	1.9
Total, western..............	357,327	56.9	1,814,955	71.5
All Species..........	627,882	100.0	2,536,799	100.0

To date, no single index is available which can adequately assess timber quality and value, since tree size, tree and log grade, and species characteristics all must be considered. Log grades, while not recognizing all end-use requirements of timber, do reflect some of them somewhat indirectly. This is accomplished by taking into account diameter, length, and amount and character of defects in individual logs.

FOREST POLICY AND TIMBER SALES

Federal

Practically all federal commercial forest land is administered by bureaus within two governmental departments. The national forests, administered by the Forest Service in the Department of Agriculture, are to a great extent the basic federal forest units and contain approximately 82 percent of the federally managed commercial forest land. The remaining 18 percent is somewhat evenly divided among the Bureau of Indian Affairs* with 7 percent, Bureau of Land Management† with 6 percent, both in the Department of the Interior, and the remaining 5 percent is identified as "other federal" (includes forest land in land utilization projects administered by the Forest Service, Defense Department, T.V.A., and other agencies).

The management and administration of practically all of the commercial forest land in federal ownership is regulated by a policy which originated when the Act of June 4, 1897,‡ became law. This

* While the Bureau of Indian Affairs is responsible for administering and managing these lands according to **sustained yield forest management** principles, the timber is the Indians' private property and is not publicly owned timber, such as that which is administered by the Forest Service and the Bureau of Land Management.

† Two-thirds of the commercial forest land administered by the Bureau of Land Management is public domain and, therefore, is subject to disposal under various public land laws. The remaining one-third, slightly over 2 million acres, comprises the O and C Lands, and is located in western Oregon. The O and C Lands get their name from the fact that the lands were granted by the government to the Oregon and California Railroad Company in 1866. When the terms of the grant were violated, Congress ordered the unsold portion returned to public ownership. Since 1946 the Bureau of Land Management has administered these lands under a forest management program.

‡ The Act of June 4, 1897, established the first provision for the administration of the federal forest reserves by placing them under the control of the Department of the Interior. At the same time the Division of Forestry in the Department of Agriculture was made responsible for providing technical advice and the U.S. Geological Survey with surveying and mapping the forest reserves. The administration of the forest reserves was transferred from the Department of the Interior to the Department of Agriculture in 1905 and the name "forest reserves" was changed to "national forests" in 1907.

act was made with specific provisions regarding the forest reserves which later became national forests. Since the enactment of the 1897 law much additional forest and land legislation has been enacted. The subsequent regulatory measures necessary to comply with this legislation are part and parcel of the rather broad and somewhat nebulous term, policy.

Presently, the forest land in federal ownership is managed and administered in keeping with a myriad of rules and regulations resulting from a host of applicable laws as related to a particular bureau within the framework of the federal government.

Gulick (1951)* has synthesized a policy statement which, even though somewhat dated, states the overall situation quite effectively:

> Led by the professional foresters and the conservationists, the United States government has now come to a very definite general policy in the management of its publicly owned forest lands. This policy as announced officially and approved by Congress includes the principles of (a) sustained yield management, designed to produce maximum continued timber production; (b) **multiple use** administration, which will recognize and make a balanced place for the other uses of forest land, particularly for water conservation, recreation, wildlife habitat, and grazing; and (c) an increasing level of forest protection and improvement. These policies apply to most federal forest lands, but do not extend to parks, military reservations, or other properties, which are generally not managed to produce timber.

Implementing this policy has been, and still is, a monumental task, particularly in regard to timber sales and those activities associated with logging operations. To further complicate the situation much controversy is taking place in respect to access roads and timberland withdrawals.

State and Local

Commercial forest land in state ownership comprises 4 percent of the national total. While most of these lands also are managed and administered with "multiple use" as the objective, the regulatory measures, particularly those involved with timber sales, are not considered to be as stringent as those associated with national forest timber sales. Over two-thirds of the commercial forest land in state ownership is located in the North and supports slightly less than one-fifth of the total sawtimber volume, while the West with one-quarter of the commercial forest land in state ownership supports slightly less than four-fifths of the total sawtimber volume.

* These citations are given in full in the References at the end of the book.

Presently fourteen states (Idaho, Louisiana, Maryland, Massachusetts, Minnesota, Mississippi, Nevada, New Hampshire, New Mexico, New York, Oregon, Vermont, Virginia, and Washington) have enacted legislation which has resulted in regulating cutting practices on private forest land. Timber harvesting in these States must be done according to certain regulatory practices, which form an important part of the State's forest policy. To be effective, regulatory measures of this type require field inspection together with a penalty system for non-compliance. Control of this sort is of particular importance in those instances where there may be a breach between what the legal directives stipulate and what actually is being done.

Commercial forest land included under the local category involves county and municipal holdings which comprise less than 2 percent of the national total. The bulk of the land is located in the North, chiefly in Minnesota and Wisconsin, where many counties have acquired large areas of land through tax delinquency and purchase. These holdings have an important place in the future timber supply of these two states. The geographical location of certain county and practically all of the municipal forest land provides society with recreation, watershed protection, and wildlife habitat. The benefits of these uses, individually or collectively, may far outweigh the tangible value associated with timber harvest and, therefore, all but prohibit this type of activity even though logging is an important activity on certain municipal watersheds in the West.

Private

Depending upon the agreement between the land owner and the buyer, timber may be removed from private timberland under a very wide range of cutting practices. At one extreme, one finds the "logger's choice" type of operation where all decisions are made by the logger, including which trees are to be cut and how the logging is to be done. In contrast, at the other extreme, the timber is harvested according to a comprehensive logging plan where the logger is bound by contract to remove specified timber using prescribed logging methods.

Considering the extremely large number of private owners of commercial forest land (4.5 million individual ownerships of less than 500 acres) it is not surprising that a policy involving timber sales for this type of ownership is nonexistent. It would appear that whatever agreement is acceptable to both the buyer and seller, in effect constitutes the "policy" for that particular timber sale. Owners of commercial forest land may dispose of the timber in any way they choose, except in those fourteen states where forest practice laws are in effect.

Various public agencies, particularly the Extension Service and the Division of State and Private Forestry of the Forest Service, have attempted to educate private timberland owners to accept certain self-imposed regulatory measures when timber is harvested from their land. While the results of these efforts have not been overwhelming, they have been mutually beneficial.

TIMBER RESOURCES OF CANADA

Canada's timber resources are of great importance to both Canada and the United States. Canadian forests cover an estimated 1,095 million acres, of which approximately 614 million acres are classed as productive forest land. It is interesting to note that this is considerably in excess of the 509 million acres classed as commercial forest land in the United States.

Exports of wood production (veneer, paper, plywood, woodpulp, and lumber) to the United States and other countries continue to strengthen and expand Canada's economy. The Canadian forest products industry leads all other industries in both the Canadian economy and in the value of exports.

The net merchantable volume of standing timber on the productive forest area in Canada amounts to an estimated 726 billion cubic feet. Of the total timber volume over four-fifths is softwood, as shown in this tabulation.

Species	Percent of Total
Spruce	34
Balsam fir	13
Jack and lodgepole pines	14
Hemlock	9
Cedar	5
Douglas-fir	5
Other softwoods	2
Total softwoods	82
Hardwoods	18
All species	100

The four pie charts in Fig. 1–7 present information about and relationships between land areas, ownership and control, and productive forest area by board forest types and by provinces. While Canada's timber resources are widely distributed among the provinces, special note must be given to British Columbia. It has approximately 52 percent of the total timber volume, about 60 percent of the total softwood volume and 64 percent of the sawtimber volume. This concentration of the larger sawtimber sizes has led to large and increasing volume of timber shipments from British Columbia to the United States.

STUMPAGE PRICE

Stumpage is a rather unusual kind of resource. While the economic worth of stumpage is based upon its utility after it is cut and removed from the land, stumpage is part of the land prior to being logged or cut.

Stumpage is the raw material from which lumber is derived after felling, skidding, transporting and milling. Hence, stumpage price is a reflection of lumber price after the various processing costs have been deducted. This relationship between stumpage price and lumber price is termed the principle of derived demand. An absolute variation between the price of the raw material (stumpage) and the finished or end product (lumber) will provoke proportionately sharper price movements of the former (stumpage). Since lumber prices frequently are ten times greater than stumpage prices, a 1 percent increase in lumber price may result in a 10 percent jump in stumpage prices. In this respect, the principle of derived demand tends to provide a reason for the wider range of stumpage price movements as against lumber price movements (Weintraub, 1959).

Should the end product be logs, the market price of logs is reduced by those costs to prepare and transport the severed stumpage from its "in place" condition to a location where it may be processed to satisfy industry demands.

Variations in quality, volume, and distance from the location of eventual use among tracts of stumpage preclude assigning a universal price to stumpage. As no two tracts of stumpage are exactly the same, the price of stumpage must be determined for each tract. The various items of value and cost associated with a particular

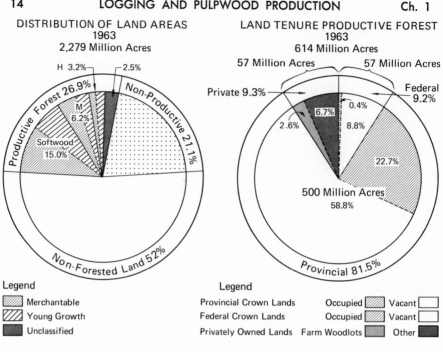

DISTRIBUTION OF LAND AREAS
1963
2,279 Million Acres

LAND TENURE PRODUCTIVE FOREST
1963
614 Million Acres

57 Million Acres   57 Million Acres

Legend

▨	Merchantable
▨	Young Growth
▨	Unclassified

Legend

Provincial Crown Lands   Occupied ▨ Vacant ☐
Federal Crown Lands   Occupied ▨ Vacant ☐
Privately Owned Lands   Farm Woodlots ▨   Other ■

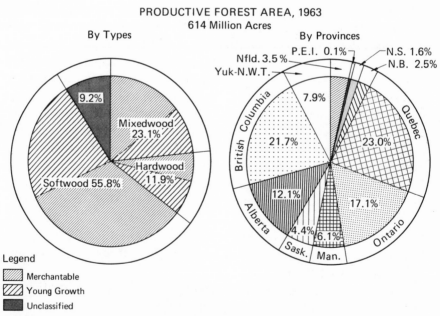

PRODUCTIVE FOREST AREA, 1963
614 Million Acres

By Types   By Provinces

Legend

▨	Merchantable
▨	Young Growth
▨	Unclassified

Fig. 1–7. Relationship of Canada's productive forest area to land distribution, tenure, forest types, and distribution by provinces. (Source: Canadian Forestry Statistics, Dominion Bureau of Statistics.)

tract of stumpage are, to a large extent, the basis for stumpage price determination.

Pertinent Items Which Influence and Determine Stumpage Values

All timber sales, both public and private, involve the determination of stumpage value to a greater or lesser degree. Some of the items which tend to establish the value of stumpage are quite apparent while others may be somewhat obscure. In general, if one will introduce and satisfy the demands of four words—what, where, when and how—when considering stumpage values he will be taking into account most of the items which either influence or determine stumpage value. Included are:

Species. Those tree species which possess certain desirable characteristics and qualities which tend to satisfy a specific purpose generally are more valuable than those which do not have this advantage.

Size. A higher percentage of better lumber grades is generally associated with large-sized trees. Logging costs usually are lower when stands of timber containing larger trees are logged.

Quality. Generally trees which have long clear boles with no visible defect have the highest value. Trees of this grade contain logs which may produce veneer, plywood, or a high percentage of the better grades of lumber.

Stand per Acre. Usually logging costs are lower and stumpage values higher when dense stands of timber with a high yield per acre are logged. This is in contrast to the value-cost relationship which exists in the case of timber stands, of the same species, which have a lower degree of stocking, in which case logging costs would be higher and stumpage value lower.

Accessibility and Terrain. In general, many miles separate timber stands, which provide the logs, and mills and plants which utilize the logs. The stumpage value of timber is rather closely related to the logging costs and the milling or conversion costs involved. Difficult terrain conditions on a logging chance usually result in higher felling, bucking, and skidding costs. This is in contrast to those localities where terrain conditions are not as severe and the various logging activities may be performed with less effort. Excessive road building and development costs, necessary to provide access to timber are, in many instances, included with logging costs. The more difficult and remote logging chances usually involve a higher logging cost which results in a lower stumpage value, while the easier and more accessible chances generally have lower logging costs and correspondingly higher stumpage values.

Logging Methods. While the conventional method of logging may be most efficient in a locality or region, a different logging method may be prescribed if other values besides logs are involved. Agitation by the public for the preservation of recreational values, watershed protection, soil stabilization, and providing adequate habitat for fish and wildlife may prohibit the use of the conventional method and demand that logging methods be used which are less destructive. Should the less destructive logging methods be more costly there will be an adverse effect upon stumpage values.

Markets. The value of logs, lumber, or other forest products are determined to a marked degree by the usual pressures of supply and demand. The best timber may have little or no value unless someone desires to purchase it. The existence of a strong, active market for logs, lumber, and other forest products has a very beneficial effect upon stumpage values. Of prime importance in this regard is the stabilizing effect provided by orderly marketing as against the uncertainty when a market, as such, is nonexistent. Markets are affected by weather, particularly in those regions where the climate is such that a log scarcity may develop at one time of the year and a flooded log market at other times.

Type of Sale. Timber may be sold on a lump-sum basis or on a unit-of-volume basis. In the case of the lump sum type of sale the owner would agree to accept a certain price for all of the timber on a certain piece of land. Many owners, particularly those unfamiliar with timber values, have regretted becoming involved with a sale of this type. However, should the owner be familiar with timber prices and values he could very well enter into such an agreement and be paid a reasonable price for his timber. Normally the unit-of-volume type of sale is considered to be more satisfactory, at least from the seller's standpoint. The transaction in this case involves payment on the basis of the number of board feet, cubic feet, or **cords** which are actually removed from the land. A further breakdown by species is also made in those instances where there is a marked difference in value between species. Practically all timber sales involving public timber are made on the unit-of-volume by species basis. The type of sale may have a considerable influence upon stumpage values.

Conditions of Sale. A complete and thorough understanding of the entire contents of a timber sale agreement is absolutely necessary. While the price and volumes are of major importance, certain less significant items may prove to have a marked effect upon stumpage values. Items such as log branding or painting, access, property line marking, **scaling,** time and method of payment, starting and termination date, insurance, slash disposal, limits of merchantability and cull log determination may very well influence the stumpage value on a particular timber sale. Sale conditions, as spelled out in the timber sale agreements which involve the public timber managing agencies, very definitely have an influence upon stumpage values.

STUMPAGE APPRAISAL

The main objective of appraising stumpage is to estimate the monetary value of standing timber, which is to be converted into commercial forest products, as accurately as is reasonably possible. Stumpage appraisal procedure varies considerably depending on who owns or manages the timber. The private owner may establish a value for his timber merely by accepting what other owners in his general locality are receiving for comparable timber. In contrast the Federal timber managing organizations are bound, by legislation and policy, to follow a definite procedure when appraising stumpage. Regardless of the procedure involved, be it simple or complicated, the elements of logging cost, milling cost, selling price, and profit and risk must be considered and evaluated. The main difference between the various methods of stumpage appraisal is the manner in which these elements are determined and manipulated. The basic relationship between the elements may be expressed as follows. Selling price − (Logging costs + Milling costs) = Stumpage + Margin for profit and risk. Oftentimes the combined total of stumpage and margin for profit and risk is identified as either conversion value or conversion return, and may be expressed by means of simple equations; such as:

$$S = R - C - M$$

where S = stumpage value
     R = selling value
     C = operating costs
     M = margin for profit and risk

The term "conversion value" identifies the difference between selling values and costs:

$$\text{Conversion value} = R - C$$

Conversion value is divided into stumpage and a margin for profit and risk. Therefore:

$$\text{Conversion value} = S + M$$

and

$$R - C = \text{Conversion value} = S + M$$

The costs of production, logging and milling, should include the best estimate of all costs relative to producing lumber according to certain specifications. While the basic relationship may be somewhat obscure in the more involved stumpage appraisal methods it still exists in principle.

The two most commonly used methods of stumpage appraisal are the overturn method and investment method, which are traditional in American forest management. The overturn method is applicable to those timber sales which will not require large capital expenditures, where a single end product such as logs will be produced, where there exists a stable log or lumber market which will allow the output to be disposed of promptly as in the case of a gyppo or contract operation, and where the major part of the investment and consequent operating risk is associated with working capital. In contrast, the investment method is more suitable to those timber sales which involve an extended stumpage supply, costly and extensive improvements, and complex conversion processes.

Federal

Acquiring stumpage from the federal timber managing agencies, Forest Service, Bureau of Indian Affairs, and Bureau of Land Management, is usually associated with a timber appraisal as a basis for a timber sale. Based, in part, upon the present volume and condition of sawtimber, the age and distribution of pole size timber, productivity of the soil, and markets, each agency assigns a volume of timber to be logged from a certain administrative district. This assignment is usually on a yearly basis and is identified as allowable annual cut, which is the goal or quota to be achieved by a particular district. The degree of timber sale activity, therefore, is closely associated with the magnitude of the allowable annual cut a district can accommodate, in keeping with the principles and application of sustained yield forest management as set forth in the agency's forest policy. Following the determination of the allowable annual cut figure, timber sales are planned for those areas within a district where timber harvesting is justified as dictated by the policy of the managing organization.

Usually the actual sale of timber is accomplished by means of oral or sealed bids made by interested parties. In this regard, legislation directs the selling agency to establish a minimum price at which the bidding may commence. The minimum price is determined by the preparation of a timber appraisal for a particular sale area within an administrative district. While there are certain differences between the three major federal timber selling agencies in their methods of developing timber appraisals, they do have one point in common. The appraised price of stumpage is a residual which is equal to the difference between the selling price of lumber or logs and the combined total of all estimated costs plus an allowance to the purchaser of the timber for profit and risk.

Each agency attempts to establish the "fair market value" for their timber by means of their particular appraisal process. In general, fair market value refers to the price agreed upon by a willing buyer and a willing seller, neither being under compulsion to enter into the transaction. While this generalization of fair market value is valid, one must also be aware of the presence and extent to which "irrational" factors affect and oftentimes distort actual market prices.

Except in those instances where a relatively small amount of timber is involved, the federal agencies are required to advertise the fact that a timber sale is contemplated. The announcement includes all pertinent information relative to the particular sale. In many instances an inspection or "look see" trip into the sale area is also publicized so that prospective purchasers may have the opportunity to examine the area and discuss those items which need clarification with agency personnel on location. Examination of a sale area allows each prospective bidder to make his own estimate or timber appraisal which, when compared to the minimum acceptable bid, will indicate how high he can afford to bid.

Most successful logging contractors have the ability to examine and assign estimated costs to the more significant cost items included in a timber appraisal. While it is possible for any element in a stumpage appraisal to vary, the more important variables which affect stumpage value are: (1) timber quality, (2) timber quantity, (3) truck haul, and (4) development cost. Other examples of cost elements such as felling, bucking, skidding, loading, sawing, and planning do not vary greatly within a general area for similar species, topography, and size of timber.

Unless a logging operator is expert in timber appraisal procedure he would be well advised to secure the services of a consulting forester or forest engineer to make an independent appraisal as a basis for bidding on a public timber sale.

U.S. Forest Service. Since the other federal timber-selling agencies tend to follow, in a general way, the system of timber appraisal used by the U.S. Forest Service, more attention is given to timber appraisal principles, guidelines, and costs as related to the appraisal of Forest Service timber.

Much of the information presented in this section is related directly to Chapter 2420, Timber Appraisal, of the *Forest Service Manual*. In certain instances information is presented which is worded similarly to that which is in the Manual.

This presentation is limited to directing attention to certain key items which are involved with appraising Forest Service timber. Should the reader desire to examine the subject more intensively he is advised to consult the Manual.

The stated objective in developing Forest Service Stumpage appraisals is to establish fair market value. In determining this fair market value or appraised value the operator-of-average-efficiency concept is used to establish market value. This procedure is expected to provide a sufficient number of interested purchasers to harvest the allowable cut according to multiple use and sustained yield principles. An important consideration in achieving this objective involves providing an adequate margin for profit and risk. The margin must be sufficient to maintain operations over an extended period of time so that a stable market for National Forest timber will be available.

According to Forest Service policy its stumpage appraisals are based on costs, returns, and profit margins which reflect average operator efficiency. There is no guarantee that purchasers will experience the profit margins, or the costs and returns used in determining fair market value. Prospective purchasers are expected to determine their own estimates of cost, return, and profit margin in keeping with the terms of the timber sale contract. These estimates should, of course, be made in respect to the operating facilities (equipment) available to a prospective purchaser, together with an honest appraisal of his operational efficiency. The prime objective of this pricing policy is to obtain a fair market value and not to penalize efficiency or subsidize inefficiency.

Usually stumpage appraisals are based on an analysis of average returns, costs, and profit margins, so that stumpage values can be determined. Returns are expressed in terms of average product selling values, as reflected by the earliest stage of processing where a free market value can be established, and where production cost, associated with the products on which the returns are based, can be determined.

Stumpage Valuation Principles, U.S. Forest Service. The Organic Administration Act of June 4, 1897, states that National Forest timber may be sold at "not less than the appraised value." In fulfilling this requirement, the objective is to determine market value for national forest stumpage by means of appraisal methods. The Forest Service considers appraised value, sometimes called "fair market value" (p. 19), to be one and the same for their purposes. When determining the appraised value of stumpage, the Forest Service relies to a large degree upon costs and returns which are applicable to typical forest industry operations of average efficiency. While prices received for national forest stumpage must also be considered, the many factors affecting these prices are also reviewed. The Manual lists some of the more important factors which may have an influence on prices. They are:

1. Local conditions of supply and demand for national forest stumpage.
2. Pre-emptive bidding.
3. Entrance or bonus bidding.
4. Income tax considerations.
5. Supplies of private timber.
6. Purchases by operators having superior merchandising or manufacturing facilities.

Market value or appraised value is not necessarily the highest price received for similar timber by the competitive bidding process. Sale procedure directs that each sale which has an appraised value in excess of $2,000 must be advertised for at least 30 days and then sold to the highest qualified bidder. Under this system the final bid price or selling price may be in excess of the appraised value. Considering that the appraised value is based upon an operation of average efficiency, and directed toward a value which at least will generate sufficient interest among enough purchasers to permit the allowable annual cut to be harvested, it is understandable why the selling price will, more often than not, exceed the appraised value.

The Forest Service employs a two-part system when making what it terms analytical appraisals. Initially, the conversion return is determined by deducting costs from sales returns. This tends to adjust the differences between tracts, such as timber quality and accessibility. The conversion return value is then divided into stumpage and profit margin. The proportions of the division of conversion value for a particular timber tract reflect what average purchasers are experiencing on tracts which exhibit similar operating conditions.

The dividing mechanism (profit ratio) is obtained by analyzing current timber sales where there is a reasonable balance between alternate purchasers and alternate timber supplies. This procedure is expected to produce a profit margin and residual stumpage which will meet two criteria:

1. A profit margin sufficient to maintain operations over an extended period of time.
2. An appraised stumpage equal to current market value considering lumber market trends.

The operator-of-average-efficiency concept is used as a basis for establishing all cost and selling values used in timber appraisals. The effect of the highly efficient purchaser, who may consistently overbid Forest Service timber sale offerings, is factored out so that profit ratios from current transactions will more truly represent the operator of average efficiency.

Analytical appraisal procedure, in keeping with the aforementioned principles should produce appraised market-value prices for stumpage which:

1. Are based upon long-term profit considerations.
2. Are sensitive to lumber market changes.

3. Are appropriate for the average efficient purchaser.
4. Are responsive to the actual prices being paid for stumpage, thus giving the United States a fair return for its stumpage.
5. Reflect all conditions of sale under which national forest timber sales are made.

The Forest Service applies a fundamental rule when it appraises timber. This rule requires the appraiser to include consideration of all significant factors which affect stumpage price. In order to function more effectively, appraisers have been provided with certain appraisal guidelines so that a systematic approach may be followed. Should close observance of the guidelines prevent proper consideration of all factors, the appraiser may present his findings in some alternative fashion which does properly evaluate them. However, in those instances where departure from established appraisal guidelines is taken, the appraiser must show justification and also be able to explain his course of action.

Conversion Value, U.S.F.S. The difference between the estimated selling value and estimated operating costs is conversion value (sometimes referred to as conversion return) and consists of two parts: the margin allowed to the operator for profit and risk and the appraised stumpage value. The margin for profit and risk (sometimes referred to as profit margin) in Forest Service appraisals includes the elements of profit to the purchaser, interest on borrowed capital, income tax payments, and margin for risk. The name implies that two main considerations are involved; an average return to the operator for services and the use of capital required to operate the business over a long term in a manner consistent with average efficiency, and a further allowance to compensate for those risk circumstances which may cause subnormal returns to be realized.

Profit Margin and Profit Ratio, U.S.F.S. In Forest Service appraisal procedure a profit ratio is used to express the margin allowed for profit and risk. The ratio is applied to a value which represents total costs which includes stumpage. However, as stumpage value is unknown at this point a selling value ratio is used. This ratio is the result of solving the expression $\left(\dfrac{P}{1+P}\right)$, where P represents the profit ratio assigned to a particular timber sale. The margin for profit and risk is determined by multiplying the selling value by the selling value ratio. The complement of the selling value ratio is the operating ratio $\left(\dfrac{1}{1+P}\right)$. Stumpage value may be determined directly by using the operating ratio.

An example will serve to indicate how the operating ratio is developed. The basic formula for the determination of stumpage value will be employed to show how the operating ratio is obtained:

$$R - C = S + P(C + S) \qquad \text{or} \qquad R - C = S + M$$

where R = selling value
   C = operating costs
   S = stumpage value
   P = profit ratio
   M = margin for profit and risk

Solving for S:

$$R - C = S + PC + PS$$
$$R - C - PC = S + PS$$
$$R - C(1 + P) = S(1 + P)$$
$$\frac{R - C(1 + P)}{(1 + P)} = S$$
$$R\left(\frac{1}{1 + P}\right) - C = S$$

Stumpage is considered as a cost in the Forest Service profit ratio method of measuring profit margin. The basic stumpage value formula on page 18 serves to show how this is accomplished. The margin for profit and risk is determined by multiplying the sum of operating costs (C) and stumpage (S) by the profit ratio.

In the above example the expression $\left(\dfrac{1}{1+P}\right)$ is the operating ratio and its complement $\left(\dfrac{P}{1+P}\right)$ the selling value ratio. Assuming that a profit ratio of 0.11 is considered adequate for a particular timber sale within a certain national forest the operating ratio is $\left(\dfrac{1}{1.11}\right)$ or 0.901 and the selling ratio is $\left(\dfrac{0.11}{1.11}\right)$ or 0.099. Relating these ratios to a specific case where lumber selling value is $60 per M bd. ft., (log scale) and operating costs are $40 per M bd. ft., (log scale) excluding stumpage. Conversion return is the difference between $60 and $40 or $20, which is divided between margin for profit and risk, and stumpage.

R × Selling value ratio = Margin for profit and risk
$60 × .099 = $5.94 per M bd. ft.
(R × Operating Ratio) − Operating Costs = Stumpage
($60 × .901) − $40 = $14.06 per M bd. ft.

Bureau of Indian Affairs. Timber sales which involve Indian forest land are unique in that a federal organization, the Bureau of Indian Affairs, is charged with the administration and management of land which, in essence, is in private ownership. Indian lands are the Indians' private property and are held in trust for them by the federal government. These lands are managed with due consideration given to the best interests of the owners according to Congressional legislation, which includes the concept of sustained yield forest management. The Bureau of Indian Affairs has the responsibility of applying and enforcing the statutes of the law.

The bureau, recognizing how closely Indian forest lands are interrelated with the general welfare of the Indian people, prepares forest management plans and timber sales which are compatible with the Bureau's overall program of social and economic betterment for the Indian. Management plans and certain policies involving timber sales must be more flexible than is ordinarily required or considered ideal from a purely technical standpoint, and must allow for adjustments to meet the immediate pecuniary needs of the individual Indians and the educational and industrial advancement of the tribes.

Basically there are two types of Indian ownership from which stumpage may be acquired—allotted lands and tribal lands. Allotted lands are those lands which are owned by one or more Indians, while tribal lands are owned in the name of a particular tribe. In either case, the bureau is requested, by the individual owner or tribal council, to prepare a timber sale for a particular tract of timber. The bureau, using stumpage appraisal procedures which are quite comparable to those employed by the Forest Service, determines stumpage rates. The individual or tribal council, as the case might be, must approve the timber sale conditions and stumpage rates before the bureau may offer the timber for sale. Following approval, the bureau prepares a timber sale contract and advertises the sale in the appropriate manner. Normally sales involving stumpage value in excess of $500 are sold on a bid basis and the payment based upon the volume as determined by a log rule or other method as stipulated in the contract. Compliance with contract provisions is a further responsibility of the bureau.

The responsibility of the bureau to enhance the welfare of the Indians, coupled with the private ownership–public agency control relationship, justifies certain deviations from the customary manner by which timber sales are prepared and administered.

Information regarding current and future timber sales involving Indian stumpage may be obtained from local Indian Agency Offices or Area Offices of the Bureau of Indian Affairs.

Bureau of Land Management. The stumpage appraisal procedure employed by the Bureau of Land Management on its **O & C lands** is unique in one particular respect—method of payment. Stumpage is sold on a lump sum basis and timber cruise volumes are used to establish an appraised price, which is the point of beginning for either sealed or oral bidding for the timber. Practically all sales are so-called cruise sales; however, stumpage may also be sold on a scale basis.

In general, BLM stumpage appraisal procedure is somewhat similar to that used by the U.S. Forest Service in all other respects. The appraisal price for stumpage is determined by establishing a **pond value** for the timber on a sale area by species. Total cost involved with moving logs from the stump to an appropriate utilization center is determined and includes the following areas of cost: felling and bucking, yarding (rig-up, swinging, and loading), slash disposal, transportation, road construction, amortization (mainline), fire protection, and other allowances associated with a particular timber sale. Conversion return is the difference between the pond value and the cost of making and moving logs to a utilization center. A conversion factor is then applied to the conversion return and thereby establishes stumpage value and margin for profit and risk. For example: pond value (at utilization center) of a certain species is $55.00 per M bd. ft., cost to utilization center is $35.00 per M bd. ft., and conversion return then is $20.00 per M bd. ft. Applying a conversion factor of 60 percent to the conversion return of $20.00 provides a $12.00 stumpage value and the balance of $8.00 would be the allowance for profit and risk.

State

Each state has its own policy regarding stumpage appraisals and timber sales, and a department is assigned the responsibility of administering and managing the commercial forest land within its boundaries. The economic importance of the forest products industry within any particular state is reflected in the procedure by which the respective forestry departments develop stumpage appraisals.

In those states where the U.S. Forest Service and other federal agencies administer and manage large areas of forest land, the state's stumpage appraisal procedure tends to follow that of the federal agencies. Timber sale activity provides the means whereby logging costs may be gathered and subsequently used to provide a sound basis for appraising stumpage.

The pressures of recreation and allied activities may have a marked effect upon the extent of timber sale activity within certain high-population states, such as in the New England and Middle Atlantic areas, where consideration is given to the effect a timber sale may have upon the recreationist, vacationist, and outdoor enthusiast. While population pressures presently are of extreme importance in certain localities, planning for future population buildup in other localities with accompanying recreational overtones is taking place.

Private

While the opportunity to purchase commercial forest land from federal, state and local timber-managing agencies is practically nonexistent, such is not the case with privately owned land in this class. An owner may elect to sell stumpage, land, and timber or just land depending upon circumstances.

The value of stumpage on land in private ownership is determined in much the same manner as is done in the case of public timber, without the necessity of adhering to a policy implemented by legislation. The entire basis for a transaction involves a price, contingent upon certain conditions, which is the result of negotiations between buyer and seller. It behooves each party to bargain to the best of his ability, giving due consideration to markets, costs, and profit prior to entering into an agreement. Here, as in timber sales which involve publicly owned timber, the final lumber or log value is reduced by the costs of production necessary to convert standing timber to either lumber or logs, as the case might be. The residual, or that which is left for profit and stumpage, is divided between the owner and logger. One of the several different methods whereby the residual may be distributed is identified as a percentage cutting contract. In this type of agreement the logger and stumpage owner may agree that the residual should be split 60–40, which refers to a situation where the logger would receive 60 percent of the residual for profit and risk and the owner 40 percent for stumpage. The apportionment of the split depends, in part at least, upon the bargaining ability of the parties involved.

Generally a private owner of commercial forest land is not as familiar with markets, prices, costs, and problems associated with logging as a stumpage or timber buyer would be. In most cases, this situation provides the buyer with an advantage when dickering for price. The seller may have certain preconceived notions regarding timber values which may or may not be valid. Many owners fail to realize that there can be a great difference in value between the type of tree growth present on woodland and that which is generally associated with commercial forest land.

A cursory examination by a qualified person, such as a county forester, farm forester, association or consulting forester, who is familiar with local conditions, usually will suffice to determine whether a timber stand has sufficient value to justify a bid on the part of a buyer or an offer on the part of a seller. Should the examination indicate the feasibility of a stumpage sale, the services of a consulting forester familiar with stumpage appraisal and sales procedures may be desirable and highly beneficial from a monetary standpoint.

One important consideration involves the degree of supervision which should be exercised in connection with a timber sale. In the case of a sale which includes large volumes of high-grade timber, supervision not only is justified, but in many cases it is an absolute necessity. Adequate supervision on the part of a stumpage owner does much to discourage an operator from diverting truckloads of logs to outlets other than those agreed upon, a situation which, in many cases, results in a direct loss to the stumpage owner. Supervision is a cost item and should be considered as such. While an owner may tend to question an expenditure for supervision, particularly if he relates it to a reduced stumpage return, the resulting benefits should be considered and evaluated. Other than lessening the chance of loads of logs being diverted, adequate and effective supervision tends to prevent excessive damage to residual trees, and alleviate problems associated with logging slash, fire, damage to improvements, and trespass, to name some of the more common areas of potential aggravation.

Absentee ownerships are particularly vulnerable should the operator elect to take advantage of existing circumstances. Expenditure for supervising the activities associated with timber sales on land in this type of ownership may do much to lessen the disappointment and frustration which, only too often, follows in the wake of logging where supervision was lacking.

Even though a timber sale contract between a logger and stumpage owner may have been drawn up by a very competent authority, there is no guarantee that the logger will abide by the provisions agreed upon. However, should there be a breach of contract, corrections should be made immediately or the contract terminated. Here again, adequate supervision can serve to detect and correct such infractions before additional damage is done. This point is of particular importance when logging takes place on absentee ownerships. To be sure, certain acts on the part of a logger, who is removing timber under an agreement, may justify legal action and may be the last recourse, but one must not lose sight of the fees involved with such litigation.

Logging activities on both absentee and resident ownerships should not be undertaken without first considering the possible benefits and cost associated with effective supervision of the operation. While no hard and fast rule is applicable, persons contemplating a sale of stumpage from their holdings should give careful consideration to the relationship between expected returns and the effects of noncompliance with contract provisions on the part of the logger. If a stumpage owner is not knowledgeable regarding the complexities and potential problems generally associated with timber sales he should engage the services of a consulting forester, to either advise, inspect, or supervise the operation.

TIMBER SALE AGREEMENTS AND CONTRACTS

Following the determination of stumpage value, by whatever means specified or employed, a sale price is agreed upon or established between buyer and seller. The sale price reflects the expectations of the buyer and the desires of the seller. A contract or timber sale agreement is the instrument by which these expectations and desires may be specified to the mutual understanding of both parties —buyer and seller. While there are numerous types of agreements and contracts which may be used to consummate a sale of standing timber each contains certain considerations depending upon circumstances. The timber sale agreements used by the various public timber selling agencies usually contain many more provisions than the agreements and contracts associated with the sale of stumpage from privately owned timberland.

Basic Provisions

There are certain provisions which embody the principal considerations between the buyer and seller of either public or private timber (and usually included in any agreement which has legal significance), such as:

1. Parties involved. The names and addresses of both the purchaser and seller involved in the contract or agreement.
2. Description of the sale area. Maps of the area together with a legal description. Included should be any available information, such as adjacent ownerships, property line and corner information, and established improvements (roads, fences, and trails).
3. Timber to be removed. Method by which the timber will be designated for removal. Individual tree marking or cutting boundary establishment. Penalty for cutting trees other than those designated. Felling snags and defective or undesirable trees.
4. Felling and bucking. Designation of minimum stump height and limit of merchantability (top diameter). Penalty for excessive breakage as a result of carelessness.
5. Skidding and truck haul. Logging methods and machinery to be employed. Provisions for location of landings and haul road location, construction, and maintenance. Use of timber for skidways or other use in connection with the construction of landings, culverts, and road construction.
6. Volume determination. Method of scaling and **log rule** to be employed.
7. Identification of logs. Method by which logs will be marked: branded or painted.
8. Payment for logs. Advance payment provisions. Method of payment, check or money order payable to whom. Disposition of scale slips. Period of payment; weekly or monthly.
9. Right-of-way and access. Designation of responsibility regarding fees associated with right-of-way agreements and costs of road construction necessary to provide access.
10. Property lines. Responsibility for marking.

11. Forest practices. Compliance with the rules and regulations as set forth by the Forest Practices Act of the State in which the timber sale area is located.
12. Fire. Adequate provision for the prevention, control and suppression of forest fires; slash burning, and suspension of activities during periods of low humidity and extreme fire hazard. Usually these items are clearly defined and enforced by both State and Federal forestry agencies.
13. Insurance. Provisions for insurance protection against claims, suits, liability resulting from personal injury or property damage caused by accidents on the operation. Types of insurance, degree of coverage, and proof that coverage is, in fact, being maintained.
14. Payments of costs and expenses. Prompt payment of all expenses and costs, including federal and state taxes, involved with both logging and hauling so that claims and liens will not be imposed upon the land or logs.
15. Termination date. Penalty for failure to comply and basis for possible extension.
16. Compliance bond. Deposit to insure compliance with covenants of agreement.

Regardless of whether a timber sale agreement is a standard form, as is usually the case when public timber selling agencies are involved, or consists of a special set of provisions between two individuals, certain measures are advised. The importance of having sufficient provisions which have been clearly stated and recorded in a contract cannot be overemphasized. Costly litigation may be the result of agreeing to vague or incomplete contract provisions. Of equal importance is that both parties read and understand what they are agreeing to do as set forth in the contract. Engaging the services of a qualified person, such as a consulting forester familiar with timber sales and contracts, to review contract provisions prior to affixing one's signature to an instrument may prevent involvement with legal action. Depending upon circumstances, further assistance may be justified, such as engaging the services of specialists in legal, tax, and accounting matters as related to timber sales.

Caveat Emptor

Generally a desire to make a profit or at least remain solvent is part and parcel of the free enterprise system regardless of whether it is a one-man operation or a large, complex industrial corporation. The Latin expression—*caveat emptor*—is a most appropriate one to use at this point. While it may be used in all instances where a buyer-seller relationship exists, it has particular significance when related to timber sales. *Caveat emptor* literally translated means; let the purchaser beware, i.e., he buys at his own risk. This caution should also be exercised by all bidders in those cases where oral or sealed bidding is used to establish the price level at which ultimately a buyer is awarded a timber sale.

Normally all timber-selling agencies provide interested parties with an ample opportunity to examine and evaluate values and volumes as associated with a timber sale. All bidders are potential buyers. The main difference among bidders is reflected in the amount they are willing to pay for the timber included in a particular sale.

Bidders who do not examine the many facets of a timber sale offering are engaging in so-called "blind bidding" and are subjecting themselves to a possible rude awakening should they become the "successful bidder." No bidder is, in fact, forced to bid on a timber sale; however, if he elects to do so it should be done only after a thorough examination of what the seller has to offer together with attendant requirements.

U.S. Forest Service Timber Sale Contract

Since the summer of 1965 Forest Service timber sales have been made according to a timber sale contract which was the result of much discussion and negotiation among representatives of both industry and the Forest Service (Fig. 1–8) (Craig, 1965). Basically the contract consists of three separate parts called divisions: A—specific conditions; B—standard provisions; and C—special provisions. Also included is a map of the sale area, drawings and specifications for development, and attachments as may be provided for in division C—special provisions.

Division A (specific conditions) embodies all those items which relate to identification and agreement as pertinent to a particular timber sale. The contents of this part of the contract will therefore be different from one sale to another. In all, there are twenty-five items which set forth what is to be done; where, how, when, and by whom.

Division B (standard provisions) contains the text of the contract

FOREST SERVICE TIMBER SALE CONTRACT

CHART 1

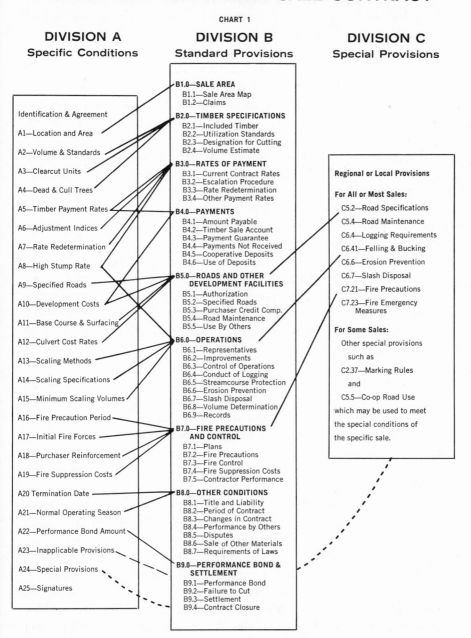

Fig. 1–8. Relationship between and among the three divisions of the Forest Service timber sale contract. (Source: *Forest Industries,* Vol. 92, No. 6, May 29, 1965, p. 31.)

and is the largest of the three divisions of the contract form. It consists of eleven pages of printed material fastened in booklet form and contains those provisions which are applicable to timber sales which involve log scaling as a basis for payment (a different set of standard provisions is used in the case of tree measurement sales). This division is organized into parts, sections, sub-sections, and items, to facilitate the application of these standard provisions to specific conditions (division A) and special provisions (division C) as required.

Division C (special provisions) includes those regional or local provisions which apply to a particular timber sale and are related to the appropriate group of standard provisions in division B. In total, special provisions applicable to a timber sale are listed on the last page of division A (specific conditions) and thereby become an integral part of the contract.

The relationship among the three divisions of the contract is shown in Fig. 1–8 and indicates the manner in which various conditions and provisions are interrelated. Once the various items germane to a particular timber sale are identified and evaluated they are compiled into a timber sale contract, consisting of a series of acceptable agreements between the Forest Service and the timber purchaser.

The contract form, while somewhat unusual in arrangement, is considered to provide a reasonable means whereby the buyer-seller relationship is made more compatible than was possible previously. Still, it is a rather complex instrument, and prospective bidders are urged to make every effort to become knowledgeable and familiar with the meaning and intent of the various conditions and provisions as related to a particular timber sale.

Since the introduction of the present contract the Forest Service has provided ample opportunity for bidders to learn how they must function to satisfy contractual agreements.

Inasmuch as the forest products industry was directly involved with the development and preparation of the present contract, it also accepted the responsibility for clarifying contract provisions. Industry leadership, realizing the somewhat unusual and unique arrangement of the contract, decided to prepare a handbook for this purpose. Seventeen forest products associations cooperated in the preparation of a purchaser's counterpart to the Forest Service Handbook on the timber sale contract, which is entitled *Buyer's*

Guide to Forest Service Timber Sale Contracts (National Forest Products Association, 1965). The guide is considered to be an indispensable supplement to the contract because it provides a very effective means whereby a bidder may more fully understand his rights and obligations.

Both the Forest Service and the forest products industry have provided means whereby bidders may become more knowledgeable regarding the present contract. Since such assistance is available for the clarification of contract provisions there should be a higher level of mutual understanding between the purchaser and the Forest Service.

SUGGESTED SUPPLEMENTARY READING

1. DANA, S. T. 1956. *Forest and range policy.* McGraw-Hill Book Company, Inc., New York, N.Y. Included are three valuable appendixes; a survey of federal policies, a complete list of major events in the development of federal policy relating to the conservation of all natural resources, and a selected bibliography.
2. FORBES, R. D., and A. B. MEYER (eds.). 1955. *Forestry handbook.* The Ronald Press Company, New York, N.Y. Section 15 includes a rather complete treatment of both the overturn and investment methods of stumpage appraisal.
3. FRITZ, EMANUEL. 1960. Timber sale contracts. *Loggers handbook,* published by Pacific Logging Congress, Portland, Oregon. The author presents a guide for preparing timber sale contracts to avoid common pitfalls and costly litigation.
4. GREELEY, W. B. 1953. *Forest policy.* McGraw-Hill Book Company, Inc. New York, N.Y. A comprehensive work on forest policy by an expert in the field of both public and private forestry.
5. U.S. Department of Agriculture, Forest Service, Forest Resource Report No. 17. 1965. *Timber trends in the United States.* U.S. Government Printing Office, Washington, D.C. This publication is the latest in a series of periodic appraisals of the timber situation in the United States made by the Forest Service.
6. U.S. Department of Agriculture, Forest Service, 1963. *Timber—story of a timber sale on a national forest.* U.S. Government Printing Office, Washington, D.C. Describes, in a very general way, the activities associated with the preparation of a timber sale on National Forest Land.
7. U.S. Department of Interior, Bureau of Land Managtment, Washington, D.C. 20402. *Timber sale procedure handbook,* Vols. 1 to 3. Includes the timber appraisal and timber sale procedures under which the B.L.M. operates.
8. U.S. Department of Agriculture, Forest Service, *Forest Service manual.* U.S. Government Printing Office, Washington, D.C. Chapter 2420, Timber Appraisal, covers in detail Forest Service policies and procedures related to stumpage appraisals.

2

Logging Planning

CONCEPTS OF LOGGING PLANNING

The concept of the subject of logging planning varies with the responsibilities of the individual concerned. From the viewpoint of the logging manager, logging planning involves the acquisition of timber to log, the markets for the logs, the selection of equipment and recruiting of labor, and, most importantly, financing the operation. When management decisions on these matters have been made, he is concerned with planning the organization of the operation, with production scheduling and with budgeting expenditure and income.

To the logging superintendent, foreman, or other production supervisor, logging planning is scheduling the interdependent activities of felling and bucking, skidding or yarding, loading and trucking, and the day-to-day distribution of machines and assignment of men. The servicing of machines and the action to be taken in emergencies such as mechanical breakdowns, accidents, and changes in the weather must also be planned.

The subject of this chapter is the logging planning with which the logging engineer, forest engineer, or industrial forester in the employ of a logging company is primarily concerned. This is the detailed planning and the layout in the field of the truck roads, the landings where logs are to be loaded on trucks, and the settings or cutting units to be skidded or yarded to the landings. In the smaller operations this planning is often done by the owner. The timber sale forester in a public agency is involved in logging planning, as the timber sale contract usually specifies where the roads are to be built and the logging methods required. A logging plan is prerequisite to the logging cost estimate which is a part of the stumpage appraisal.

Logging planning is unique in that every tract of timber differs to some degree from every other tract in timber stand, topography, soil and forestry requirements. The concomitant consideration of many factors is required in order to achieve the most economical logging operation consistent with silvicultural, protective and other forestry objectives. Analytical thinking, imagination, and judgment are demanded of the logging planner. The basic principles of logging planning and recommended planning procedures follow.

Considerations in Logging Planning

The primary objective of logging planning is economy of operation. The continued existence of a company and its ability to provide employment depend upon operating at a profit. Rising labor wage rates and more mechanization, with higher capital investment in equipment, enhance the importance of planning. Just as the superintendent or foreman is judged by his logging costs, the logging engineer is judged by his ability to develop the most economical plan possible under the given conditions. The timber sale forester planning logging on public lands should likewise be concerned with the economics of logging. The stumpage value of a timber stand increases with reduction in logging cost. The considerations in logging planning may logically be grouped under the four classifications which follow.

Physical Requirements of Logging Methods. First to be considered are the physical requirements and limitations of the available **skidding** or **yarding** methods. For example, all tractive methods are more efficient when skidding down the slopes, which requires roads at the lower elevations, such as near the bottoms of the valleys. Highlead yarding is most effective upslope, so roads at the higher elevations or on the ridge crests are indicated. There is no physical limitation to tractor skidding distance, but effective **highlead yarding** distance is limited by topography, spar tree height, and yarder line capacity. Mobile loggers are limited in efficient yarding distance by the low lead to the boom sheave. Skyline systems require adequate deflection between supports. The horsepower and size of the log trucks to be used, and the performance characteristics of the trucks, influence the determination of road standards. Road standards include maximum gradient, minimum radius of curvature, and width of road surface.

Road Pattern and Spacing. The road pattern and spacing of roads which will give the lowest combined total cost of skidding, truck hauling, and road construction and maintenance per unit of volume is the ideal to be sought. This necessitates estimation of skidding costs, road construction and maintenance costs, and trucking costs. The volume per acre to be logged is an important element in road spacing computations. Topography often determines the road pattern. The road pattern for a tractor-yarded flat-bottomed valley and two cable-yarded hillside settings is shown in Figure 2–1. The fan-shaped pattern of the cable yarding roads is also seen. The flat was logged first and was restocked with young growth when the photograph was taken.

Forestry Considerations. Forestry considerations may include the prescribed silvicultural system; the percentage of the stand to be cut in partial cuts or the maximum size of clear-cut settings; and the applicable state forest practice laws. Such laws require the leaving of a minimum volume per acre of trees under a specified diameter where partial cutting is done. State forest practice laws may require leaving seed blocks where clear-cutting is practiced, unless the land owner obligates himself to seed or plant the logged area. Protection consideration may include fire protection, soil and water resource protection, and protecting the residual stand or the uncut settings from damage. Where clear-cutting by staggered settings is practiced, the protection of the uncut settings from windthrow by providing windfirm setting boundaries is of paramount importance. Logging planning in multiple-use forests subject to heavy recreational usage requires consideration of public safety and aesthetics. Public relations considerations may necessitate special attention to protection of fishing streams from siltation and debris, preserving potential camping and picnic sites along streams, and leaving uncut forest strips to screen logged areas from the view of recreationists.

Fig. 2–1. Road pattern for tractor-yarded flat, and two cable-yarded hillside settings. (Courtesy of the Weyerhaeuser Company.)

Safety. Because of the hazards associated with logging and the relatively high accident frequency rate experienced in the logging industry, the safety of the crew is an important consideration in logging planning. **Landings** are selected with dimensions and side slope that will enable logs to be safely unhooked and loaded. **Skyline roads** are located to afford **deflection** for an adequate factor of safety. In demarcating the boundaries of clear-cut settings the hazards to the yarding crew due to topography are taken into consideration. Road standards of gradient, curvature, width, and turnout spacing affect the safety of all traveling the road. In projecting roads, long sustained descending grades are flattened at intervals to permit release and cooling of log truck brakes. It has been demonstrated that safety can be engineered into the logging operations.

The Logging Plan

The planning results in the preparation of a **logging plan.** The logging plan serves the same purpose for the development of the logging operation that an architect's blueprint does for the construction of a building. The type of plan will vary with the size of the tract of timber and the difficulty of the logging chance. For a small tract on flat ground presenting no topographic difficulties, the logging plan may be a simple idea in the mind of the logging operator. For a large tract of timber on steep, broken ground, the plan may be so complex as to require delineation on a large-scale logging plan map combining topography and timber stand data. The logging plan usually shows the truck roads, both existing and to be constructed, the landings to which logs will be yarded or skidded for loading on trucks, and the boundaries of the settings or areas from which timber will be skidded to each landing (Fig. 2–2). The skidding method planned for each setting is noted. The priority sequence in which the settings are to be logged is specified if the staggered setting system is used.

In states where forest land can be logged only under a permit from the state forester, a logging plan map must accompany the application for a permit. The map shows the location of tract, together with the legal description, the existing and proposed roads, the areas to be cut, the cutting system, and the regeneration measures proposed. The permit system helps to secure compliance with the forest practice laws, informs the state fire wardens where logging operations are being conducted, and enables the state forester to halt operations during periods of extreme fire danger by withdrawing the permit.

The detailed logging plan for industrial forest lands in the western regions is usually made by the company logging engineer. He works under the supervision of the logging manager or superintendent, who establishes the policies governing the planning. The final plan is subject to his approval. In other regions the industrial forest logging plan is made by the company forester or forest engineer. The logging plan for small tracts may be made by the land owner, by the logging operator, or by a consulting forester. The logging plan for public lands is made by timber sale foresters of the

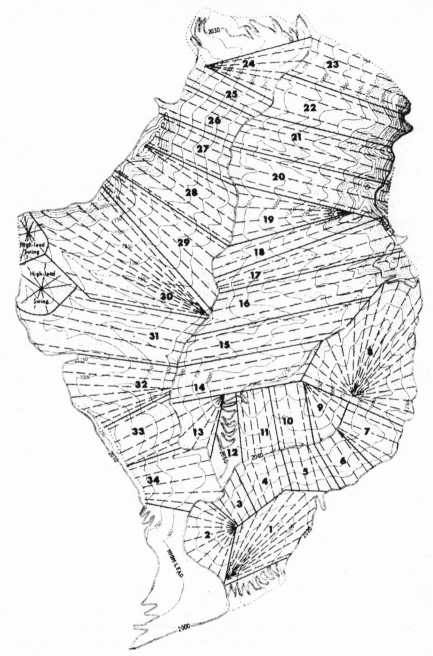

Fig. 2–2. Logging plan for skyline crane yarding. (Courtesy of the U.S. Department of Agriculture, Forest Service, Pacific Northwest Forest and Range Experiment Station.)

agency managing the forest. The logging plan is made a part of the timber sale contract.

LOGGING PLANNING DATA

The first step in logging planning is to collect all available data relating to the tract involved. These data may include the following.

Maps. (1) *Topographic maps.* United States Geological Survey quadrangles, scale 1 inch to the mile, with 40-foot or 80-foot contour interval, are usually available. Large-scale topographic maps are often made by photogrammetric methods especially for logging planning. The common scales are 400 feet to the inch with 20-foot contour interval or 1,000 feet to the inch with 50-foot contours. Sometimes U.S. Geological Survey maps are enlarged photographically to the scale of the aerial photographs, usually 1,000 feet to the inch. (2) *Timber type maps.* Such maps, showing boundaries of timber types, species composition, size class, and degree of stocking of each type, are commonly made in connection with the timber cruise or forest inventory. Small-scale, generalized-type maps are published by the Forest Survey branch of the U.S. Forest Service. Where operations research data are available for planning, site maps are needed, if site is the basis for stand factor used in classifying work places. (3) *Special logging plan maps.* The most desirable kind of logging plan map combines timber type and volume data with large-scale topography. The larger industrial forestry corporations often contract the making of such maps of their lands to consulting forestry firms specializing in photogrammetry and photo-interpretation. (4) *Geological maps.* Geological quadrangles, published by the U.S. Geological Survey, show the geological structure, parent rock and land forms, and the nature of the materials, such as unconsolidated glacial till, river run rock, or consolidated bedrock. Geological maps may also be available from state geology departments. Geological maps are especially useful guides in forest road engineering. (5) *Soil survey maps and reports.* Soil surveys of private land are made by the U.S. Soil Conservation Service and by state agencies. They are usually limited to agricultural lands, but may be helpful in planning the roads to forest lands adjacent to or intermingled with farm lands. Some industrial forest land owners have soil surveys made of their holdings. The soil survey maps and reports of National Forests are especially valuable for logging planning. Included in their interpretations are ratings of the hazards of slump, slide, potential erosion and windthrow, soil bearing strength for roads, and the value of bedrock for surfacing material.

Aerial Photographs. Most large companies have aerial photographs made of their timber land holdings. Others buy prints of photographs made by aerial photography contractors for public agencies, such as the bureau of land management, forest service, soil conservation service, state forest departments, and county assessors. Planimetric base maps are sometimes obtainable from the contractor or the agency.

Land Survey Notes. Copies of the original cadastral survey field notes are essential for determining the legal boundaries of the property to be logged. They are obtainable from the nearest office of the cadastral engineer, bureau of land management, and from the office of the county engineer. State departments of maps and surveys are sources of current information on corners recovered or restored and the coordinates of points established on the state plane coordinate system. The U.S. Coast and Geodetic Survey publishes data on triangulation stations and bench marks useful for photogrammetric control.

Route Survey Notes. If any route surveys have been made in or adjacent to the planning area, the plans and profiles of such surveys are helpful in photogrammetric control and road system planning. Among route surveys which may be available are highways, county roads, forest roads, electric power transmission lines, and pipe lines. Notes or sketches of any reconnaissance made in the logging planning area are collected.

Timber Stand Data. Data on the timber stand may include forest inventory data and timber cruises. If any adjacent areas have been logged recently, data on the log recovery, including gross or woods scale, net scale, and volumes per acre, and a comparison of the cut with the cruise, will be useful. Where tree selection is practiced, marking and enumeration of the marked trees before the road system is planned is preferable.

Cost Data. All available logging cost data, which are applicable to the tract, are assembled. If cost data are not available from time and cost studies, cost and production data on past logging operations in comparable tracts, as detailed as records provide, are collected. Construction cost data on roads built in similar terrain are helpful guides. Logging cost data compiled by public agencies for timber sale appraisal are also useful in logging planning.

Equipment Data. If the planning is done for an operating company, the logging equipment to be used should be ascertained. Data on equipment specifications and performance germane to planning include the following: (1) Tractive skidding—the size and type of tractors, the log skidding accessories to be used with them, the optimum favorable gradient which will give the fastest round trip time, tractor performance on adverse gradients, and the efficient skidding distances on various slopes for animal skidding. (2) **Cable yarding**—the line capacities; power and speed of **yarders;** the length of skylines available if skyline systems are to be used; whether tree **spars** or portable steel **towers** will be used and the height of the towers; the type of loaders and the swing radius of crane loaders for bank clearance on the landings; the size and efficient yarding distance of mobile combination yarding and loading machines; and the height of the lead block or sheave above the ground. (3) Log trucks—the specifications of the log trucks, including gradability or speed at which they can negotiate favorable and adverse grades in the various gears; turning radius; bunk width; weight; and load capacity.

In planning for a public agency timber sale, who the successful bidder will be is unknown. Therefore, the equipment common to prospective bidders is the basis for the planning. The public agency usually avoids specifying logging methods which will restrict bidding to one operator with uncommon equipment. For example, a tract of state land of steep, rocky topography could have been more economically logged by skyline yarding systems, because of a saving in the length of truck roads, and consequently in road cost. However, highlead logging was planned because the prospective bidders in the locality did not have skyline yarding equipment. Forestry considerations may outweigh equipment considerations. An example is found in the western pine regions where soil erosion hazards have led public agencies to specify cable systems in localities where tractor skidding is the prevailing method. **Skyline cranes** are required on some Douglas-fir region timber sales where avoidance of soil disturbance is paramount.

Policies. It is essential that the planner be briefed on company or agency policies germane to the logging plan. Matters of policy may include the following: (1) The cutting system to be used: If cutting is by tree selection, the marking rules or guides to be followed and the volume per acre or percentage of the stand to be cut; if clear cutting, whether by staggered settings, strips, or progressive cutting leaving seed blocks, and whether artificial or natural regeneration is planned. (2) The permissible logging methods and, in the case of a company, the logging equipment that will be made available; and the preferred external yarding or skidding distances, as developed by company experience. (3) Whether log trucks legal on the highway or off-highway trucks will be used and the reload or transfer site for off-highway vehicles. (4) The road standards including gradient, curvature, width, and surfacing for all-weather roads. (5) The volume of production desired from the tract by periods of time or, for public timber sales, the volume of timber to be sold. (6) Any special precautions for watershed, stream, or fire protection. (7) Anticipated recreational use of roads, fishing streams, and camp sites.

Preplanning Reconnaissance

A **reconnaissance** is a preliminary examination of an area or route. The preplanning reconnaissance is made in the field to familiarize the planner with the tract, and to supply information needed to supplement the data available from maps, aerial photographs, and cruises. The intensity of the reconnaissance will depend upon the value of these data. If accurate, large-scale topographic and timber-type maps have been made, less reconnaissance is required than if only small-scale maps are available. If the planning is to be done on aerial photographs, the density of the crown canopy will determine the amount of ground reconnaissance needed. Open stands require less reconnaissance than dense stands which obscure the ground. The reconnaissance is customarily made by walking over the tract in a systematic manner. The use of the helicopter for reconnaissance is increasing. In open stands on flat or rolling ground, a gridiron pattern of hand compass lines is followed. In dense stands or on rough topography, following existing trails, supplemented by walking along the ridge crests, is the quickest way to make the reconnaissance. Where access roads present the greatest problem in planning, projected road routes are followed by running **Abney level** grade lines between major control points such as benches, saddles, stream-crossing sites for bridges or large culverts, and points to be avoided such as rock outcrops and slide or slump areas. Taking the aerial photographs on the reconnaissance and correlating photo images with features on the ground will enhance skill in photointerpretation.

Equipment for the reconnaissance includes hand compass, Abney level, aneroid barometer for obtaining elevations of control points, map, aerial photographs, pocket stereoscope, and a field note book. Among the items which are observed and recorded are:

1. Conditions affecting road location: soils, rock outcrops, and gravel deposits suitable for road surfacing, bridge and large culvert sites, slide or slump areas, and major control points.
2. Conditions affecting logging methods and setting: soils, flats or benches suitable for landings, and other topographic features; indications of problems in obtaining windfirm boundaries adjoining clearcut settings.
3. Conditions affecting priority of cutting: tree maturity, health and vigor, evidences of insect or fungus attack, and windthrow.
4. Barometric elevations of potential landings and road control points. These are pin-pricked on the photographs and the elevations recorded on the back of the photo, as are bench marks or other points of known elevation.
5. Land survey corners, such as section, quarter-section, witness and meander corners, and metes and bounds survey corners, are pin-pricked on the photographs.

PLANNING PROCEDURES

For efficient use of time and effort, the planning must proceed in logical steps. Procedures have been developed through experience which attain the best and final plan expeditiously without neglecting any of the considerations in logging planning. Procedures vary with the logging methods to be used, the kind of topographic information available, and, of course, with the individual doing the planning. Following are the recommended consecutive steps in planning for each of the basic types of logging operation and for pulpwood production.

Planning for Cable Yarding

Planning the logging tracts of old-growth timber growing on steep, rough ground presents the most difficult and challenging problem to the logging planner. Following is the recommended procedure for planning cable yarding operations where clear-cutting is practiced, and large-scale logging plan maps are available. The same procedure may be used with aerial photographs by the skilled photointerpreter.

Priority Sequence. In non-uniform old-growth stands the first step is to mark the priority sequence of cutting and demarcate the boundaries of such areas on the map or photograph. Priority I areas include dead or dying stands of fire-killed, insect-infested, or diseased trees, and over-mature or decadent stands. These are the areas which should be cut first. Priority II areas are mature stands on the steeper or rougher slopes. If staggered settings are to be cut, the reserve settings should be left on the less steep slopes for easier salvage of windthrow along the cutting lines. Priority III areas are the remaining mature stands. Priority IV areas are immature stands, which would be the last to be cut. Priority areas are determined by study of the type map, the aerial photographs and cruise data and confirmed by field reconnaissance.

If the cutting system has not previously been dictated by company or agency policy, study of the priority areas will indicate the preferable system. Possible alternatives are clear-cutting by staggered settings, by strips of consecutive settings leaving uncut parallel strips, or by progressive cutting of adjacent settings leaving seed blocks. Where the staggered setting system is used (Fig. 2–3) the first cutting sequence would take the priority I areas plus enough settings in priority II areas to obtain the volume desired from the tract or to keep the initial unit road construction cost within reasonable limits. Highlead setting areas are usually 30 to 40 acres, with a maximum clear-cut unit of 60 to 80 acres. If natural regeneration of the logged settings is depended upon, the uncut settings are left as a seed source until after the logged settings are regenerated. If the logged settings are seeded or planted, the leave settings are cut after the new crop is well established. The staggered setting system also has advantages in fire protection as it keeps the **slash** areas relatively small and surrounds them with green timber. The re-

Fig. 2–3. Clear-cut staggered settings in the Douglas-fir region. Mt. Rainier in the background. (Courtesy of the Weyerhaeuser Company.)

maining priority II areas and some of the priority III areas are taken in the second cutting sequence. The third priority sequence would take the remaining merchantable timber. If natural regeneration is depended upon, the interval between cutting sequences is about 10 years in the Douglas-fir region. The staggered setting system is practiced on some industrial forest lands. However, there is a trend toward cutting long strips a setting in width, leaving uncut strips on each side of the logged strips to provide a seed source. While the slash areas are larger, the logged strip parallels the contour so that the shorter dimension is on the slope. Fire travels faster up the slope than along the contour. The first strips are laid out through the priority I and II areas. The strip system has the cost advantage in the first cutting sequence of reducing by $\frac{2}{3}$ to $\frac{1}{2}$ the length of road required by the staggered setting system. The strip system also reduces the length of the perimeter exposed to windthrow. On other private lands, planning is done to comply with the state forest practice laws. For example, in the state of Washington, eight acres in each clear-cut quarter-section (160 acres), strategically situated for seeding the cut area, must be left in seed blocks unless the land owner plans on artificial regeneration and deposits a performance bond with the state. The priority IV areas of immature timber are the first choice for seed blocks if of cone-bearing age. Second would be the more open stands, as they are more windfirm than dense stands and leave less volume. Seed blocks of mature timber preferably should be left where they can be salvaged after regeneration of the cutover area. However, often one finds seed blocks left where they cannot be reached economically. This may have been due to unimaginative planning, but the block may have been inaccessible with the logging methods which were available.

Landings. Good landings are important to safe, efficient operation. Natural flats suitable for landings are usually scarce where the topography necessitates cable yarding. Therefore, the second step in planning is to find and mark prospective landings with an "X." Natural landings are found on benches, on ridge crests, and in flat valley bottoms. The side slope on highlead landings should not exceed 20 percent. The preferred minimum landing circle radius around the spar is one and one-half the maximum log length plus the width of the road. If a landing must be excavated on a steep slope, the minimum safe radius is equal to the maximum log length.

With a pencil compass set to scale at the most efficient external yarding distance for the yarder and height of spar tree to be used, circles are drawn around each landing marked. Some of the circles will overlap and others will be too far apart. The landings which are correctly spaced are marked with the highlead landing symbol —an X in a circle. The landing spacing for mobile steel towers is less than the spacing for spar trees and settings are rectangular in shape. A transparent plastic templet is a convenience for outlining such settings. The templet width is the average landing spacing and the length is twice the external yarding distance. A hole is drilled in the templet for centering over a landing.

If any areas are suitable for tractor yarding, their boundaries are delineated. Such areas are found on slopes under 30 to 35 percent on non-erosible soils. Tractor yarding is usually cheaper than moving cable yarding on ground suitable for tractors. Landings are located at the lower margins of the tractor settings. An ideal combination, where topography permits, is to highlead the lower side and tractor yard the upper side to the same landing. Such a combination side gives greater production and lower loading cost. Tractor landings should be larger than highlead landings to afford room for dropping each turn of logs without delay. For safety from rolling logs the side slope on the tractor landing should not exceed about 10 percent. The conventional symbol for marking a tractor landing on the logging plan map is a small rectangle.

Road Projection. The third step is to project road routes to the trial landings to ascertain whether they can be reached within the grade limits of the standard of road. Methods of route projection are given in Chapter 3, *"Forest Road Engineering."* Some of the trial landings may be found to be inaccessible by road. They are marked "CD" for potential cold-decks to be swung to a landing. Some settings may have no natural landing flats, so landings would have to be excavated. This increases logging cost, exposes more soil to erosion and removes land from tree production. Landings which can be reached by road having been ascertained, the next step is to attempt to develop a systematic road pattern from the random pattern generally resulting from the initial road projection. A systematic pattern of roads generally requires less total length of roads than a random pattern. The systematic pattern also tends to reduce the area lying beyond optimum external yarding distance. Road patterns are illustrated in Fig. 2–4. Classification of logging roads by the following three systematic patterns was first proposed by Silen in 1955.

1. Parallel climbing roads connected by a single climbing road of maximum allowable gradient. The length of climbing roads is a minimum with this pattern. The contour roads parallel the main drainage on a **"grade contour"** of low gradient. The spacing of contour roads for cable yarding is twice the optimum external yarding distance.
2. Parallel climbing roads taking off on switch backs from a main road in the bottom of the valley. They may also take off in the same direction on a steeper gradient than the valley road.
3. Ridge and valley road pattern. The road follows up the valley floor until topography is favorable to climbing to the parallel ridge. The road then follows the crest of the ridge. This pattern is adapted to valleys with steep side slopes where construction of contour

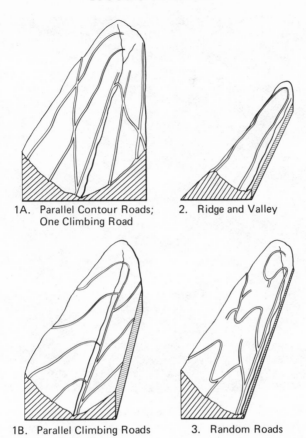

1A. Parallel Contour Roads;   2. Ridge and Valley
One Climbing Road

1B. Parallel Climbing Roads   3. Random Roads

Fig. 2–4. Road patterns. (Courtesy of the Bureau of Land Management, U.S. Department of the Interior.)

roads is uneconomical and the distance between ridge crest and valley is not more than twice the external yarding distance.

The random pattern of roads generally is the result of not planning far enough ahead. However, where the land is steep or rough or interspersed with rock cliffs or slides, and natural landings are scarce, the random pattern may be the only solution. To reach good landing sites it may be necessary to use short stub spur roads (Fig. 2–5).

The road pattern having been developed, the final landings along the roads are marked. Settings which cannot be reached with a road must be cold-decked and swung to a landing for loading. In

Fig. 2–5. Stub spur road to a highlead landing (*lower*) from the main road (*upper*). (Courtesy of the Department of Natural Resources, State of Washington.)

planning for a skyline swing, the principal consideration, after the swinging distance permitted by the equipment is obtaining adequate deflection (see "Planning for Skylines"). For swinging by wheel or crawler tractor, the swing road on a favorable gradient is projected in the same manner as a truck road.

Setting Boundaries. The final step is delineating the setting boundaries. **External yarding distance** circles are drawn around the final landings. The topography near the points where the yarding circles touch or intersect is studied to find the boundaries which will best conform to the physical requirements of the yarding equipment. The planner must visualize both the vertical and horizontal projections of the **yarding roads,** or the paths the **turn** of logs will follow. In profile, an unobstructed line of sight between the tail block and the mainline lead block on the spar tree is essential. In plan, yarding across a truck road is to be avoided because the road would be damaged and blocked to traffic during yarding and afterwards by debris. For highlead yarding parallel to the contour on steep slopes, the path of the turn is the resultant of two forces. These forces are the line pull toward the spar tree and the force of gravity tending to make the turn roll or slide sidewise down the slope. It is highly desirable that the planner have a background of practical experience as a member of a yarding crew in order to visualize the yarding operation.

If staggered setting system of clear-cuts is used, the selection of windfirm setting boundaries is of paramount importance. Losses from windthrow along the boundaries of leave settings in the Douglas-fir region have been extensive. Where the prevailing storm winds are from the southwest, the most windthrow occurs on the north and east boundaries of the clear-cut setting. Windthrow is more severe on the lee or sheltered side of a ridge than on the windward side. This is attributed to "lee flow" or the tendency of wind blowing over a ridge to flow down the slope with increased velocity. Wherever the wind funnels, as through a saddle in a ridge or a narrowing indentation cut in the trees, the velocity is accelerated. Trees growing on shallow or poorly drained soils and trees infected with root rots or butt rots are susceptible to windthrow. Consequently, windfirm setting boundaries are selected on the windward sides of ridges; on well-drained, deep soils; and in sound healthy trees. Hardwoods or mixed hardwoods and conifers and young stands are generally more windfirm than old stands of conifers. Poorly stocked or open stands are more windfirm than dense stands. Figure 2–6 illustrates the relationship between topography and windthrow.

Stream protection is another consideration in demarcating setting boundaries. Yarding across streams disturbs the stream bed, causing siltation. Slash tends to accumulate in the streams. These conditions are detrimental to both water and fish resources. Yarding should be done away from streams so far as possible. On erosible soil an uncut filter strip should be left between the cutting line and the stream. In multiple use forests, where recreational travel is anticipated, a scenic strip is left between the clear-cut boundary and the main road, and spur roads take off on a curve to obscure the logged area from the main road. Setting boundaries on steep slopes

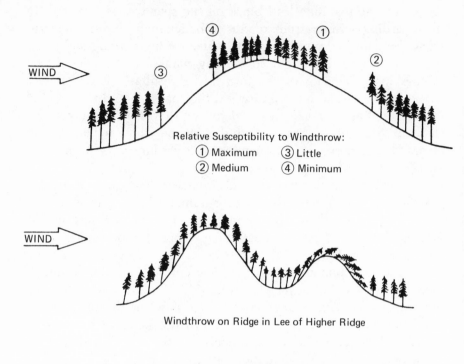

Relative Susceptibility to Windthrow:
① Maximum   ③ Little
② Medium   ④ Minimum

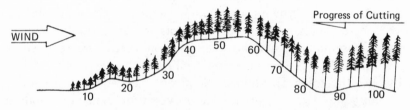

Windthrow on Ridge in Lee of Higher Ridge

Distribution of Age Classes at the End of the First Rotation,
Progressive Strip Cutting.

Fig. 2–6. Relationships between topography and windthrow. (Courtesy of the Bureau of Land Management, U.S. Department of the Interior.)

where the fallers would have difficulty in holding trees from sliding into an adjacent leave setting is to be avoided.

Delineation of the setting boundaries may indicate the desirability of changes in landing locations. Further revision of landings, roads, and setting boundaries is made until the best possible paper plan is achieved.

Field Check. The paper plan is checked in the field. The amount of checking required depends upon the accuracy of the topographic data. Plans developed from aerial photographs require more field checking than plans made from large scale topographic maps. The details to be checked are as follows: (1) *Road routes.* An Abney level grade line is run along the projected road routes to ensure that the road can be located where planned. (2) *Landings.* Proposed natural landings are checked to see that they are suitable. Sites of landings to be excavated are inspected for the presence of rock which would greatly increase the cost. (3) *Boundaries.* Setting boundaries are checked for windfirmness.

If the field check indicates the desirability of any changes in the logging plan, corrections are made on the logging plan map. The settings to be logged in the first cutting sequence are marked with Roman Numeral I, lettered alphabetically, or numbered with arabic numerals in the proposed order in which they will be logged. If the first cutting sequence involves several years' operation, the year in which it is proposed to log the setting is added. An overlay tracing of the roads, landings, and setting is drawn. This can be combined with the topographic map tracing to print final logging plan maps.

Planning for Skylines

A **skyline** is a cable stretched between supports on spar trees for yarding logs with a carriage which rides on the skyline. The various skyline systems used in North America are described in Chapter 8, Cable Yarding. The older systems drag the log turn with only the front end lifted to clear obstacles. The newer systems, termed **skyline cranes,** carry the turn suspended from the carriage clear of the ground. Single span systems are rigged between a head tree or tower at the landing and tail trees at the back end of the setting. **Multispan** systems have intermediate supports and require an open-side carriage which can ride over the supports. In the systems referred to as **"running skylines"** the carriage rides on the haulback line and the load is divided between the haulback and the mainline.

Single-Span Skylines. The most important consideration in planning for single-span skyline yarding is to select landings and setting boundaries so that the profile of the ground between them will allow adequate deflection or sag in the skyline. The ideal profile along the skyline road is concave. A skyline can be used on a uniform slope if the spar trees are tall enough to permit ground clearance with deflection large enough to enable the skyline to carry the desired payload. A convex profile requires a multispan system. Obviously, good topographic maps or aerial photos are essential for skyline planning.

The load-carrying capacity of a skyline decreases with decrease in deflection and with increase in the slope of the ground. Since the tension in a skyline is the maximum when the load is at midspan, deflection is measured vertically in feet at this point, and is expressed as a percentage of the **span,** which is the horizontal distance between supports. In situations where the skyline would have to be tightened to a small deflection, it is necessary to compute the payload capability to ascertain whether that location is feasible. The deflection required for ground clearance is found graphically (Fig. 2–7). The profile of the ground along the proposed skyline

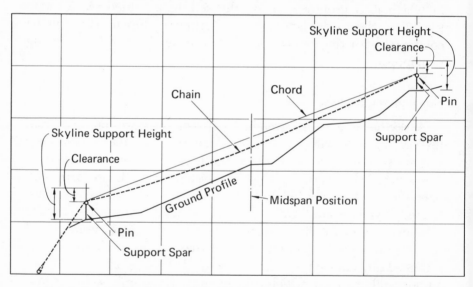

Fig. 2–7. Profile plot for determining allowable skyline deflection. (Courtesy of the U.S. Department of Agriculture, Forest Service, Pacific Northwest Forest and Range Experiment Station.)

road is plotted on cross-section paper at a scale of 100 feet to the inch, both horizontally and vertically. A light fine-link or beaded chain is used to simulate the skyline, as it will assume the same catenary as the cable. It is supported by push-pins at the spar tree locations at elevations equal to the skyline support height minus the required ground clearance. The ground clearance needed will vary with the system. A small weight is hung on the chain at the critical point on the profile, which is found by trial when sliding the weight along the chain, and the chain is adjusted so it just clears the point. The weight is moved to mid-span and the deflection measured.

The computation of the payload capability is greatly facilitated by using the *Skyline Tension and Deflection Handbook* (Lysons, 1967). This booklet gives, in both graphical and tabular form, the data required to compute tension due to the weight of the cable and tension due to the load at mid-span, for the range of deflections and slopes which are apt to be found. Convenient forms for both single-span and multi-span computations are given, as well as clear instructions and examples. The safe working tension of the skyline is obtained by dividing the breaking strength by the desired factor of safety, usually 3. The tension due to cable weight is computed and subtracted from the safe working tension. This remaining tension capability is divided by tension per kip (1,000 pounds) of load to obtain gross load capability. The weight of the carriage and rigging is subtracted and the remainder is the payload capability. Skyline tensions may also be computed by using the formulas given in the handbooks published by the wire rope manufacturers.

If the payload capability is insufficient to carry the largest log to be yarded, or the desired weight of turn of logs, the proposed skyline road cannot be used. It may be possible to change the location of the spar tree to obtain more deflection. The tail tree may be moved closer to the landing and the tension due to cable weight reduced by the shorter span. Or it may be moved farther away if by so doing a higher elevation can be reached to increase the deflection. If yarding down a steep slope, moving the head tree out from the toe of the slope will give more deflection. Where the landing is in the bottom of a narrow valley, some skyline systems permit the head spar to be located above the landing on the opposite side of the valley from the tail tree.

In selecting sites for cold decks which are to be swung with a skyline system, the resulting profile and its effect on skyline deflection must be considered. Locating the cold deck spar tree well back from the edge of the bench gives more decking area. However, if the ground drops off steeply on the lower side in the direction of the swing, the cold deck may have to be placed near the edge of the bench for clearance of the skyline.

Running Skylines. Running skyline systems usually are operated by mobile yarders and are used to yard parallel strips perpendicular to the truck road. The major problem is to locate the truck roads so that adequate deflection can be obtained. Roads on the crests of ridges, on the outer edges of benches, or on the lower edges of the gentler slopes are favored. The effective maximum yarding distance is usually about 700 to 800 feet, but will vary with the ground profile and the consequent deflection. Forms and instructions for computing tensions in running skylines are given in *Single-Span Running Skylines* (Lysons and Mann, 1967).

Multispan Skylines. Planning for multispan skyline systems does not present the problems in obtaining deflection associated with long single spans because intermediate supports can be placed at points where ground clearance could not otherwise be obtained. However, the design of the skyline layout and the computation of payload capacity is more complicated. Since the skyline is not clamped at the intermediate supports, and is free to slide over the intermediate support jack, the deflection in all spans changes as the carriage moves along the skyline. To design the system, the weight must be moved along the entire length of the profile plot and the height of the skyline adjusted so that ground clearance is obtained at all points along each loaded span (Fig. 2–8). Unless the critical span which limits payload capability is obvious, tensions must be calculated for all spans. The tension due to the cable weight in the unloaded spans above the loaded span must be included in the computation. Helpful guidance in planning for skyline crane yarding systems, both single span and multispan, and in estimating production costs, will be found in Research Paper PNW-25 (Binkley, 1965).

For accurately estimating log weight for use in designing skyline locations, a "density index number" system has been developed by the office of Forest Engineering Research, Pacific Northwest Forest and Range Experiment Station, Forest Service. The density index number (D.I.N.) for a species growing in a given locality is obtained by weighing and measuring sample truck loads of logs. D.I.N. equals log weight with bark divided by volume in cubic feet inside

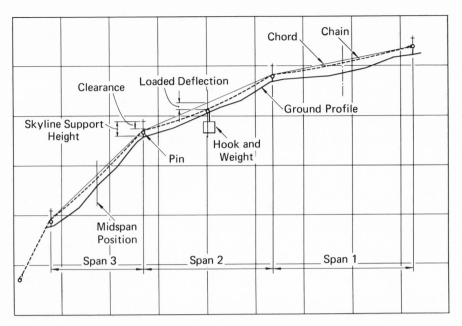

Fig. 2–8. Multispan profile plot for determining allowable deflection. (Courtesy of the U.S. Department of Agriculture, Forest Service, Pacific Northwest Forest and Range Experiment Station.)

bark. It is anticipated that tables of weights of a wide range of log sizes by density index numbers will be published. In addition to use in skyline design, the tables will be useful in instructing the cutting crew in the log lengths to buck in order not to exceed the payload capacity of the skyline. The same Forest Engineering Research agency has developed an electronic computer program for skyline design which will greatly reduce the time required for planning for yarding with skylines. The program reads the ground profile from a topographic map or an aerial photograph, plots a graph of the profile and skyline deflection, and computes the tension and payload capability.

Planning for Mobile Loggers

The efficient skidding distance of mobile loggers, **jammers**, or other combination cable skidding and loading machines is limited by the relatively low elevation of the lead sheave on the end of the boom. Planning for such operations is generally a matter of pro-

jecting parallel contour roads at a spacing which will give the
desired skidding distance. For the Idaho jammer, the customary
skidding distance below the road is about 400 feet. On the upper
side, skidding is limited to the distance the jammer can cast or
throw the tongs, as pulling tongs and line uphill is laborious and
inefficient. The road spacing is usually about 500 feet. The Big
Stick Highlead, used for pulpwood production in the mountains of
the Tennessee Valley, is limited by drum capacities to 400-foot
skidding distance, so the road spacing is not more than 800-foot
slope distance. For mobile loggers the yarding distance on the
lower side of the road is usually about 500 feet. Although these
machines have a haulback line to pull the butt rigging and mainline
out, the lack of lift above the elevation of the lead sheave limits
efficient yarding above the road to 100 or 200 feet. Road spacing is
thus 600 or 700 feet. On an experimental operation for the salvage
of dead and down timber in an old-growth stand with a mobile
logger, rock-surfaced roads were spaced 1,000 to 1,500 feet apart.
Paralleling earth roads of lower standard were built between them,
giving a spacing of 500 to 750 feet. Yarding to the earth roads was
done only in the summer dry season. Landing spacing averaged
about one-half the external yarding distance. A description of this
operation and analysis of the costs is given in Carow (1959). A
paper company logging steep ground in western Washington yards
up to 800 feet with Washington Trakloaders.

Planning for Balloon Yarding

Balloon yarding distance is limited only by the line capacities
of the yarder drums. Due to the high line speeds of balloon yarders,
yarding distance is an unimportant factor in production. Balloons
have been used for yarding up to one mile, and even longer distances, from tidewater to timberline, are being considered in Alaska.
Since the log turn is flown through the air, the ground profile along
the yarding road is of little concern. However, the location of **corner
blocks** and **tail blocks** — to afford clearance for the **haulback line**
running between them — may require engineering planning. Balloon
yarding is usually done downslope, and the truck road is located at
the toe of a long slope or along the bottom of the main valley. The
crest of the ridge which forms the divide between major drainages
is usually the upper boundary of the setting. **Bedding ground** sites

Fig. 2–9.　Raven logging balloon tied down at bedding ground. (Courtesy of Raven Industries, Inc.)

for tying down the balloon for replenishment of helium gas loss, or during wind storms, must be planned (Fig. 2–9).

Planning for Tractor Skidding

The planning of logging operations in the pine regions, where tractor skidding predominates, is customarily done on aerial photographs. In the more open pine stands, enough of the ground can be seen to permit truck roads to be projected accurately. The trial plan is best drawn on a transparent overlay of one of a stereo pair of photographs under a mirror stereoscope. The first step is to demarcate the areas suitable for tractor skidding from areas where terrain or soils require other methods. The tractor skidding areas are stratified by volume per acre to be cut as this is an important variable in road spacing. Patches of immature timber to be left undisturbed are marked. Since partial cutting is generally practiced in pine forests, the volume per acre to be cut will depend upon the policy of the land owner. In ponderosa pine where the marking of trees to be cut is generally based on risk of bark beetle attack, and sometimes on mistletoe infestation, the health and vigor of the stand will determine the amount to be cut. Preferred practice is to mark the trees to be cut before the roads are located. If marking is deferred until after the roads are located, the volume per acre previously cut in comparable stands is used as a planning guide.

The second step is to determine the **optimum road spacing.** This is the spacing, or distance between parallel truck roads, which will give the lowest cost per unit of volume of skidding plus road construction. At the optimum road spacing unit, variable skidding cost will equal unit road construction cost. This can be demonstrated by plotting a graph such as is illustrated in Fig. 2–10. The vertical axis is cost in dollars per unit of log volume measurement, which may be cords, cubic feet or board feet. The horizontal axis is road spacing in "stations" of 100 feet. The variable skidding cost plot is an ascending straight line. It is termed a variable cost because it varies with the average log volume. The larger the logs the larger the turn volume, and, consequently, the lower is the skidding cost per unit of volume. Variable skidding costs are based on time studies which reveal the effect of log volume on production and cost. In the example given in Fig. 2–10, if roads are spaced 10 stations apart, the external skidding distance will be 5 stations and the average skidding distance 2.5 stations; with a variable skidding cost of $0.40 per unit the skidding cost is $1.00. The cost per unit of the truck road at increasing road spacing is a descending curve. Points on the curve are obtained by dividing the cost of one station of road by the volume of timber on the area served. In the example, at a road spacing of 10 stations, each station of road serves a strip 5 stations long or 1.15 acres on each side of the road. At 12 M bd. ft. per acre the total volume served per station of road is 27.6 M bd. ft.

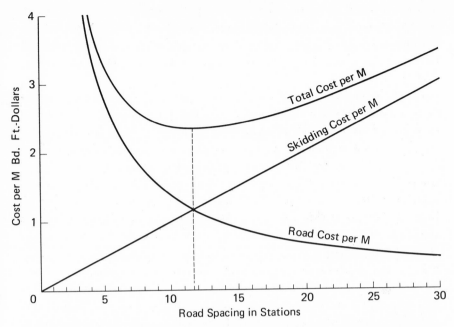

Fig. 2–10. Chart for finding optimum road spacing. Skidding direct to the road, from both sides, for loading by mobile loader. Variable skidding cost is $0.40 per M board feet per station; road construction cost $38.00 per station. Volume per acre cut is 12 M bd. ft. Break-even point of 11.8 stations = optimum spacing.

Dividing the road cost of $38.00 per station by the volume served gives a unit road cost of $1.38; adding unit skidding and road costs gives a total unit cost curve. The low point of the curve will be at the road spacing where skidding and road cost curves intersect. In the example, this point is at the road spacing of 11.8 stations, which is the optimum spacing for the given conditions.

Optimum road spacing may be calculated by a number of methods. The most comprehensive is the ratio method developed by Wallace and published in the *Journal of Forestry* (Wallace, 1957). The reference is available in every forestry library which maintains files of this journal. Ratios are given for a wide range of conditions of landing spacing and landing costs, as well as skidding and road costs for both one-way and two-way skidding. The optimum combination of landing spacing and road spacing may be calculated by the Wallace ratio. As the road spacing increases, average skidding distance and variable skidding cost increases. Unit road costs decrease with wider spacing, as each segment of road serves a larger acreage of land. The data needed for road spacing computations are volume to be cut per acre, skidding cost per station (100 feet), and road construction cost per station. If skidding is to landings, rather than to the nearest point alongside the road, the cost of making landings is required. Skidding costs are usually based on time study data. As skidding cost varies with log size, the average log volume is obtained from cruise data. Road construction costs are estimated from previous experience in similar terrain. Optimum road spacing computations are of the most value when planning for crawler tractor skidding on flat or gentle topography where roads can be built anywhere. The faster travel speed of the wheel tractor reduces the importance of skidding distance as a production factor in wheel skidding. Where roads and landings are determined by topography, the optimum road spacing computation is only a general guide to the average spacing to be sought. In some operations the determining factor in skidding distance and consequent road spacing is balancing tractor production with loader capacity. For example, one company, which skids to landings with crawler tractors, found that the balance is achieved when external skidding distance is 400 feet per tractor. When three tractors per loader are to be used, a maximum skidding distance of 1,200 feet is planned. Where logs are to be windrowed along the edge of the truck road for loading by a rubber-tired loader — after enough logs have been skidded to enable the loader to operate full capacity — optimum road spacing should be computed.

The third step is to project a systematic parallel road pattern. The first road will ordinarily be located along the main valley bottom or at the lower edge of a long slope. On flat ground the best location for the main road will first be determined, and spur roads projected at optimum spacing perpendicular to the main road. On sloping ground contour roads, approximately parallel to the first road, with the average distance between roads at the computed optimum spacing, are projected. Finally, the parallel roads are connected by a climbing road. If the topography is characterized by narrow branch valleys, the road system may consist of a series of spur roads following up the branch valley bottoms from a main road in the principal valley.

If the logging plan is predicated on the use of landings, the fourth step is to mark the location of the landings at the optimum landing spacing. The boundaries of the area from which logs are to be skidded to each landing are delineated. Finally a field check is made to ensure that the truck roads can be built as projected and that the landings planned are satisfactory.

While tractor logging is more flexible than moving-cable logging, planning for operating flexibility — to meet changing conditions of weather or market demand — may demand special attention. For example, in the California pine region some companies log white fir and Douglas-fir in the spring and summer for the dimension lumber market, and log pine in the autumn — when the pine is less susceptible to blue stain — for winter sawing when pine lumber prices may be higher. Planning the operation so that logging can begin at the back end of a spur road and progress toward the main road will minimize interference of skidding with trucking. This practice is particularly desirable with earth spur roads which are also used in part as tractor roads.

While the routes traveled by the skidding tractor are ordinarily left to the judgment of the tractor driver, under some conditions it is advisable to plan the tractor road system and mark the roads in the field. Advance planning of skid roads is desirable to minimize soil disturbance and concentration of runoff on erosible soils, and to protect scattered patches of young growth. Where climbing roads are to be bulldozed on steep side slopes, their locations should be planned.

Aesthetics

Where aesthetics is to be considered in logging planning for forests used by the public for recreation, the planner endeavors to present as natural a scene as is feasible. Clear-cut settings may be screened by leaving strips of uncut timber bordering the main roads which will be travelled by recreationists, or the roads and settings may be located so that topography will assist the screening. Screening is illustrated in Fig. 2–11. The uncut strip must be wide enough to be windfirm. Landings are located on stub spur roads which leave the main roads on curves so that the logged setting is obscured from view. Where clear-cut settings will be visible from a distance, the use of irregularly curved boundaries gives a less artificial appearance than does straight squared-off cutting lines. The upper and lower boundaries may be naturally curved by following

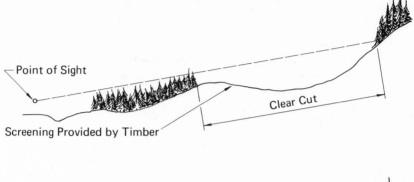

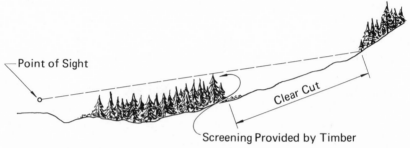

Fig. 2–11. Clear-cuts in the back foreground portion should be screened, at least in part, from the viewing area whenever possible. Take advantage of topographic features and timber screening existing in the foreground area. (Courtesy of the U.S. Department of Agriculture, Forest Service.)

the contour. On steep slopes, clear-cut settings may imitate the paths of slides or avalanches, which are often a natural part of the mountain forest scene. Where yarding is done with skylines, long narrow strips may be cut perpendicular to the contour. For other cable yarding methods, triangular-shaped settings are preferable. Shapes of clear-cut settings advocated in the *Forest Service Manual* are illustrated in Fig. 2–12.

Where the timber on both sides of a road is to be logged, for aesthetics the lower side should be logged first, leaving the upper side until the lower side is reforested.

Where partial cutting is practiced, considerations of aesthetics may necessitate making lighter cuts at shorter time intervals than would otherwise be made. Monotonous stretches of similar forest scenery may be avoided by varying the degree of cutting. Landscape architects advocate leaving "special interest" trees, such as wolf trees, or malformed trees which would ordinarily be marked

UNDESIRABLE

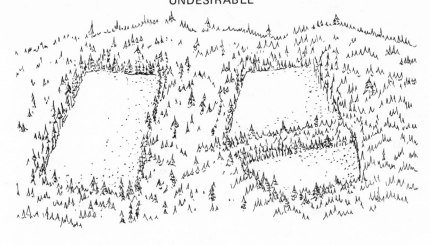

GOOD

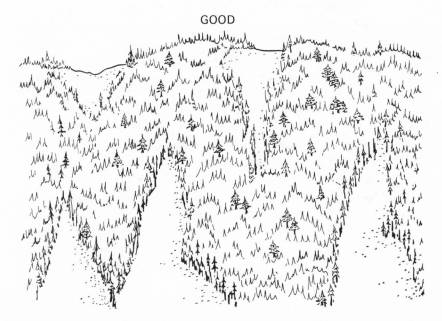

Fig. 2–12.　Clear-cut boundaries designed in the shape of avalanche paths appear much more natural than squared-off straight-sided boundaries. (Courtesy of the U.S. Department of Agriculture, Forest Service.)

for cutting. Disposal of all unsightly slash or debris following road construction and logging is, of course, an important part of the aesthetics program.

In the Forest Service the landscape architect joins with the road engineer in planning to make the forest roads aesthetically more pleasing to the recreationists. The road may be located to reach points affording scenic vistas, such as valleys, lakes, or interesting rock formations. Moving vistas are provided by openings cleared at a narrow angle with the road, and stationary vistas are created at a wide angle from parking turnouts. The tunnel effect of straight lines of trees bordering the edge of the road right-of-way clearing is relieved by irregular cutting lines and openings. Similar treatment is applied to the timber edge at the top of cut banks to deemphasize the artificial appearance of the cut slope.

Allied with aesthetics is planning to avoid water pollution from road construction and logging. Roads are located far enough away from main streams so that ravelling of embankments will not cause stream siltation. Landings are selected so that the logs will be yarded away from streams and not across them. Tractor landings are placed where skid trails leading to them will not cross streams. In watershed forests where increasing the quantity of water is desired, clearcuts are laid out in narrow strips in an east-west direction. Such strips will enable more snow to reach the ground, and provide shade to lower the rate of melting. The width of strip recommended in the Forest Service Manual varies from one-half tree height on south slopes, and twice tree height on flats, to five times tree height on north slopes. Partial cutting to reduce stand density may also be used to increase water quantity. The cutting creates openings for snow accumulation, and reduces the amount of interception, evaporation, and transpiration of water. The uncut trees shade the openings. Heavier cutting to aid water supply may well conflict with lighter cutting desirable for aesthetic reasons.

Planning Pulpwood Operations

In planning pulpwood production from large tracts of flat or gently sloping ground, there is no topographical limitation on road location. The main problem confronting the planner is determining the plan which will give the lowest total production cost. If data are available from time and cost studies, or from operations research, on production standards and unit costs of the various machines or methods to be used, and construction costs of various standards of roads, the optimum plan can be determined mathematically. Items to be computed may include optimum spacing and standards of main and spur truck roads and skid roads; optimum spacing of location of landings, skidways and slasher sites, if used; and the break-even point between skidding and hauling. If the data necessary for operations research solutions of the problems are not available, then the planner must proceed by trial, making alternative plans and estimating costs until the most economical plan is found. Plans for production on flat ground are often made in the field by the producer or the foreman, based on experience and customary practice. Whether the optimum plan is achieved depends upon his judgment.

Rules of thumb developed from operating experience are the guides used for planning skidding distance and road spacing. For example, in the operations of one company in the Northeastern region, branch roads are laid out perpendicular to the main road at a spacing of 500 to 600 feet for horse skidding and 1,300 to 2,000 feet for tractor skidding. In the Lake States region where 8-foot wood is cut and piled during the summer, and forwarded during freezing weather, trails 8 feet in width are cleared perpendicular to the truck road at a spacing of 60 to 90 feet.

For planning pulpwood production on steep or broken ground where cable systems are to be used, the procedures are similar to those given for planning for cable yarding. Planning for animal skidding is just as important as planning for machines. This fact may not always be recognized by pulpwood producers because of the smaller capital investment in animals. However, when one considers the physical limitations of animals and the relatively lower productivity of the teamster as compared with the machine operator, the value of planning is evident. The planner should know the size of the animals and the skidding accessories to be used and the relationships between variable skidding time, distance, and slope. For example, it makes a difference whether 1,000-pound farm horses or 1,500-pound woods horses, trained for skidding, are to be used. The spacing of haul roads and of landings or skidways and the boundaries of the cutting area for each skidway as related to skid trails are subject to optimum determination.

Planning by the Pulpwood Producer. Advance planning to achieve the most economical operation is as important for the pulpwood producer who cuts small, scattered tracts of timber, as it is for the large logging company. The following consecutive steps in planning are suggested to ensure that nothing be overlooked which would affect the subsequent operation. It is assumed that the producer has familiarized himself with the tract before entering into an agreement to purchase the timber, or to contract the operation, and has obtained the information on timber distribution and volume necessary for planning.

1. Get the legal description of the boundaries of the tract to be cut and the names of adjacent landowners from county courthouse records.
2. Collect all available maps and aerial photographs. Maps which may be available include U.S. Geological Survey topographic quadrangles, Forest Survey timber-type maps, and county road maps. Among possible sources of aerial photographs are the state forest department, federal agencies, such as the Soil Conservation Service and the Forest Service, and pulp companies owning forest land in the vicinity of the tract.

   One does not have to be an expert photo-interpreter in order to make use of aerial photographs in planning. By taking them into the field and correlating features on the ground with their appearance on the photographs, the ability to obtain helpful information on timber and terrain may be developed.
3. Locate and clearly mark the boundaries of the tract to avoid trespass on adjacent ownerships. Colored plastic flagging tape, obtainable from suppliers of forestry or surveying equipment, is the most convenient material for marking lines.
4. Drive the existing roads to find the best access to the tract and the route to the concentration yard or other pulpwood market.
5. On the best map available, sketch any haul roads or prehauling trails required to be built, and landings or truck loading sites. Check the plan on the ground and make any changes indicated. Among the factors to consider in making the plan are topography, soil types, varying densities of timber within the tract, cost of road and trail construction, and the economical skidding or prehauling distance of the equipment to be used.
6. Mark the center line of roads and trails to be built. This may be done by blazes on trees, by stakes, or by plastic tape of a different color than that used for marking the boundaries. Obtain rights-of-way if a road must cross other ownerships.

7. Plan the sequence of operations. This includes the roads and trails to be built in advance of cutting, where to start operations, the order in which the areas served by the various landings are to be cut, and the moving of the equipment. The plan should be flexible to accommodate changes in weather, in order to attain the objective of sustained average daily production.

LOGGING PLAN FIELD LAYOUT

Layout for Cable Yarding

The layout in the field of the logging plan for a cable yarding operation, which has been projected on a map or aerial photograph, generally involves locating the truck roads, marking spar trees, or landings, and marking and traversing setting boundaries. Establishing the projected road on the ground requires identifying controlling points, running an Abney level grade line between them at the computed gradient, and tagging the final grade line. The grade line is followed with a preliminary line (P-line) traverse for a designed road or a direct location survey (see Chapter 3, Forest Road Engineering). At the landing site the road should be located so as to allow room for the log turns to be landed and safely unhooked. If a growing tree is to be used as a spar, the road is located at one-half the subgrade width from the tree. On a steep side slope the width to be excavated for a landing is noted on the road plan-and-profile sheet, as the landing should be built when the road is constructed. When locating climbing roads, particular care must be taken to reduce the gradient when approaching a landing in order to allow room for a vertical curve. The maximum road gradient on the landing depends upon the loading method. If the truck has to shuttle back and forth during loading, the gradient should not exceed 5 percent. For drainage of water from the landing, the minimum gradient is 1 percent. A turn-around is located at a convenient place so the empty log trucks can reverse direction and back up to the landing.

If a tree growing on the landing is to be used for a spar tree, it is marked with a big "X" chopped in the bark. This conveys to the cutting crew the message: "Do not cut this tree." The largest, tallest, straight sound tree of suitable species is selected for the spar. In over-mature timber subject to conk rot, one or two additional trees are marked, in case the first tree proves to be defective when topped. If no tree suitable for a spar is found on the landing site, a tree termed a **dummy** is marked to be rigged to provide the lead to raise a spar which is dragged to the landing, usually by a crawler tractor. Or a suitable tree growing on the landing may be "jumped" to a better position for yarding by being topped, guyed, cut off at the stump, and moved horizontally in an erect position. If a steel tower is to be used, the landing is marked with a flagged stake.

Finally, the setting boundaries are established. Because of the importance of windfirm boundaries where the staggered setting system of clear-cutting is used, a boundary reconnaissance is first made. Using map or photograph as a guide and hand compass and pacing for control, the projected boundary is followed. It is flagged at intervals with plastic tape for guidance in running the final cutting line. The setting boundary is revised as indicated by the requirements of the yarding system and to obtain windfirm margins of the adjacent settings. Then the cutting line is blazed or tagged and traversed for plotting on the map and computing the acreage. Public timber sale units are usually traversed by staff compass and chain. On private lands, a hand compass and pacing traverse is made or sometimes the boundaries are sketched on a map using topographic features as a guide.

Skyline Layout. Skyline systems require more field engineering than do other cable yarding systems. Tail trees are selected and marked, with consideration given to how the skyline tailhold can be anchored. The skyline tension is greatest at the upper support, and if available anchor stumps are small, a system of tiebacks to additional stumps has to be designed. Where deflection appears to be critical, a profile of the ground along the proposed skyline road is made with Abney level and tape. The payload capability is computed, and, if it is inadequate, the position of the tail changed to increase deflection. Multispan systems especially require careful engineering (Fig. 2–13). The skyline road should always be profiled on the ground and alternative positions of the intermediate spars and resultant skyline tensions computed. The intermediate spar trees are marked, or the location staked and dummy trees marked, if spars must be raised. The cutting line on each edge of the strip to be yarded is flagged. On slopes transverse to the skyline, the distance to the cutting line will vary with the slope. For example, if a strip width 200 feet each

Fig. 2–13. Hoisting the log turn to a Skagit Skycar carriage on a multi-span skyline. (Courtesy of the Skagit Corporation.)

side of the skyline on a flat traverse slope is yarded with a skyline crane, on a 70 percent transverse slope the strip width should be limited to 100 feet slope distance on the upper side and increased to 300 feet on the lower side. The maximum size of log turn which can be safely carried on the skyline should be communicated to the foreman of the yarding crew, and to the cutting crew if the payload capability requires bucking large trees into shorter lengths than usual. The head rigger should be informed of the height to rig up the unloaded skyline in order to provide minimum clearance for the log turn. Changes in unloaded deflection due to loading may be obtained from graphs in the *Skyline tension and deflection handbook* (Lysons, 1967).

Layout for Tractor Skidding

Where tractor skidding is to be done downslope from above the road, the truck road is located near the lower edge of the landing to leave as much room as possible on the upper side. For efficient operation the tractor should never have to wait because of lack of room on the landing to unhook the log turn. Where skidding is to be done from both sides of the road, the road is located through the middle of the landing. The boundaries between areas to be logged to different landings are flagged. Where skidding is done to the nearest point on the road, and logs windrowed for loading by mobile loader, turn-arounds for the log trucks are marked during road location and built when the road is constructed. The turnout spacing may vary from 300 feet along roads with steep grades or sharp curves, to 500 feet or more where the driver has good visibility for backing the truck.

If tractor roads are located, the objective is the shortest skidding route to the landing which will avoid damage to the residual stand. On steep slopes which necessitate bulldozing of tractor roads, the optimum gradient which will give the fastest round trip travel time, consistent with alignment which avoids sharp bends, is followed. Marking tractor roads with plastic flagging tape is preferable to blazing trees.

LOGGING COST ESTIMATING

The final step in logging planning is the preparation of an estimate of the cost of building the roads and logging the tract. If alternate plans are made, comparative cost estimates are needed to determine which plan would give the lowest production cost. Most companies require expenditures to be budgeted well in advance, which necessitates cost estimating and also production scheduling. The contractor who is going to bid on, or negotiate, a logging contract first makes a cost estimate, as does the company offering the contract. The decision on whether to log a tract of timber with company equipment and crew, or to contract the logging, is often based on the cost estimates. When a change in equipment or methods is contemplated, estimates of the operating cost with the new system are made. The logging cost estimate is an essential part of the stumpage value appraisal which is made by the timber sale staff of a public forest agency, and by the prospective bidders. Costs are expressed in dollars per unit of merchantable volume by dividing dollars to be expended by the volume which will be produced. The volume unit used is that by which the product is sold and may be board feet or cubic feet for logs, lineal feet for poles and piling, and cords or cubic feet for pulpwood. Since production is affected by logging conditions, which may vary widely between timber tracts, logging cost data which will take into account the effect of the measurable variables are required for cost estimating.

Variables Affecting Production

The production obtained by a given machine or method, and consequently the unit cost of operation, is affected by variations in timber, terrain, and other factors. Following are the variables which commonly affect production for the consecutive steps involved in logging or pulpwood operations.

1. Planning, engineering and layout: The quality of the information available from maps and aerial photographs and the consequent amount of field reconnaissance needed; detail of timber cruise data; terrain conditions (topography, brush, etc.) and the accessibility of the tract; weather during the field work season.

2. Road construction: Length and standard of road; number and size of stumps; type of soil; volume of earthwork and type and volume of rock excavation; length of haul of rock or gravel surfacing material; and drainage structures (ditches, culverts and bridges) required.

3. Road maintenance: Gradient, curvature, and type of surfacing material. Maintenance costs increase with steepness of grade and sharpness of curve, and loss of surfacing rock from dusting or erosion.

4. Felling, bucking and limbing: Average tree diameter and height; average number of bucking cuts per tree; volume per acre; terrain conditions; and walking distances. The percentage of loss in scale from defect and breakage is an important variable in old-growth timber.

5. Skidding or yarding: Average log volume; average skidding or yarding distance; volume per acre; volume per setting; direction up or down the slope; and terrain conditions; and expenditure required for landing preparation. For tractive skidding the ground conditions of type of soil, moisture and snow cover affect production.

6. Loading: Average log size; payload capacity of log trucks; and balance between loader capacity and truck availability. If "hot" loading is done coincident with skidding or yarding, loader output is dependent upon skidding or yarding production. If "cold" loading is done from decks or windrows, the distance the loader moves between concentrations of logs, and the volume in the decks, are production factors.

7. Hauling: Average log volume and consequent payload volume; round trip travel time, which is determined by gradient, curvature, road surface, and turnout spacing on single lane roads and haul distance. Weight restrictions and speed limits on public roads affect log trucking costs. The difference between gross and net scale is an important factor in unit cost of log trucking.

8. Unloading and subsequent transportation: Average log volume is the principal variable affecting the unit cost of unloading or dumping, scaling, booming rafting, and towing, and loading railroad cars, barges, or ships. Transportation cost is proportioinal to length of haul.

Two important variables affecting production which have not been listed are labor quality and weather. Labor quality involves skill, efficiency, motivation, and incentives. Labor accounts for a considerable percentage of the cost of logging. For example, in a study of six second-growth thinning operations in the Douglas-fir region, the labor cost percentage was found to be as follows: Felling and bucking, 84 percent; tractor skidding, 38 to 49 percent; loading, 26 to 44 percent. An analysis of logging costs for fifteen cutting units in old growth showed labor costs to be 70 percent for highlead yarding and 64 percent for mixed methods which included tractor, highlead and skyline yarding. It has been observed that in small contractor operations, where the crew members are partners, or are paid on an incentive basis, the production rate is higher than it is in large company operations employing union labor. Adjusting cost estimates for labor quality requires experience on the part of the estimator.

The weather anticipated during the season of the year when roads will be built, or when the tract will be logged, may require adjustment of estimated costs. Labor efficiency decreases in inclement weather and winter storms may necessitate cessation of operations. Shutdowns during the fire danger season, either by order of the state forester or to comply with the requirements of the operator's fire insurance policy when the relative humidity drops below a specified percent, may reduce anticipated production. In some localities floods from heavy rainfall or melting snow may be expected to halt operations. A margin for risk due to the foregoing contingencies may have to be added to the cost estimate.

The overhead costs of a logging company or an individual operator must be added in order to estimate the total cost of logging a tract of timber. Overhead costs include managerial and supervisory salaries, office expense, and expense items which are usually paid annually, such as taxes, insurance, and license fees. Unit overhead costs are obtained by dividing annual overhead expense by annual volume of production. Other costs which may have to be considered are fire protection, slash burning in clear-cuts, and piling and burning of debris in partial cuts. The cost of moving equipment at the start of operations on a new tract and of moving log production machines between landings is usually charged to the activity in which the machine is engaged.

Sources of Cost Data

Firms engaged in the business of logging maintain monthly cost accounts which show the average unit cost of production for each activity. Such cost accounts are indispensable for business management. However, they do not disclose variations from the average due to differences in logging conditions. They are valid for estimating the costs of logging other tracts only if operating conditions are similar to those present during the period covered by the cost account. Special cost studies which show the effect of measurable variables on costs are made for estimating purposes. The time study method is used by public forest agencies, for use in timber sale appraisal, and by companies which employ operations research staff or consultants. The gross data collection method is used by some companies.

Time Study Method. In the time study method the clocked time required to perform each element of the operation, the volume produced, and the operating conditions are measured in the field and recorded. For example, in making a tractor skidding time study the following data would be recorded: the time for each turn, the number and volume of logs in each turn, the distance the turn was skidded, the machine and crew, the weather, and the terrain conditions, such as slope percent and skidding direction on the slope, soil, brush, and snow. Usually the turn time will be subdivided into outhaul, hooking, inhaul, and unhooking time, and delay time and cause will be noted. If the volume cut per acre is not available from cruise data, the area from which the volume of production was obtained is measured. Time studies are made over as wide a range of logging conditions as possible. To ascertain seasonal variations in production, time studies are made at various seasons of the year.

The field data are analyzed in the office and multiple regression curves plotted for turn time, skidding distance, and log volume for each terrain class. Since a multiple log turn is composed of logs of different sizes, a least squares analysis is made to find the effect of log size on production time. Times in minutes are converted to cost per unit of volume by multiplying by the machine rate per minute. A machine rate is the cost per unit of time of owning and operating a machine. The resulting unit costs are shown in the form of graphs or tables. A graph of skidding costs by log volume for an average skidding distance of 400 feet, published in Adams (1967), is shown in Figure 2–14. Since it is not practical to make compilations for all the variables affecting costs, only those which have the most influence are compiled. They usually include log volume for all log-making and log-handling activities, tree size for felling, and the distance the log is moved for skidding or yarding and trucking. Sometimes data are compiled by terrain class, but more often a description of the conditions where the time study was made is appended. Application of cost study data requires experienced judgment on the part of the estimator to make suitable adjustments. The volume per acre to be cut is an important variable which affects felling and bucking cost, and activities in which unit cost varies with the setting volume, such as moving in equipment, landing preparation, and road construction.

In making felling, bucking, and limbing time studies the following data are recorded: sawing time for felling and for bucking, limbing time, walking time between trees, and delay time and cause; tree diameter, height, log scale and number of logs; length of bole limbed,

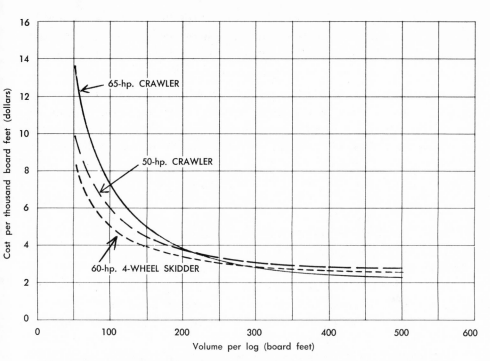

Fig. 2–14. Direct skidding-cycle cost per thousand board feet (for distance = 400 feet), based on log volume by type of skidding machine. (Courtesy of the U.S. Department of Agriculture Forest Service, Pacific Northwest Forest and Range Experiment Station.)

type of limbs and limbing methods; volume cut per acre; and terrain factor relevant to production. In making loading time studies the data recorded are: loading time per load; delay time and cause, such as waiting for logs or waiting for trucks; and number of logs and log scale on each truck load. The equipment, method, and crew is noted in all time studies.

Machine Rate. The **machine rate** is composed of fixed or ownership costs, and variable or operating costs. Fixed costs include depreciation, interest, taxes, and insurance. Repair and maintenance costs may be included in fixed costs as a percentage of depreciation cost, or under variable costs if reliable data are available from owner experience. Annual costs are expressed in costs per hour by dividing by the number of operating hours a year. Variable costs include labor and supplies, such as fuel, lubricating oil, wire rope and other consumable items and are calculated on an hourly basis. The total of fixed and variable costs per minute is the machine rate applied to time study data. Table 2–1 is an example of machine rate calculations for a crawler tractor with rubber-tired arch operating 1,800 hours a year. It is reproduced by permission from Adams (1967).

Gross Data Method. The gross data collection method of logging cost analysis requires reporting output and operating conditions along with the daily time sheet on which each supervisor reports the hours worked by the members of his crew. When enough data are accumulated, the effect of the variable conditions on costs are analyzed. A paper company in the northeastern region used the gross data technique to obtain data for production control, cost control, and forecasting. The camp clerk compiles a weekly report for each crew which shows daily production and cost and coded timber type and terrain conditions, lost time, and weather. The report is reviewed by a company logging engineer and forwarded to a central tabulating department where the information is transferred to computer punch cards. The data on the cards are analyzed periodically and a summary of labor productivity prepared. Duplicate copies of the weekly report forms are available to superintendents, foremen, and logging engineers for current use in production control. A detailed explanation of the technique and sample report form and summary are given in Donnelly (1962).

The gross data method is used by a logging company in the Douglas-fir region for establishing piece work and contract rates, and for stumpage appraisal and budgeting. The daily time sheet submitted by each crew supervisor to the foreman has spaces for reporting where his crew worked, the machines used, and the output, as well as the job and hours worked for each member of the crew. As an example, the hooktender of a yarding and loading side reports setting location by road and landing number, identification numbers of machines used, the man-hours spent in each activity associated with yarding, lost time and cause, and the output in number of logs. Since the logs from each setting are branded or color-marked, log volumes are obtained when the logs are scaled at destination. The supervisor's reports are reviewed and approved by the foreman and turned in to the timekeeper. Terrain conditions and timber stand on the setting are obtained from large-scale logging-plan maps and cruise reports when the data are analyzed by the company forester.

For estimating the cost of logging a setting, a basic point value of

TABLE 2–1. Machine Rate, 65-hp. Crawler Tractor with Arch, Based on 8-hour Day, 1965[1]

Item	Amount	Per Year	Per Hour[2]
		Dollars	
List price, including freight[3]	20,800.00	—	—
Rubber-tired arch, used	750.00	—	—
Residual value, 10 percent of initial price	2,155.00	—	—
Amount to be depreciated	19,395.00	—	—
Average investment for imputed interest[4] $\dfrac{I+R}{2} = \dfrac{\$23,705}{2}$	11,852.50	—	—
Average investment for insurance and taxes[4] $\dfrac{I+R+D}{2} = \dfrac{\$27,584}{2}$	13,792.00	—	—
Fixed costs:			
Depreciation, 5 years $\dfrac{\$19,395}{5}$	—	3,879.00	—
Imputed interest, 6 percent of average investment	—	711.15	—
Insurance, 1 percent of average investment	—	137.92	—
Taxes, 2 percent of average investment	—	275.84	—
Repairs and maintenance, 90 percent of depreciation	—	3,491.10	—
Imputed interest on initial choker investment, 6 percent × $10.50 × 10	—	6.30	—
	—	8,501.31	4.72
Variable costs:			
Fuel, 3.15 gallons per hour × $0.16	—	—	.50
Engine oil, 0.05 gallon per hour × $0.65	—	—	.03
Lubrication, 0.1 lb. per hour × $0.15	—	—	.02
Filters, estimated	—	—	.01
Wire rope: 3/4-inch drum line, 65 feet × $0.456 = $29.64 + $4.65 hardware = $34.29 every 6 months	—	68.58	—
1/2-inch chokers, 6 per year at $6.70	—	40.20	—
Replacement hooks, 3 per year at $3.80	—	11.40	—
	—	120.18	.07
Labor, at $3.075 per hour × 1.25 payroll overhead	—	—	3.84
Crew transportation	—	—	.49
Total per hour	—	—	9.68
Total per minute	0.161	—	—

[1] Horsepower is net engine horsepower.
[2] At 225 days × 8 hours = 1,800 hours per year.
[3] Includes blade, canopy, and winch.
[4] I = initial cost; R = residual value; D = depreciation.

81

relative difficulty is assigned to each factor affecting costs. Terrain factors used are soil, elevation, gradient, character of topography, and underbrush. Ground cover factors are tree height, diameter and volume, stand density in volume and number of trees per acre, windfalls, and defect. Area and yarding distance is noted for each strip which will be assigned to a cutting crew. The point values are totaled and converted to unit cost. The point values remain constant for given conditions. The equivalent costs change with changes in wage rates, in expenditures for supplies, and in ownership costs of machines.

Cost Estimate Example. Following is an example of logging cost estimate for a 160-acre tract in the ponderosa pine region, using time study cost data. Unit costs are given in dollars per 1,000 board feet.

1. Stand conditions: Volume per acre marked for cutting 7,000 board feet. Total tract volume 1,120,000 board feet net scale. Average log 350 board feet gross scale. Net merchantable volume 90 percent of gross volume.
2. Terrain conditions: Moderate slope, light underbrush. Soil suitable for tractor skidding and bulldozer road construction. Patches of young growth to be avoided in skidding.
3. Operating plan: Fell, buck and limb by contract. Road spacing

Direct Costs	Gross Scale	Net Scale	Indirect Costs	Net Scale
Felling, bucking & limbing	$3.60	$ 4.00	Road construction— 1 mile, $5,280.00	$ 4.71
Skidding	4.23	4.70	Road maintenance— 5 mile at 10¢ per M per mile	.50
Loading	2.19	2.43	Engineering	.54
Move in expense $300.00		.27		
Cost on trucks		11.40	Total road cost	5.75
Truck haul—35 miles	8.10	9.00	Administration, supervision, general expense	2.85
Total Direct Cost at Mill		20.40	Total Indirect Costs	8.60
			GRAND TOTAL COST	29.00

$\frac{1}{4}$ mile. Skid one way downslope to nearest point on truck road. Average skidding distance 660 feet. Increase 20 percent for slope and tractor weave to 790 feet. Operating time to skid average log an average distance, 11.6 minutes per thousand board feet. Load with rubber-tired mobile crane. Truck haul 5 miles off-highway, 30 miles highway, by contract at state-regulated common carrier rates.

4. Machine rates: 160 horsepower crawler skidding tractor $0.366 per minute. Mobile loader $0.382 per minute. Use two tractors per loader.

SUGGESTED SUPPLEMENTARY READING

1. BINKLEY, VIRGIL W. and LYSONS, HILTON H. 1968. *Planning single-span skylines.* Pacific Northwest Forest and Range Experiment Station, U.S. Department of Agriculture Forest Service. Research Paper PNW-66. Portland, Oregon. Gives detailed instructions for designing single-span skyline layouts.
2. CARSON, WARD W.; STUDIER, DONALD D.; MANN, CHARLES N., and LYSONS, HILTON H. 1970. *Running skyline design with desk top computers and plotters.* P.N.W. Research paper. Pacific Northwest Forest and Range Experiment Station. U.S. Department of Agriculture Forest Service, Portland, Oregon. A computer program designed by forest engineering research engineers for doing all the calculations involved in running skyline design. Included is automatic plotting of skyline road profiles from a topographic map, and the catenary of the skyline.
3. CARSON, WARD W., DONALD D. STUDIER, WILLIAM M. THOMAS. 1970. *Digitizing topog data for a skyline design program.* Forest Service Research Note PNW-132. Pacific Northwest Forest and Range Experiment Station. U.S. Department of Agriculture Forest Service, Portland, Oregon.
4. LARSSON, GERHARD. *Studies on forest road planning.* 1959. Transactions of the Royal Institute of Technology, No. 147. Stockholm, Sweden. This work examines road spacings, road standards, and methods of transportation in respect to achieving minimum cost per unit of volume of timber.
5. MANN, CHARLES N. 1969. *Mechanics of running skylines.* Pacific Northwest Forest and Range Experiment Station, U.S. Department of Agriculture Forest Service. Research Paper PNW-75. Gives instructions for designing running skyline layouts.
6. PEARCE, J. KENNETH. 1960. *Forest engineering handbook.* 220 pp., U.S. Department of the Interior, Bureau of Land Management, State Office, Portland, Oregon. Covers all aspects of logging planning in the Douglas-fir and ponderosa-pine regions.
7. ROLSTON, K. S. Logging planning and layout. Pp. 157–175. *Proceedings Forest Engineering Seminar II.,* American Pulpwood Association, March 1968. An analysis of the factors involved in planning the pulpwood production operation and their application.

3

Forest Road Engineering

The location and construction of a network of roads are prerequisite to logging and other multiple use forest management activities on most forest lands. Construction of these roads requires large capital investment, especially if they are all-weather roads in mountainous terrain. Log truck roads must be located in order to serve the skidding or yarding operations as well as the log trucking operation. The goal of the logging road engineer is to develop a road system which will minimize the combined costs of skidding, road construction and maintenance, and truck transportation.

The paramount importance of roads to all forest management activities—administration, protection and recreation, as well as timber harvesting—justifies emphasis on engineering. This does not imply that the precision instruments and methods of highway engineering are required. Forest road engineers have developed surveying techniques which are much less time-consuming and considerably less expensive, yet are adequate. Since complete coverage of the subject would be of book length, this chapter will be limited to the fundamentals of road alinement, and to the survey and design methods most commonly used by forest road engineers in the logging industry. References are given to recommended sources of more complete information on the subject. This chapter is included primarily for the benefit of foresters, and other readers concerned with truck roads, who do not take formal courses in the subject.

ROAD STANDARDS

Before a road survey is undertaken, the standards for the road must be known. The classification of the road and the standards for the class are usually established by the forest land owner or the public forest management agency. Among the factors influencing the determination of standards are road construction and maintenance costs, log trucking costs, and the volume of timber on which the cost of the road will be amortized. The anticipated traffic density is often a determining factor. On multiple-use forests, the traffic due to recreationists, local farmers, stock men, and miners, must be considered in addition to log trucks and logging service and administrative vehicles. The road standards generally set limiting specifications for gradient, curvature, surface width, bank slope ratios, and drainage structures, including ditches, culverts and bridges. Turnout spacing on single lane roads, surfacing materials and depths, and the width of the subgrade—or roadbed before surfacing—may also be specified.

The road standards vary with the classes of roads. Forest roads are classified first as permanent or temporary. Temporary roads are to be abandoned after the logging of the adjacent timber is completed. They are often seeded or planted to bring the land occupied by the road back into forest production. The only limitations on curvature are the turning radius of log truck or equipment hauler, and on gradient the state safety code requirement, usually a maximum of 20 percent. Roads also may be classified as all-weather roads when surfaced with gravel, crushed rock or asphalt, or as summer roads which are unsurfaced and useable only in dry weather.

Roads classified as permanent are those which are planned to be maintained for traffic for many years. They may be further classified as primary or **main** roads and secondary roads. The main road is constructed to the highest standard justified by the density of the traffic anticipated and the volume of timber to be hauled. The route of a main road usually follows the valley of a major drainage. The secondary road system consists of branch roads and spurs. The branch roads connect the spurs with the main road. They usually follow valleys tributary to the main drainage and are constructed to intermediate standards. The **spur** roads are generally short roads to

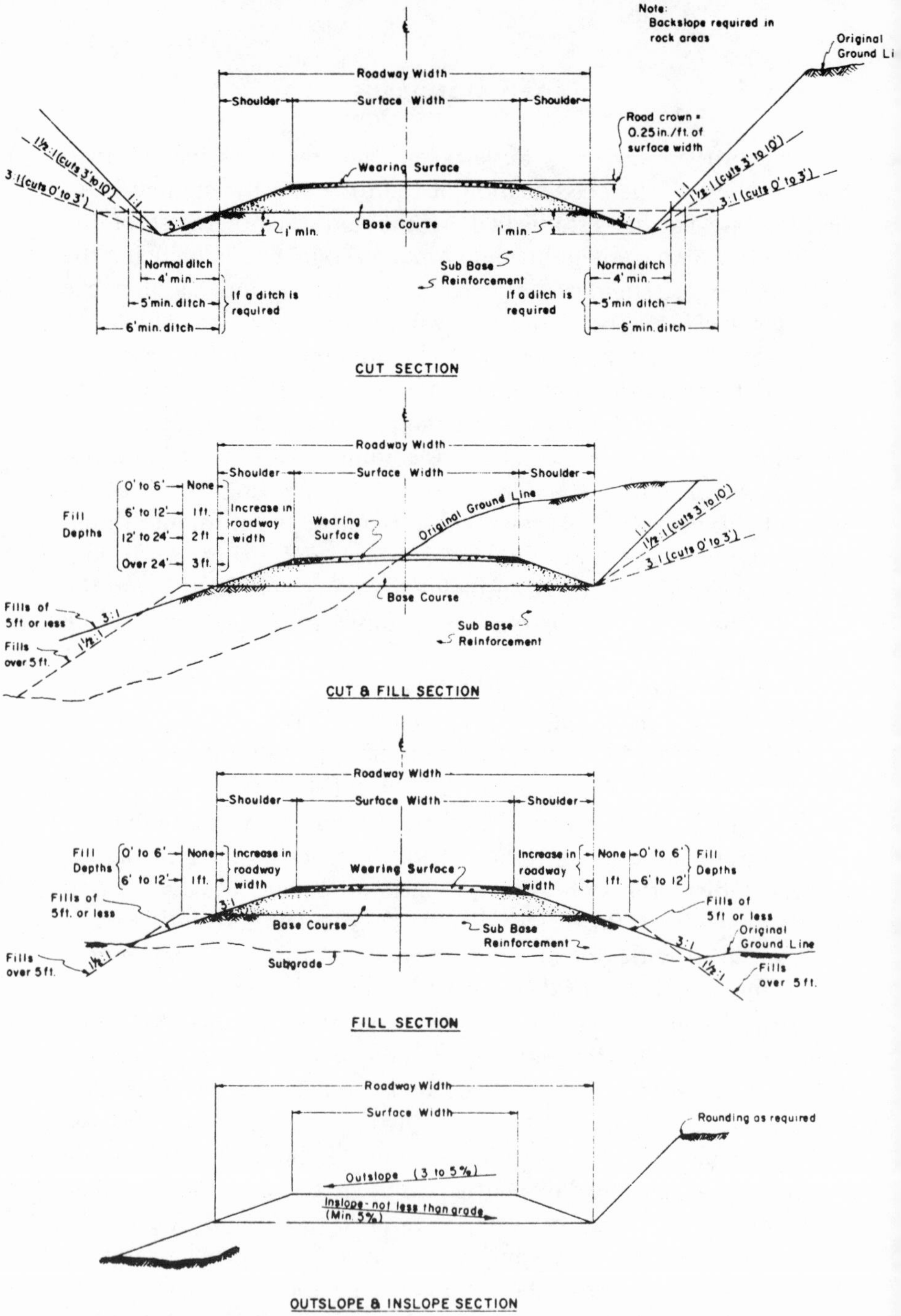

Fig. 3–1. Typical road sections. (Courtesy of the Bureau of Land Management, U.S. Department of the Interior.)

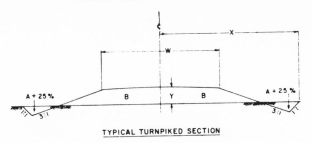

TYPICAL TURNPIKED SECTION

Fig. 3–1. *Continued*

landings. They are constructed to the lowest standard of the permanent system. They are maintained only if required for fire protection or administrative use, or for subsequent logging within a period of time which will justify maintenance of the road. Examples of road standards adopted by public forest agencies and by private companies follow.

Federal Forest Road Standards

Both the United States Forest Service and the United States Bureau of Land Management designate forest roads by a combination of a capacity code letter—S for single lane, D for two lane—and a duty code letter—L for light, N for normal, I.E. for trucks of legal weight and dimensions on the state highway, and H for heavy or off-highway trucks. The surface width of the single-lane road is twelve feet for normal duty, and for duty code H log truck bunk width plus four feet. The roadbed or subgrade width depends upon the thickness of the surfacing material. Since the shoulder slopes are 3:1 (three feet horizontally to one foot vertically), the roadbed width is equal to the surface width plus six times the surfacing thickness in feet. The recommended **back slope** in cut sections in common earth vary with the natural ground slope, as follows: $1\frac{1}{2}:1$ up to 30 percent; 1:1 from 30 to 55 percent $\frac{3}{4}:1$ over 55 percent. In hardpan the back slope is $\frac{1}{2}:1$ and in solid rock $\frac{1}{4}:1$. The slope between the edge of the roadway and a ditch is 3:1 for maintenance with a road grader. The minimum ditch depth is one foot. Fill bank slopes vary with the soil but are generally $1\frac{1}{2}:1$. Typical road sections are shown in Fig. 3–1. **Turnouts** are required on all blind curves on single-lane roads with additional turnouts as needed for a minimum turnout spacing of 750 feet.

The Bureau of Land Management *Roads Handbook* provides a geometric design guide for alinement and gradient of logging roads (Table 3–1).

TABLE 3–1. Design Guide

Design Speed mph	Alinement, Radius in feet, Single Lane Curves		Climbing Gradient in percent		
	Open Curves	Blind Curves	2-axle	3-axle	4- or 5-axle
10	50	80	11.3	8.0	5.3
15	70	200	8.5	5.4	3.4
20	125	400	6.3	4.0	2.4
25	200	800	5.0	3.0	1.7

For Class SN, which is the most common class of logging roads, Forest Service engineers appear to favor limiting gradients to 4 percent adverse (against the loaded truck) and 7 percent favorable, although 5 percent adverse and 10 percent favorable are permissible. A minimum curve radius of one hundred feet is preferred with fifty feet the smallest radius permitted.

Private Road Standards

The standards of private roads vary greatly among logging companies. They are influenced by the size and horsepower of the log trucks, by the differences in elevation to be gained, by the yarding equipment to be moved over the road, and by the volume of timber to be hauled over a given road. In mountainous terrain maximum favorable gradients of 10 percent on primary roads and 15 percent on secondary roads are common on private roads. Where log trucking is to be done by contract haulers at rates established by a state regulatory agency, the rate structure influences the gradient standard. For example, in the state of Washington there is one rate per mile for 0 to 6 percent; a higher rate for 6 to 12 percent; and a still higher rate for 12 to 18 percent. This often results in maximum gradient standards of 12 percent on primary roads and 18 percent on secondary roads. The maximum permitted by the state safety standards is 20 percent. Adverse gradient standards depend upon the horsepower-weight ratio and the consequent gradeability of the log trucks used. The recent trend toward using engines of 300 to 350 horsepower in trucks legal on the highway, and up to 475 horsepower in off-highway trucks, has resulted in raising the adverse gradient limits from the 5 or 6 percent previously used to 9 percent or more.

For log truck and trailer combinations the minimum radius of curve that can be traveled by highway vehicles is fifty feet and by off-highway sixty-two feet or more. Where mobile steel yarding towers are used the minimum radius in through cuts or open curves is determined by the curve the tower can travel without scraping the bank.

Typical standards of secondary roads set by one paper company (which uses off-highway trucks with ten-foot bunks, carrying twelve to fifteen thousand board feet) are as follows: surface width fourteen feet, subgrade width nineteen feet; maximum favorable grades, 10 percent preferred, 12 percent acceptable, and 18 percent if necessary to reach a landing; maximum adverse grade 7 percent, with 8 percent permitted up to one thousand feet distance; minimum radius curve, eighty feet.

Private roads which are continuously maintained are generally constructed with back slopes in cut sections steeper than those prescribed by public agencies. Back slopes in common earth are often $\frac{3}{4}$:1 on moderate ground slopes and $\frac{1}{2}$:1 on steep slopes. The steeper back slopes result in reduction in excavation cost, in the area of cut bank exposed to ravelling and the land area taken out of timber production for the road. The slope of the shoulder and the ditch bank on private roads may be as steep as $1\frac{1}{2}$:1. This permits a saving in roadbed width of three feet for each one foot thickness of surfacing material, as compared with the public forest road.

Annual Cost Formula

In determining the more economical of two or more classes of permanent road the following annual cost formula is a useful guide:

$$A = R + I + M + T$$

where A is annual cost in dollars per mile; R is the annual cost of road construction, for the amortization period; I is the annual interest on the investment in the road; M is the annual road maintenance cost; and T is the average log trucking cost per mile multiplied by the annual volume to be hauled. All costs are on a per-mile basis. A comparison of the value of A for the various classes of road under consideration will reveal which class would be the most economical. The Forest Service *Logging Road Handbook* (Byrne, 1960) is recommended for computing the round trip travel time as affected by curvature, gradient, surface, and turnout spacing on single-lane roads. Applying the cost per unit of time, computed from operating cost records available or from the tables given in the preceding reference, to the travel time will give the average log trucking cost per mile.

For example, a lower standard road to be constructed in a West Coast logging operation is estimated to cost $10,000 per mile to construct and $500 per mile annually to maintain. The estimated trucking cost is 35¢ per thousand board feet per mile. A higher standard road would cost $17,000 per mile, but maintenance would be reduced to $400 per mile and trucking cost to 25¢ per thousand board feet.

Assuming an amortization period of 25 years, an average annual interest rate of 3 percent, an average annual volume of logs to be hauled of 10,000 thousand board feet, the annual cost A would be as shown in Table 3–2.

TABLE 3–2. Annual Cost A

	Lower Standard	Higher Standard
R	400	680
I	300	510
M	500	400
T	3,500	2,500
A	4,700	4,090

The higher standard road would give an annual saving of $610 per mile. For any volume of annual production of less than 3,900 thousand board feet the lower standard road would be the more economical. If the higher standard road involved more mileage because of lesser gradients, then the annual cost per mile multiplied by the number of miles of each road would be compared.

ROUTE PROJECTION AND RECONNAISSANCE

One succinct dictionary definition of "route" is "a way to go." Road **route projection** is the laying out on a topographic map or aerial photo of the alternative ways a road could go. **Reconnaissance** is reconnoitering the projected route in the field to determine its feasibility. If maps or photos are not available, then the route must be found by field reconnaissance. Selection of the best route requires concomitant consideration of many factors, which are presented in the ensuing section.

Considerations in Route Selection

The primary consideration in selecting the route for a road for the transportation of logs is cost. The annual cost of maintenance and the truck travel time and consequent log trucking cost must be considered as well as initial construction cost. In choosing between alternative routes the total combined cost per unit of log volume measurement of the road amortization, maintenance, and trucking will usually be the decisive factor.

In mountainous terrain the route will often be determined by the topography. Since the main road usually follows a major drainage, the alternatives of a valley bottom route — possibly involving bridges and stream disturbance, and high road maintenance costs due to flood damage — must be weighed against a hillside route with heavier excavation and poorer alinement due to crossing side valleys and ridges. Edaphic considerations—geology and soils—are important in both construction and maintenance. A route in coarse-grained soils is preferable to one in fine-grained soils. Where the subsurface strata are inclined, the side of a valley where the cut bank slopes parallel the strata would be more subject to slides than the other side where cut banks would cross the strata. Aspect is related to maintenance. A road on a slope with southerly aspect in the northern hemisphere will dry faster after rain, be less subject to damage from traffic, and require less maintenance. A ridge crest route will have construction and drainage cost advantages, but may involve segments of adverse gradient and a longer haul. A route through rock outcrops will increase construction cost. Slide and slump areas and poorly drained soils present maintenance problems and are to be avoided if possible.

In multiple-purpose forests consideration is given to the needs of traffic generated by other uses than logging. Recreational use may dictate the valley bottom route for access to camping or picnic grounds, fishing streams, and natural swimming pools. The route may skirt the edges of flats suitable for recreational development instead of crossing them. In the valley of a meandering stream, alinement and distance might be sacrificed to avoid channel changes and siltation. In the interests of amicable public relations the needs of farmers or miners along the route who want access to the road may be considered.

The checkerboard pattern of forest land ownership found in many regions makes avoidance of **right-of-way** problems a prime consideration. The most obvious and direct route may not be feasible because of adverse ownerships encountered. Condemnation of rights-of-way involves lengthy and costly legal procedures, and is generally resorted to only if no other route can be found. Even public agencies with the right of eminent domain are inclined to condemn a right-of-way only as a last resort.

Since logging planning and secondary roads are interdependent, the reader is referred to Chapter 2, Logging Planning, for further considerations in route selection of secondary roads.

Route Projection

Before starting to make a route projection all available information on the area through which the road is to be built is assembled. This may include U.S. Geological Survey topographical and geological quadrangles, soil survey maps and reports and land ownership plats. Some companies and public agencies have large-scale topographic maps of their forest lands, with scales ranging from 400 feet to the inch, with contour interval of 20 feet, to 1,000 feet to the inch and contour interval of 50 feet. For secondary road projection timber type maps and cruise data are obtained. The various kinds of maps and their sources are described in Chapter 2. Aerial photo coverage of most forest lands is now available, and sometimes planimetric base maps accompany them. In the absence of large-scale topographic maps, aerial photos are the mainstays of route projection.

Control Points. The first step in route projection is to identify and mark major control points. Terminal control points are the takeoff from an existing road, and the point which it is desired to reach with the road. Intermediate control points may include saddles or passes in ridges, points where streams narrow for possible fill or bridge sites, points above or below cliffs or rock outcrops, slides and swamps, and points where existing roads, railroads, or electric power lines can be crossed. For multiple-use roads potential recreation sites are marked. For secondary roads, benches or flats suitable for road junctions or landings are important control points.

The second step is to determine the elevations of the control points, measure the distance between them, and compute the percent grade. On topographic maps, elevations are interpolated from the contours. Methods of obtaining differences in elevation and distances between points on an aerial photo are given in textbooks and handbooks on photogrammetry. If the computed grade exceeds the maximum grade allowed by the road standard that route is abandoned and another solution is sought.

Grade Contour. The third step is to plot a **grade contour** between usable control points. To plot on a topographic map, a pair of drawing dividers are set at the distance, to map scale, for one contour interval at the computed grade percent. For example, if difference in elevation is 80 feet and the distance 1000 feet, the grade is 8 percent, and the dividers are set for 20/.08 or 250 feet for a contour interval of 20 feet. If the map scale is 400 feet to the inch the dividers are set at 250/400 or 0.625 inch. Starting at the contour nearest one control point, successive contours are stepped off to the next control point. If this trial grade contour does not reach the elevation of the next control point, the distance and grade is recomputed, the dividers reset, and the new grade line stepped off. For methods of plotting a grade contour on aerial photos the reader is referred to a photogrammetry text. The final grade contour is plotted by drawing lines connecting the points ticked on the contour lines with the dividers. The plotting continues until all the consecutive segments of a route are plotted. A notation of the grade percent and distance of each segment between control points is made on the map or photo. If there are possible alternative routes they are plotted in the same manner.

Cost Comparison of Alternative Routes. An estimate of the more economical of two alternative routes may be made by tabulation of the elements affecting cost. For construction cost comparison, samplings of the side slopes in each segment are measured, and the total length of each side slope class, usually by 10 percent class intervals, is compiled. Any heavy through cuts or long fills requiring long hauls of earth or borrowing are considered separately from the side-hill cross-sections. As an example of the effect of side slope on the volume of excavation, and consequently of construction cost of the common single-lane road with ditch, the volume of excavation on a 30-percent side slope will be doubled on a 50-percent slope, will be six times as much on a 70-percent slope, and ten times as much on an 80-percent slope. Comparative volumes of excavation may be obtained from Calder's Table 18 (Calder, 1957) or similar tables in route survey textbooks. The clearing width and cost is also increased on steeper-side slopes. A tabulation of the number of draws or creeks crossed will give a comparison of the number of culverts required. Stream crossings requiring exceptionally large culverts or bridges are noted. The other construction cost elements will be directly proportional to the lengths of the alternative routes.

Maintenance costs on alternative routes generally will be in proportion to their lengths. However, unit maintenance costs of gravel or crushed rock surfaced roads will be increased on steeper grades and on segments with sharper curvature.

To estimate truck travel time the total length of each **favorable** and **adverse grade** by 1 percent intervals is tabulated. Alinement is classified as good, fair or poor, based on the average radius divided by the number of curves per mile (poor is less than twenty, fair is twenty to fifty, and good is more than fifty). Travel time in minutes per round trip mile may be readily obtained from tables or graphs in the *Logging Road Handbook* (Byrne, 1960). If a gradeability chart is available for a specific truck — representative of those which will be used on the road — travel time as affected by gradient may be computed from the chart. The effect of curvature on truck speed is given in graphs in the *Logging Road Handbook* and in tables in the *Forest Engineering Handbook* (Pearce, 1960).

Tabulation of the elements affecting road construction and maintenance cost, and computation of truck travel time may be sufficient to determine the most desirable of alternative routes. However, if the higher cost road would give faster travel time, it is necessary to calculate hauling costs to determine the most economical route. Travel time may be converted to cost by multiplying by the cost per minute of truck operation from company records or from cost data given in the *Logging Road Handbook*. In states where log trucking rates are set by a state agency, the state classification and rates are often used to estimate hauling costs. The route which gives the lowest combined annual cost of road amortization, maintenance, and trucking, for the anticipated volume of annual production, will usually be the preferred route.

Reconnaissance

The initial road route reconnaissance may be extensive in nature if no large scale maps or aerial photos suitable for route projection are available, or intensive if the route has been projected. The extensive reconnaissance is made to obtain the information on which to base the selection of the route. It generally covers a relatively wide belt of land. The field procedure depends upon what aids are available, such as small-scale maps or aerial photos on which the crown cover is too dense for route projection. Sometimes the ground reconnaissance is preceded by flying over the area by helicopter or light airplane. On the ground possible major control points are found, their elevations are obtained with an aneroid barometer, and the distances between them paced for estimating the gradient. Side slopes are read with an Abney level and a hand compass is used for direction. Some forest engineers keep notes on 10x10 cross section paper on which form lines are drawn to build up a sketch map. The watch time of each barometric reading is noted for later correction of elevations from the record of atmospheric pressure changes during the day kept by a recording barograph in the office. The reconnaissance proceeds in a systematic manner until all alternative routes have been investigated. The extensive reconnaissance is followed by an intensive reconnaissance of the selected route.

Grade Line. The intensive reconnaissance traces the projected or selected route on the ground. It checks the feasibility of the route and establishes a tagged or flagged **grade line** for the survey line to follow. The map or photos showing the projected route are taken into the field. The customary procedure is to walk from the first control point to the next control point. The distance is paced so the gradient can be corrected if the trial line ends above or below the second control point. Minor control points, which did not show on the map or photos, may be encountered requiring changes in the gradient. At the projected stream crossing the possibility of a better crossing upstream or downstream, without exceeding the gradient's limit, is investigated. The final grade line is run from the second back to the first control point. This procedure is repeated between successive control points until the final grade line has been established along the entire route to be surveyed.

The grade line is marked with plastic **flagging** tape or red tags at intervisible intervals. The practice of blazing the grade line on trees should be discouraged. Only the final location survey line should be blazed for the guidance of the right-of-way felling crew. If the grade line should be subsequently changed, as sometimes occurs, the previous line is erased by removing flagging or tags, or by using different colors on the new line. Tree blazes cannot be removed. The flagging tape is tied to trees or branches on line. The tags of red Bristol card, about $1\frac{1}{2}$x5 inches, are quickly set by driving the blade of a hand axe vertically into a tree, opening the slit by twisting the blade horizontally, and inserting the end of the tag in the slit. When the axe blade is removed the slit closes to grip the tag.

A grade line can be run by one man by setting the tags at his eye height above the ground, and sighting back at the tag with his Abney level, to place his feet on grade. In the interests of safety a two-man party is customary. The Abney man goes ahead, and finds his grade point by sighting back at his eye height on the other man, and marks it by scuffing the ground. In dense brush the rear man holds a flashlight at the sighting point. He then proceeds toward the grade point marked by the Abney man, setting tags or tying flagging at intervisible intervals to mark the line between grade points.

In running a grade line at or near the maximum permissible gradient for the road standard, care must be exercised to avoid tagging a grade line longer than the final location line, which would result in exceeding the permissible gradient. In rounding sharp-nosed ridges or narrow valleys the curve that will be located must be visualized and the grade line shortened accordingly. At a road junction point between roads on different gradients, allowance for grade separation and a vertical curve must be made. In flattening the gradient for a landing, allowance for a vertical curve is required. To check that the curve standard is not exceeded, a hand compass may be used to turn deflection angles. For detailed explanation of techniques of running grade line so that the location gradient and curvature will not exceed the desired maximum, the reader is referred to Chapter 420 of the *Forest Engineering Handbook*.

The importance of the reconnaissance in forest road engineering cannot be over-emphasized. The major decisions affecting the location of the road are made at this time. To make major changes during the road survey is difficult and more expensive. Reconnaissance is hard work, demanding mental as well as physical effort, and tenacity to persist until the best solutions of all the problems encountered have been found.

ROAD LOCATION SURVEYS

A variety of methods are used in making location surveys for forest roads. They vary with the class of road, the topographic difficulties encountered, and the experience of the locator. The methods used by logging companies vary with the personal preferences of the logging engineer. The public forest agencies publish detailed handbooks or manuals which specify the methods to be used by their engineers. Road location methods range from simply walking a route and blazing a line for a bulldozer operator to follow, to transit-traversed preliminary lines and highway design by electronic computers, with balanced sections of earthwork in **excavation** and **embankment**. Only a few of the simpler methods, which are commonly used in logging industry, will be given here.

Elements of Curves

The alinement of the center line of a road consists of a series of straight lines, termed tangents, connected by curves. Simple circular curves are used on forest roads. The spiraled transition curves and compound curves of highways are seldom used. Figure 3–2 illustrates the elements of the curve, the initials of the elements as commonly used in the field note book and on road plan drawings and specimen values which are computed. The magnitude of the curve may be designated by radius or by degree. The **degree of curve** is the angle subtended by a 100-foot arc or by a 100-foot chord. The arc definition is more commonly used for forest roads. Since the radius (R) of a $1°$ curve is 5730 feet (to the nearest foot), $R = 5730/D$ or $D =$

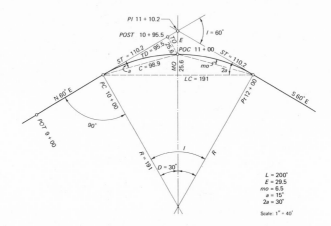

Fig. 3–2. Circular curve elements and specimen values.

PI = Point of intersection of two tangents. Specimen station 11 + 10.2

I = The intersection angle of the tangents, which in the specimen curve = 60°. The Greek letter delta (Δ) is often used instead of I.

D = Degree of curve = 30°.

R = Radius of curve = 5730/30 = 191 feet.

L = Length of curve = 100 I/D = 6000/30 = 200 feet.

E = External, the distance from the PI to the mid-point of the curve = $R \times$ exsec $I/2$ = 191 × 0.1547 = 29.5 feet.

ST = Semi-tangent = $R \times \tan I/2$ = 191 × 0.577 = 110.2 feet.

LC = Long chord = $2R \times \sin I/2$ = 382 × 0.500 = 191 feet.

MO = Middle ordinate for L = R vers $I/2$ = 191 × 10.134 = 25.6 feet.

mo = Middle ordinate for 100-foot arc = R vers $D/2$ = 191 × 0.034 = 6.5 feet.

C = Chord for 100-foot arc = $2R \sin D/2$ = 382 × 0.259 = 98.9

a = deflection angle between ST, and c to POC = $D/2$ = 15°.

Distances along the center line of a road are measured in stations of 100 feet. Following are the computations of the stationing of the curve:

PC = Point of curvature = $PI - ST$ = 11 + 10.2 − 110 2 = 10 + 00

PT = Point of tangency = $PC + L$ = 10 + 00 + 200 = 12 + 00

POC = Point on curve = 10 + 00 + 100 = 11 + 00

TD = Tangent distance to $POST$ = $C \cos a$ = 98.9 × 0.966 = 95.5 feet.

$POST$ = Point on semi-tangent = $PC + TD$ = 10 + 00 + 95.5 = 10 + 95.5

TO = Tangent offset = $C \times \sin a$ = 98.9 × 0.2585 = 25.6 feet.

POT = Point on tangent between curves.

5730/R. For the curve illustrated in Fig. 3–2 the nomenclature is given, and the formulas for computing curve data using trigonometric functions, and the computations, which can be made by anyone conversant with tables of trigonometric functions, without access to curve tables in route survey books.

Staking Curves. The points of the curve may be established on the ground in a number of ways. *PC* and *PT* may be set by measuring *ST* from the *PI* along the bearings of the tangents. *PT* may be set from *PC* by measuring *LC* on a deflection angle of one-half *I* from the back tangent. The mid-point of the curve, station 11 + 00 in the example, may be set by measuring *E* from the *PI*, along a line which bisects the included angle between the tangents, and *C* from *PC*; by simultaneously measuring *TO* from *POST* and *C* from *PC*; by *MO* from the mid-point of *LC* on a line perpendicular to *LC*; and by turning deflection angle "*a*" from the line of *ST*. If station +50 stakes are to be set, station 10 + 50 may be set by measuring the *TO* for a 50-foot arc, which is equal to the middle ordinate of a 100-foot arc, along a perpendicular from the line of the *ST* which intersects a measurement of the **sub-chord** length, which is 49.9 feet in the specimen curve, from the *PC*. It may also be set by turning a deflection angle of $D/4$ and measuring the sub-chord length from *PC*. Station 12 + 50 may be set in a similar manner back from the *PT*. For any arc length the deflection angle in minutes is 0.3 minute per foot per degree of curve. The sub-chord length is 2 R times the sine of one-half the deflection angle.

The computation of *ST* and *E* is expedited by using a table of functions of a 1° curve, which may be found in all route survey books. Dividing the function for a 1° curve for the given intersection angle by the degree of curve gives the value of *ST* or *E*. Conversely, dividing by the desired *ST* or *E* will give the degree of curve. For field use, Calder's *Forest Road Engineering Tables* are recommended. They are of pocket size, printed on waterproof paper, and contain tables of tangent offsets, middle ordinates, and chord lengths which obviate computation of these elements. Functions of a 1° curve are given for *LC* and *MO* as well as *ST* and *E*.

Direct Location

The direct location method, whereby the curves are computed and staked in the field as the alinement is traversed, is commonly used on spurs, and on branch roads which do not involve difficult location problems. The tagged grade line is followed with the traverse of a series of tangents with staff compass and 100-foot steel tape. The head chainman lines himself by eye with two tangents demarcated by the tagged grade line and sets the *PI*. The distance from the last *PT* to the *PI* is chained, stakes being set at each station and sometimes at each +50 foot. The compassman sets up at the *PI*, reads the back and forward bearings of the tangents and computes the *I*. The degree of curve which will best fit the conditions is determined, and the curve data and stationing computed. The *PC* is set by measuring from the nearest stake previously set. The stakes along the semi-tangent are offset to the curve. The *PT* is set by chaining the *ST* length along the line of the forward tangent.

The degree of curve is determined by one of the following methods: (1) from the external distance measured to the point desired for the mid-point of the curve. The *E* for a 1° curve divided by the desired *E* gives the degree of curve. Usually the nearest even degree of curve, or the nearest 5°, is selected. (2) From the allowable *ST* which will not overlap the last *PT*. It is preferable to have at least a truck and trailer length of tangent between curves in opposite directions. The *PI* station minus the last *PT* station plus the desired tangent length *PC* and *PT* gives the allowable *ST*. The *ST* for a 1° curve divided by the allowable *ST* gives the degree of curve. (3) The last method is from the maximum degree of curve, or the minimum radius, permitted by the road standard. The sharpest curve would raise construction costs unduly. "Broken back" curves, with short tangents between curves in the same direction, are undesirable. It is better to compound the curve by making the end of the first curve a *PCC* (for point of compound curvature), beginning the second curve.

The fastest method of staking a curve is by tangent offsets. By trial the chainman finds the point on the semi-tangent where, by extending his arms along the line of the semi-tangent, and swinging his arms until the palms touch, his hands point to the *POC* which is the chord length from the *PC*. Simultaneous measurement of chord and *TO* establishes the *POC*. The first half of the curve is staked forward from the *PC*, and the second half back from the *PT*. The distance between the last two stakes set from each end of the curve is measured as a check on both the staking and the curve computations. Where the *PC* or *PT* is not a full station or a +50, the *TO* may be the formula $TO = c^2/2R$, or read directly from Calders' Tables. Staking only the midpoint of the curve by measuring *E* from the *PI*, along the line of the bisector of the interior angle between tangents, may define the curve sufficiently for construction.

On a long curve where the tangent offsets are too long to be conveniently measured, or where the *PI* is inaccessible, the curve may be staked by deflection bearings, computed from deflection angles. The compassman sets up at the *PC*, and alines the stake held by the chainman simultaneously with the chaining of the distance. To avoid clearing a wide swath in thick brush, the compass is set up on successive *POC*'s to prolong the curve. At each setup the bearings must be recomputed. Applying the deflection angle used to set the *POC* to the bearing between *PC* and *POC* will give the bearing of the tangent to the curve at the *POC*. Applying the deflection angle of $D/4$ for 50 feet or $D/2$ for 100 feet to this bearing will give the bearings to the next stakes. At the *PT* the forward tangent is established by applying to the bearing from the last *POC* the deflection angle which would be required to set that *POC* from the *PT*. The deflection bearing method of staking the curve in Fig. 3–2 is illustrated in Fig. 3–3.

A long curve may be staked by the middle ordinate method with

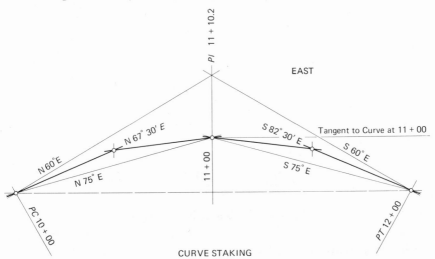

Fig. 3–3.   Deflection bearing method: compass at *PC* 10 + 00 and *POC* 11 + 00.

a minimum of line brushing. The first stake on the curve is set by the tangent offset method. The middle ordinate for the arc to the second stake is measured inside the curve, perpendicular to an estimated tangent to the curve at the first stake, and a temporary stake or a range pole set. The second *POC* is set in line with the *MO* point and the *PC*, at the chord distance for the arc to that point. At the second *POC* the *MO* is measured, and the third *POC* set in line with the *MO* point and the first *POC*. If the *PI* is inaccessible, the line of the forward tangent is established by one of two methods: by measuring the *TO* for the distance between the last *POC* and the *PT* to the line of the *ST*. Or by prolonging the curve beyond the *PT* by using the *MO* for the chord, measured from the *PT*, and measuring the same distance in the opposite direction from the prolonged point to the tangent. Sometimes both methods are used to give three points from which to aline the forward tangent. The middle ordinate method is illustrated in Fig. 3–4.

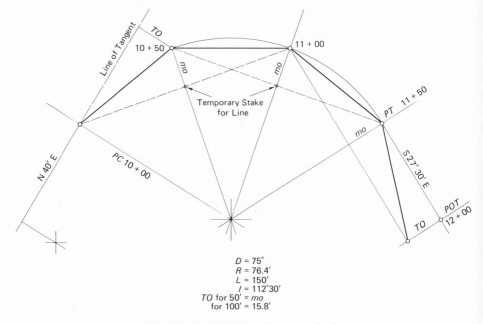

$$D = 75°$$
$$R = 76.4'$$
$$L = 150'$$
$$I = 112°30'$$
$$TO \text{ for } 50' = mo$$
$$\text{for } 100' = 15.8'$$

Fig. 3–4. Middle ordinate method.

Preliminary Traverse for Road Design

For main roads and for branch roads through difficult terrain involving heavy construction, or where the road is to be designed for balanced quantities in excavation and embankment, a **preliminary line**, commonly abbreviated **P-line**, is run. The P-line is a random traverse following the tagged grade line. The road is designed in the office from the plotted P-line and accessory topographic information. The P-line is a traverse of a series of straight lines, following as close to the tagged grade line as practicable. Forward and back bearings are read at each angle point with a staff compass, and distances are measured with a 200-foot tape. Stakes are set at angle points, and at breaks in the ground profile. Differences in elevation between each stake are obtained with an Abney level, by double-Abney leveling, or with hand level and rod. Side slopes on both sides of the line are read at each stake with the Abney. Traverse notes are recorded on the left-hand page of the field book, and a form line sketch of the topographic features together with notes on stream widths and bearings, rock outcrops, poorly drained areas and other features which would affect road location, on the right-hand page. Field notes are entered from the bottom of the page up to accord with the sketch. For high standard roads, or where a right-of-way plat will be required by the land owner, the P-line is run as a transit traverse, and elevations are obtained with a tripod-mounted level and rod. In this case stakes are set at every station.

Where road construction cost justifies the extra surveying expense, a contour map of a strip of land along the P-line may be made. The traverse is plotted on detail drawing paper at the scale of 100 feet to the inch. This copy is termed the **hardshell.** A field copy is made, either on tracing paper or by pricking the angle points through onto another sheet of drawing paper. At each staked point a line is drawn perpendicular to the traverse line, and the elevation of the point is marked. In the field 5-foot contour points are located on the perpendicular lines with hand level and rod, and the distance from the stake to the contour measured with a tape on a reel. These points are plotted to scale on the map sheet, and the contours drawn halfway to the adjacent lines. A faster method, but one requiring more skill on the part of the topographer, is to obtain elevations at breaks in the ground profile along the perpendicular line with an Abney level and slope chaining. The contour points are interpolated. The width of the strip mapped is that which, in the judgment of the topographer, will include the possible lateral limits of the final location line. The steeper the slope, the narrower is the width of topography required. All the features mentioned in the preceding paragraph which might affect the road location are noted on the strip map. The subject of road design will be covered in a later section of this chapter.

Location Line Staking of Designed Road

After the road has been designed in the office, a tracing of the *P*-line and the **location** or **L-line**, with distances between them at key points scaled and marked, and curve data computed, is taken into the field. Long tangents are alined by measuring two or more offsets from the *P*-line. Intermediate stakes are set by chaining forward from a point of known station. If there is any discrepancy between the computed station and the chained station the latter is the one marked on the stakes. Field notes are kept on the *L*-line points and stationing. For short curves the *PI* is set by measuring the scaled distance from the two nearest *P*-line stakes, much as one would establish a point by swinging two arcs with a drawing compass. The *PC* and *PT* are set by measuring the *ST* distance along the lines of the tangents. The stakes on the curve are set by tangent offset method described above in "Direct Location." For long curves, or where the *PI* is inaccessible, the *PC* and the *PT* of the curve is set from the *P*-line stakes nearest to them, and each half of the curve staked forward from the *PC* and back from the *PT* by the middle ordinate or the deflection bearing method.

Elevations of the *L*-line are taken by tripod level and rod, hand level and rod, or by the double-Abney method, depending upon the class of road and the precision desired. Hubs or pegs, driven flush with the ground, are set at each stake, and the leveling rod is held on the peg. The pegs facilitate restaking the line after clearing and grubbing, as many of them will survive, whereas the stakes will be knocked out. Often the staking of the *L*-line is deferred until after the clearing and grubbing operation has been completed. Key points, such as *PI*'s of short curves and *PC*'s and *PT*'s of long curves, are located and referenced to trees or stakes outside the upper clearing line. The center line points are reset by measuring back from the reference points. Referencing is covered in a later section of this chapter.

The center line stakes are marked on the front side with the road identification which may be a letter or number, and the station number, preceded by P or L for preliminary or location line. The front of the stake is the side seen as one walks from the direction of the beginning of the line. The vertical cut or fill to the grade elevation is marked on the back of the stake. For the example given in "Slope Staking" below, the marks would be C $3^{\underline{0}}$ and F $6^{\underline{0}}$.

Corner Ties

Where the L-line crosses a section line or a land ownership line, a stake is set on that line, and a tie is made to the nearest corner. The corner may be a section corner, one-quarter section corner, or other subdivision corner, or ownership corner. A tie is the bearing and distance from the L-line to the corner. It serves to establish the position of the road with respect to legal land subdivision. If a right-of-way plat is to be made, ties are made to the corners on each side of the L-line.

A **right-of-way plat** is a map drawn to scale showing the L-line data, the width of the right-of-way, which is the strip of land to be occupied by the road, the acreage of right-of-way in each legal subdivision (quarter-quarter section or lot) and the ties. The plat may be required by the land owner as a condition of granting an easement for the use of the right-of-way. The plat provides the data necessary to draft the legal document granting the easement. Sometimes the bearings and distances between all corners on the section lines are required to be shown on the plat, necessitating a survey of the entire perimeter of the section. The tie is usually made with staff compass and tape in a cardinal direction. For a right-of-way plat a transit traverse tie may be required.

Slope Staking

A **slope stake** marks the point where a bank slope intersects the surface of the ground. The operation of locating the slope stake points is termed **cross-sectioning.** Cross-sectioning is essential where a construction contractor is paid on the basis of cubic yards excavated. The cross-section field notes give the data for computing end areas of the excavation and embankment prisms. The value of the slope stakes is noted in the guidance of the construction machine operator. On side hill sections only the upper slope stake, and the "grade out" point where the subgrade intersects the ground surface on the lower side, are set. On through cuts or fills, slope stakes are set on both sides to mark the tops of the cut banks or the toes of the fills. Cross-sections are taken perpendicular to the center line on tangents, and perpendicular to the tangent to the curve at the *POC* on curves.

The conventional method of setting slope stakes is with a hand level and rod, and a steel tape on a reel. The horizontal distance to the slope stake point equals the half-base of the road bed plus the bank slope ratio times the difference in elevation between the subgrade and the point. For example, given a 16-foot width subgrade and a 3-foot width ditch, the half-base for the upper side = 8 + 3 = 11 feet. With a center line cut of 3.0 feet and a cut bank slope ratio of ¾ to 1, on a 50 percent side slope the difference in elevation would be 13.6 feet. Then the horizontal distance = 11 + (¾ × 13.6) = 21.2 feet. The slope stake fraction C 13.6/21.2 is marked on the front of the stake, and entered in the field book. The slope stake point is found by trial. With experience the slope staking party develops the ability to estimate closely where the point will be so that usually one trial reading and one final reading will locate the point. If the slope between center line and slope stake is not uniform, the distance to the point where the slope changes and its difference in elevation from the subgrade are noted. The grade-out point for the example would be 3 feet in elevation below the ground at the center line stake, and 6 feet out on a 50 percent slope. The complete cross-section notes for the example as entered in the field book are:

L	CL	R
0	C3.0	C13.6
6.0	0	21.2

As an example of cross-sectioning a fill of 6.0 feet in height on the same width road, bank slope ratio of 1½ to 1, and a ground side slope of 20 percent, the half-base is 8 feet, since there is no ditch. On the lower side the difference in elevation between grade and slope stake point would be 10.9 feet. Then the horizontal distance = 8 + (1½ × 10.9) = 24.3 feet. On the upper side the respective figures would be 3.4 and 13.1 feet. Following are the field notes for this cross-section:

L	CL	R
F10.9	F6.0	F3.4
24.3	0	13.1

On steep slopes where **pegging up** with the hand level and rod would be required, a faster method is to use Calder's slope staking tables, Abney level, and slope chaining. For the first example above, from Calder's Table 14 the slope distance is 23.7 feet and the difference in elevation 13.6 feet. From Table 17 the grade-out on the lower side is 6.0 feet. For the second example, from Calder's Table 16 the slope distance is 24.8 feet and the difference in elevation 10.9 feet. The field procedure is as follows: the Abney man reads the slope percent and finds the distance in the table while the chainman is proceeding up the slope with the tape. The Abney man holds the tape at this distance, and takes a check reading on the chainman at the point. If the slope reading differs, the distance is corrected. The Abney man is also the notekeeper. For the conventional method the party consists of a level man-notekeeper and rodman.

Referencing

Since most of the survey stakes within the clearing limits will be obliterated during the clearing operation, it is essential that they be **referenced** so they can be replaced readily before the earth-moving operation. Where slope stakes are set, the stake is offset 5 feet or 10 feet beyond the clearing line, which is usually 5 feet beyond the slope stake point. A blaze on a tree in line with the slope stake and a center line stake enables the slope stake to be reset by measuring 10 feet back in line with the blaze. The center line stake may be reset by measuring a distance equal to the denominator of the slope-stake fraction marked. A right-of-way clearing line is blazed or flagged connecting the points 5 feet from the slope stakes. The center line is also blazed for the guidance of the right-of-way felling and bucking crew.

If slope stakes are not set, the method and intensity of referencing varies widely. Some engineers reference every stake, others reference only the *PI* of a short curve, or the *PC* and *PT* of a long curve. If *L*-line staking is deferred until after right-of-way clearing, the angle points of the P-line are referenced. Some use a staff compass and take bearing and distance to a tree beyond the clearing line. The station, bearing, and distance are marked on a large blaze or on a stake nailed to the tree. Others use only a tape and measure the distance to two trees at such an angle that simultaneous measurement of the two distances will accurately restore the center line station point. In open forest stands two stakes are driven 10 feet apart in line with the center line stake, and the distance and station marked on the near stake. All reference trees or stakes are also marked with the initials RP for reference point. Reference notes on distance, bearing, tree species and diameter are entered in the field book, so that the reference points can be used even after the marks on them have become obscured by weathering over a long period of time.

The use of a square of aluminum foil nailed to reference trees, instead of blazes, is recommended. Reference data can be indented on the foil with a ball point pen and will not become obscured by pitch as will crayon marks on blazes. The shiny foil is highly visible, even through dense brush.

Reference trees are also used as grade references where slope stakes are not set. A horizontal notch is chopped at the bottom of the RP blaze or foil square. The difference in elevation between the notch and the center line stake ground elevation is obtained by hand level and rod or Abney level and slope chaining. The cut or fill at the center line stake is added to or subtracted from the difference in elevation, and the difference in elevation between the notch and the grade elevation marked on the RP and noted in the field book. Another method of grade referencing is to locate with hand level and rod, on a tree beyond the lower clearing line, the elevation of the subgrade at the center line point nearest the tree. The elevation is marked with a stake nailed horizontally, or with a strip of aluminum foil. This enables the construction crew to determine easily when they have excavated or filled to grade. Driving a hub or peg flush with the ground at each center line stake aids in restoring the center line, as most of the hubs will survive the clearing operation.

Bridge Site Survey

Where a bridge of a permanent type is to be designed a special bridge site survey is made. The proposed center line is run with transit and tape, setting stakes every 20 feet and at grade breaks of the ground profile. The center line extends far enough beyond the banks of the stream to include possible approach fills. If the stream is too wide to chain across, the width is obtained by triangulation from a base line, approximately as long as the stream is wide, laid out along one bank. Pegs are driven at each center line stake and elevations obtained with tripod level and rod. Bench marks are established on each side of the stream outside the area to be cleared.

The center line is plotted on a scale of 20 feet to the inch. A tracing or hard copy on 10×10 cross-section paper, with the elevation at each stake marked, is taken to the field for mapping topography with a 2-foot contour interval. Contour points are located with hand level and rod and tape, along lines perpendicular to the center line, and plotted. Usually a strip 200 feet in width is mapped. The contour points are connected by contour lines, drawn in the field as the topographer interprets them. All identifiable water lines, exposed rock masses or large boulders, and piles of driftwood are mapped. If the stream is shallow enough to wade, elevations of the stream bed are taken along the center line, and along parallel lines to each side. In deep water soundings are made from a raft or boat and the position at the instant of each sounding obtained by transit and long tape or by triangulation with two transits from a base line. The stream bed is contoured by interpolation of contours from the elevation points.

The depth to bedrock at proposed abutment sites is obtained with a **refraction seismograph,** or by digging test pits. The soils are sampled and the location of concrete aggregate and suitable approach-fill material noted. A copy of the site map and a profile of the center line on a scale of 10 feet to the inch, both horizontal and vertical, is furnished to the bridge design engineer.

If large multiplate or concrete **culverts** are a feasible alternative to a bridge, a profile of the stream bed for 500 feet on each side of the center line is made. The stream bed is cross-sectioned for the wetted perimeter, and the stream velocity is measured, during flood stage if possible.

ROAD DESIGN

The P-line traverse is plotted on the hardshell to the scale of 1 inch equals 100 feet. The quickest method of plotting is to use a drafting machine with protractor head and scales on two radial arms. One arm is used for drawing perpendiculars at points at which side slopes were read. A more accurate method, as it avoids cumulative errors, is to compute latitudes and departures and plot by coordinates. All information on the right-hand page of the field book which would affect the road location, such as rock outcrops, swamps, etc., is transferred to the hardshell. Water courses are mapped and stream widths marked. The side slope is marked at the end of each perpendicular line. Ties to corners are plotted and property lines drawn. The points at which elevations were taken are plotted on pencil profile paper to the horizontal scale of 1 inch equals 100 feet and vertical scale of 1 inch equals 10 feet. The elevation points are connected with a freehand line to give a profile of the ground along the P-line. Control points are marked, the gradient between them computed and a tentative grade line drawn on the profile.

Graphic Method

The next step is to determine the horizontal distance between each staked P-line point and the desirable center-line point for the L-line. This may be computed by the contour offset method, or determined graphically. The graphic method is recommended for use by the less experienced designer. One method is to plot the ground cross-section at each elevation point on 10 × 10 squared paper (10 squares to the inch) to the scale of 1 inch equals 10 feet. Using the heavy lines as even 10-foot elevations, the elevation of the P-line point is plotted, and the side slopes from this point drawn with a protractor. **Templets** the shape of the finished road cross-sections are cut to scale from transparent plastic. The side-hill section templet shows the cut bank slope, ditch, subgrade, and embankment slope. Templets for through cut section and all embankment or "turnpike" section are also made. Placing the appropriate templet over the plotted ground cross-section with the subgrade at the grade elevation read from the profile shows the end areas in excavation or embankment. The templet is shifted horizontally to give the desired center-line cut or fill, and the distance between P-line and templet center line read by counting the 1-foot squares horizontally.

On side-hill sections the horizontal shift of the templet is determined by how much of the subgrade should be on solid ground.

Where the lower side slope is 66⅔ percent or more, a 1½ to 1 fill slope will not catch and a full bench is required. On lesser side slopes some logging companies prefer to have the subgrade intersect the ground line at least 4 feet from the center line on the down-hill side, so the log truck wheel tracks will be over solid ground. Where balanced sections are desired on moderate slopes the templet is shifted until end areas in excavation and embankment are approximately equal.

Miller Design Aids. A faster graphic method, since it obviates drawing each ground cross-section, is to use the design aids devised by Roswell K. Miller (Miller, 1964). The background design aid is plotted on 10 × 10 paper to a scale of 1 inch equals 20 feet. Lines for the range of upper side slopes are drawn by 10 percent increments radiating from the center point, which represents the P-line elevation point. Similar lines are drawn for the lower slopes up to 70 percent. The overlay aid is drawn on transparent plastic to the same scale. It is similar to the templet, except that all cross-sections for one subgrade width are drawn on one overlay, and the overlay is not cut out along the lines of the finished road and bank slopes. The overlay is placed upon the background aid, and shifted to the position desired with respect to the side slopes at each P-line point. The offset from the center-line is read on the background scale. For detailed instructions on the use of Miller's design aids see Miller (1964), which will be found in any library maintaining files of the *Journal of Forestry.*

Location Line Design

To design the L-line alinement, the desirable center-line points located by the graphic method are plotted on the perpendicular lines on the hardshell by scaling the offset. The ground elevation read from the plotted cross-section, or computed by applying the vertical intercept on the background aid to the P-line elevation, is marked at the offset point. Tangents are drawn with a straight-edge through each series of points which are approximately in line. Some of the points will be off the line, so the designer uses his best judgment in the points he selects to aline tangents. For example, on a steep slope the higher of a series of points are selected, since alining the lower points would give insufficient cut or possibly even fill at the stations of the higher points. Curves are fitted to the tangents graphically, either by swinging arcs with a drawing compass or with curves cut to scale from transparent plastic. The curve which will best fit the center-line point lying inside the semitangents, without overlapping adjacent curves, is found by trial.

A profile of the trial L-line is plotted to check the design. The ground elevations of L-line points which do not coincide with the previously plotted center-line points are obtained by scaling the distance between points, multiplying by the slope percent, and applying this difference in elevation to the known elevation. A trial grade line is fitted to the profile. If there is an appreciable difference in length between the P-line and the L-line, new L-line stationing is obtained by stepping off 100-foot station points with drawing dividers, and scaling the plus distance to elevation points. Plan and profile are studied together to find segments where the design could be improved. The alinement is corrected and the new ground line plotted on the profile. This procedure is continued until the best design has been found. Design is completed for a segment between major control points before proceeding with the design of the next segment, so the latter will not be affected if changes are made later in the preceding segment.

Balanced Design. Where balanced earthwork in excavation and embankment is desired, the ground cross-sections plotted for the templet method can be used for obtaining end areas. The templet is placed in the correct position on the cross-section. The bank slope, ditch, and subgrade lines are drawn by moving the pencil point along the cutout part of the templet. End areas are measured by planimeter or computed. The average of two end areas in square feet multiplied by the distance between them in feet, divided by 27, gives cubic yards. Tables which expedite yardage computations are given in route survey textbooks. Where a section of earthwork terminates in a point, as when going from cut to fill, the volume is computed as a pyramid, with the end area as the base and the estimated distance to where cut changes to fill as the altitude, and divided by 3 times 27 or 81 to get cubic yards. Excavation and embankment are, of course, computed separately, and embankment yardage is adjusted for shrinkage of earth or swell of rock. Shrinkage and swell varies with the material and the construction method. Local experience is the best guide to follow. Common earth shrinkage averages about 25 to 30 percent with bulldozer construction. Rock swell varies with the size of the fragments and may be from

25 to 45 percent or more. Quantities usually may be balanced by raising or lowering the grade line on the profile. Sometimes changes in alinement are required to obtain a balance between excavation and embankment.

Computed Offset Method. For design by the computed contour offset method it is preferable that station and +50 stakes be set in running the traverse. The P-line traverse and profile is plotted as previously described, and a grade line fitted. The ground elevation, grade elevation, and side slopes at each P-line station are tabulated on a design form. The P-line cut or fill is computed from the difference in elevations. The contour offset is computed by dividing the cut or fill by the slope percent. The contour offset points are plotted on the hardshell to outline a grade contour. An L-line is fitted, using the grade contour as a guide. The offsets from each P-line station to the L-line are measured perpendicular to the P-line and the difference in elevation, DE, computed by multiplying the offset by the slope percent. The L-line cut or fill is computed by the following formula: L-line C or F = P-line C or F ± DE. The sign of DE is positive if the L-line is above the P-line, and negative if below. To keep the L-line stationing close to that of the P-line, an equation is used at the PT of the curve, where the length of the curve is shorter than the distance by the P-line. The PT station back equals the PC station plus L. The station ahead equals the P-line station on the offset perpendicular to the P-line. All of the above data are tabulated in appropriate columns of a design form. Where a vertical curve is required the vertical curve ordinates and the corrected C or F is entered. The L-line is staked by offsetting the P-line stakes the distance shown on the design form.

Road design on a contour map strip begins with plotting a grade contour between control points in the same manner as described under "Route Projection." An L-line is fitted to the grade contour, stations are stepped off with dividers, and ground elevations at each station read by interpolating between contours. A profile is plotted and desirable changes in alinement and profile made until the best location has been achieved. Cross-sections for end area and yardage computations may be plotted by measuring the map distance to contours above and below the center line. Where the side slope is uniform the slope percent is rapidly obtained by noting the difference in elevation between divider points set at 1 inch or 100 feet to scale.

For more detailed instructions on road design methods and design forms and field note forms, the reader is referred to Chapter 710 of the *Forest Engineering Handbook*. Manuals on highway design with electronic computers are published by the United States Forest Service, the Bureau of Public Roads, and state highway departments.

Vertical Curves

Where the straight grade lines on the L-line profile intersect, a transition from one grade to another is effected by a parabolic vertical curve. The length of the vertical curve chosen depends upon the desired traffic speed, the algebraic difference in grades, and the construction cost. Figure 3–5 gives the length of vertical curve re-

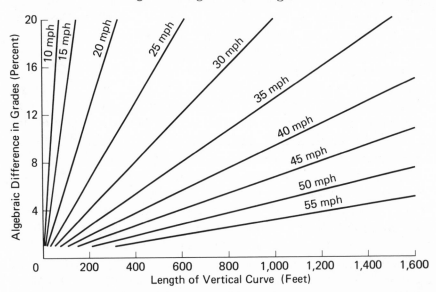

Fig. 3–5. Chart for determining length of vertical curve required for desired speed, for given algebraic difference in grade. (Courtesy of the Bureau of Land Management, U.S. Department of the Interior.)

quired for a desired speed for algebraic differences in grade from 1 to 20 percent. Following is a simple method of computing vertical curves:

Read from the profile the elevations of the beginning of vertical curve, A; the end, C; and the vertex, V. The elevation of mid-point B on a line connecting A and C is $\frac{1}{2}(A + C)$. The elevation of M, the mid-point on the vertical curve is $B + \frac{1}{2}(V - B)$. Since the equation of the parabola is $y = KX^2$, the ordinate y from the grade line to the vertical curve is K times the fraction of the distance from the beginning or end to the vertex, where $K = V - M$. At half the distance the ordinate is $(\frac{1}{2})^2 K = \frac{1}{4}K$. At $\frac{1}{3}$ the distance the ordinate is $(\frac{1}{3})^2 K = \frac{1}{9}K$, at the $\frac{2}{3}$ point $(\frac{2}{3})^2 K = \frac{4}{9}K$. In the example given in Fig. 3–6, with an algebraic difference in grades of 10 percent,

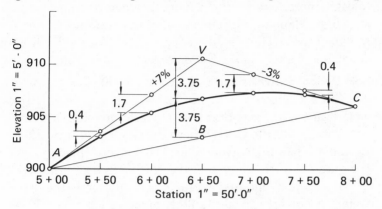

Fig. 3–6. Parabolic vertical curve.

a 300-foot vertical curve is required to give a speed of 25 mph. Then the elevation of $B = \frac{1}{2}(900 + 906) = 903$. The elevation at station $6 + 50$, $M = 903 + \frac{1}{2}(910.5 - 903) = 903 + 3.75 = 906.75$. $K = 910.5 - 906.75 = 3.75$. The offset at station $5 + 50 = \frac{1}{9} \times 3.75 = 0.4$ and the curve elevation $= 903.5 - 0.4 = 903.1$. The offset at station $6 + 00 = \frac{4}{9} \times 3.75 = 1.7$ and the elevation $= 907.0 - 1.7 = 905.3$. The offset at $7 + 00$ is the same as at $6 + 00$ and at $7 + 50$ the same as at $5 + 50$.

Drainage Structure Design

Road drainage structures include ditches, culverts, and bridges. The ditch carries runoff and seepage to the **culvert.** The culvert carries ditch and stream water laterally under the road. The bridge crosses the stream with flow too large for a culvert, or where conditions are unsuitable for a fill. The crown of the surface of the road might also be considered a drainage structure as it carries rainfall off the road to avoid saturation of the subgrade and consequent loss of bearing strength.

The shape and dimensions of the ditch are usually standardized in a given operation or locality. Public forest agencies specify a triangular ditch with a 3 to 1 slope from the bottom of the ditch to the edge of the road so that the ditch can be maintained with a motor grader. The back slope is the same as that of the cut bank, and the minimum depth is 1 foot. Logging companies prefer narrower and deeper ditches on hillside roads. If a ditch segment will have to carry more than rainfall runoff, as when draining a swampy area or springs, Manning's formula is used to determine the required cross-sectional area. The ditch gradient should never be less than 0.5 percent, and a minimum of 2 or 3 percent is preferable.

Culvert Design. Culverts are required under all fills across draws or water courses which would act as dams if accumulated water were not drained. Lateral drainage culverts are also required on long segments of hillside roads where the ditch would overflow before reaching a natural culvert site. The culvert spacing depends upon the gradient of the ditch and the erosion index of the soil. The steeper the gradient and the more erosive the soil, the closer is the spacing. A culvert spacing table is given in Chapter 650 of the *Forest Engineering Handbook*. Lateral drainage culverts are of the minimum permissible diameter. Public agencies specify minimums of 15 to 18 inches and logging companies use minimums of 6 or 8 inches.

The determination of the culvert diameter required is often the most perplexing problem facing the road designer. Culverts are expensive, and the cost of a culvert of ample size must be weighed against the consequences of failure if the culvert is too small to carry the peak runoff, which may be expected only once in a period of years. Experience with the culverts in roads previously built in the same watershed, or in a locality where conditions are comparable, is the best guide. The handbooks published by the pipe culvert manufacturers give various methods of computing culvert diameter. Some of the methods commonly used in forest road design will be discussed briefly here.

If the culvert is to carry the flow of a stream which can be measured at flood stage, the required pipe diameter may be obtained from tables in the handbooks distributed by the manufacturers of culvert pipe. The flow of a large stream is best obtained by measuring the velocity with a portable current meter, and the cross-sectional area with level rod and tape. The flow in cubic feet per second equals velocity in feet per second multiplied by area in square feet. Flow of a small stream is measured with a **weir**, a notched plank set in the crest of a temporary dam. The depth of the water flow through the notch is measured, and the flow through a triangular weir computed from the formula $Q = 2.5H^{5/2}$ where Q is volume in cfs. and H is the depth in the notch in feet. An approximation of stream flow may be obtained by measuring the cross-section and the slope of the stream bed, and estimating the roughness of the channel. Flow is computed by the Chezy formula, $Q = CA\sqrt{RS}$, where A is the cross-sectional area to highwater mark, R is hydraulic radius or A divided by wetted perimeter of the cross-section, and S is the slope of the stream bed in percent. C is a coefficient for roughness, ranging from 30 to 40 for rough, obstructed channels and 40 to 60 for stony earth channels, to 60 to 80 for clean earth.

Where stream flow is not measured, the Talbot formula is most commonly used for estimating required culvert size. The original formula for a maximum intensity of rainfall of 4 inches per hour is $A = CM^{3/4}$ where A is the required cross-sectional area of culvert in square feet and M is the drainage area in acres. For other rainfall intensities A is proportional to 4, e.g., for 1-inch intensity the area is $\frac{1}{4}$ the A given by the formula. The value of C varies with topography and steepness of slope, ranging from 1.0 for steep, rocky ground to 0.05 for flat land. Fig. 725-1 (*Forest Engineering Handbook*) gives culvert diameters for C values of 1.0, 0.8, and 0.6 and rainfall intensities of 1, 1.25, and 1.5 inches, the range in western Oregon mountains. Rainfall intensity is obtained from weather bureau or local records, and drainage area from maps or aerial photos. The United States Geological Survey is a source of runoff data in localities where they maintain stream gauging stations.

The length of the culvert is found graphically by plotting the cross-section at the culvert site. A convenient homemade graphic "Culvert Length Finder" is described in Miller (1964). Ordinarily the culvert is laid perpendicular to the road center line on the stream bed gradient. The end of the culvert should project 1 to 2 feet beyond the toe of the fill. If the thread of the stream is at an angle to the center line, the line of the culvert is skewed, and the length found by multiplying the distance scaled on the cross-section by the cosine of the angle. If the stream gradient is so steep as to require an excessively long culvert, the culvert may be skewed to the nearest edge of the fill to discharge on solid ground. Discharging a skewed culvert onto natural ground above the stream has the added advantage of reducing stream sedimentation, as some of the sediment will be trapped or filtered out before the water returns to the stream. A "cannon" installation discharging on the side of a high fill may be used to reduce culvert length, but this requires protecting the fill bank from erosion with rock rip-rap or discharging into a flume of culvert pipe halves laid on the fill bank.

BRIDGES

A bridge is required where the stream flow cannot be carried by a culvert or where flood conditions would endanger a fill. On streams traveled by anadromous fish, bridges are preferred to culverts. Some logging companies consider the maximum economic pipe culvert diameter to be 8 or 9 feet. If a larger culvert would be required they build a bridge. Forest road bridges are classified as permanent or temporary, although the terms are relative. Permanent types include glued-laminated creosoted timber beams or trusses, steel girders, pre-fabricated pre-stressed concrete girders, and reinforced concrete arches poured in place. The last type appears to be the most popular with public forest agencies. The permanent types are usually designed by a structural engineer who specializes in bridge design.

The temporary types include sawn **stringer** bridges for short spans, up to about 32 feet in length, and log stringer bridges for longer spans. Spans as long as 120 feet have been built of Douglas-fir, but maximum spans of 70 or 80 feet are more common. The ends of the stringers rest on abutments of piling, timber-framed bent, logs or concrete, depending upon the ground conditions. The **deck** is of heavy plank laid crosswise for public travel with outside guard rails, or lengthwise on sawn timber cross beams to form two wheel tracks for logging vehicle traffic. A pole guard rail is bolted to the inside of each wheel track. Since the individual log stringers in a bridge may differ in stiffness and consequently in the amount they deflect under loading, equalizer beams are desirable. The equalizer is a heavy sawn timber beam or a log bolted crosswise of the bridge to the bottom side of the stringers. Equalizer beams are placed at one-third span points on the longer spans, or at mid-point on the shorter spans. If the timber available for stringers is not large enough in diameter to carry the design load with a single tier of logs, two tiers of logs, one above the other, are sometimes used. The load capacity of two log stringers may be increased by keying the top log to the bottom log so they cannot slide in a horizontal plane when deflecting under load. The strength of a keyed beam is up to 3.2 times that of one of the members, whereas the strength unkeyed is only twice that of one beam. Other means of strengthening a bridge used by loggers include suspending the bridge by wire ropes running from trees or towers near each end of the bridge to a cross-log at mid-span, or supporting the stringers by inclined bents from the abutment footings to one-third points of the span. The life of the log stringer depends upon the durability of the species used. Douglas-fir, peeled of bark, will last 12 to 15 years. The deck plank subject to traffic wear must be replaced at intervals averaging about 4 years.

The graduate logging engineer or forest engineer is competent to design a timber bridge. The timber available for stringers is investigated for size and structural grade. The bridge is designed to carry the live load which will cause the maximum bending moment. This may be a truck and lowboy machinery trailer loaded with the heaviest item of logging equipment to be transported, a portable steel tower, or a mobile loader. In localities of heavy snowfall the snow load may be the maximum live load. The bending strength of a log stringer is computed for the diameter inside the sapwood.

For crossing wide streams which are shallow during the logging season, the low water bridge may offer an economical solution to the bridge problem. It is designed with the deck sloping downward toward the upstream side, so that water flowing over the bridge at flood stage will create a vertical component of force to help hold the bridge in place. Low concrete abutments are grouted into bedrock and bridged with concrete or sawn timber stringers. The timber stringers are removed before flood season, stacked on the bank above high water level, and replaced on the abutments when flood season is passed. The composite beam of sawn timber and concrete which serves as the deck may also be used.

PLAN AND PROFILE

The **plan and profile** of the final location is the blueprint for the construction of the road. Standard plan and profile tracing sheets have profile engraving printed on the lower half. The plan shows L-line stationing, bearings of tangents, curve data, corners and corner ties and land subdivision or property lines. The P-line traverse is also shown on the plan for designed roads. Classification of material, as common or rock, and the acreages of right-of-way to be cleared are sometimes shown. Lines showing the edges of the right-of-way are sometimes drawn on the plan, especially if the road crosses other ownerships and right-of-way acquisition is involved. The profile is of the L-line only, drawn to the same horizontal scale as the plan. If the L-line has been staked and levels run, the staked ground elevations are plotted. Otherwise the elevations are computed from the P-line offsets. The profile shows the ground line, grade line, vertical curves, and the location and dimensions of each culvert. On single-lane roads **turnouts** to be constructed are shown graphically. If the design is to be checked and approved by the designer's superior, the side slopes are shown along the top line of the profile. If earthwork has been computed, cubic yards by segments of excavation and of embankment are entered on the profile. This information is useful to the construction supervisor in planning earth moving. The standard sheet has lines on the margin for the names or initials of the individuals responsible for surveys, design, drafting and computing.

CONSTRUCTION COST ESTIMATE

The final step in the engineering of a forest road is the compilation of a construction cost estimate. Companies require estimates for budgeting the construction funds and for determining whether the road as located is of economically feasible standards. The cost of roads to be built in connection with a timber sale is an important element in the appraisal of stumpage value by a public agency. The estimator assembles all available records of costs of roads previously built, and of machine rates and production data of construction equipment to be used. Following are the usual road construction operations in the chronological order in which they are performed and the cost elements on which the estimate is based:

1. Felling and bucking right-of-way timber. Based on a 100 percent cruise of timber and cutting cost per 1,000 board feet.

2. Yarding and decking right-of-way timber. Based on the cruise, the average yarding distance to decking sites where the logs can be piled for loading after the road is built, and yarding cost per 1,000 board feet.

Logging companies customarily charge items 1 and 2 to logging and not directly to road construction.

3. Clearing and grubbing. Based on the number and size of stumps from the right-of-way cruise and cost records of blasting stumps, if blasting is required, and grubbing costs per acre. If brush piling and burning is to be done, which is sometimes required by public agencies, the cost per acre is added.

4. Excavation or earthwork. If volumes have been computed, the material is classified and the appropriate rate per cubic yard applied. If balanced sections requiring end haul beyond the normal free-haul distance are designed, the cost of over-haul per station-yard is added. If there is not enough yardage in excavation to complete a fill the cost of borrow per yard is added. If yardage is not computed the cost of excavation per station is estimated, based on bulldozer or power shovel production records and machine rate. Where rock is encountered, drilling and shooting or ripping cost is added.

5. Culverts. Based on cost per lineal foot for each size of culvert. Pipe culvert cost includes purchase price, trucking, bed preparation, and installation.

6. Trimming subgrade and ditching. Cost per station based on motor grader performance and machine rate.

7. Surfacing. Cost per cubic yard of digging gravel or quarrying and crushing rock, of hauling, and of spreading.

8. Bridges. Cost per lineal foot and total cost.

9. Costs for the entire road project are totaled, and the average cost per station and per mile computed.

SOILS ENGINEERING

Constructing permanent forest roads for multiple use, and minimizing maintenance activities, necessitates consideration of the engineering properties and characteristics of forest soils. Soil is the construction material of the **subgrade.** Coarse-grained soil (gravel) or rock, the parent material of the soil, is the construction material of the pavement structure (surfacing) of all-weather roads. It is as important to be familiar with the properties of these materials as it is to be familiar with the properties of materials used in designing a bridge or a building. The properties of soils as they relate to the erosion and infiltration of water affect the design of drainage structures. Consequently, soil is an important consideration in every step of forest road engineering: reconnaissance, location, design, and construction. Soil also has an important effect upon the degree of road maintenance which may be necessary on a road.

In light of the importance which is attributed to soil, insofar as forest road construction is concerned, soils engineering has been the most neglected phase of forest road engineering. Surfacing failures, subgrade failures, slides and washouts, resulting in high maintenance or replacement costs and slowed or interrupted traffic, tend to emphasize the magnitude of this neglect.

It is assumed that the reader has had introductory courses in geology and forest soils, which are included in the curricula of most forestry schools.

Necessity for Soils Investigation

Investigation of soils along a road route by some type of soil survey provides pertinent information which may be used:

1. To avoid construction problem areas, such as potential landslides, poorly drained areas, and soils of low-load bearing capacity.
2. To locate suitable fill materials and surfacing rock.
3. For guidance in road prism design, such as relative cut and fill slope ratios, and stability of embankments.
4. To determine the bearing strength of the subgrade and the pavement structure design.
5. To assist in the design of adequate drainage facilities. For example, culvert spacing depends on soil as well as gradient.
6. To help plan erosion control measures.
7. For control of road building materials during construction.
8. For logging planning, silviculture, and other forest management purposes.
9. For estimating road maintenance costs.

Engineering Properties and Characteristics of Soil

Soils have basic engineering properties that determine their characteristics as materials of construction. The most important factors affecting the properties are soil grain size, gradation of sizes, and moisture content. The basic soil properties of importance to the forest road engineer, such as internal friction, cohesion, capillarity, elasticity and compressibility, and permeability are defined and discussed in the *PCA Soil Primer* (Portland Cement Association, 1956).

Classification and Identification of Soils

Several systems of soil classification have been developed. The silviculturist makes use of a system which relates the soil to site quality for tree growth. The Bureau of Public Roads and state highway departments use a classification system which is based on grouping soils by load-carrying capacity. The *Unified Soil Classification System* developed by the Corps of Engineers, United States Army (1953), is recommended as the most practical for use by forest road engineers.

The system is based on identification of soils according to their textured and plasticity qualities and their grouping with respect to their performance as engineering construction materials. The soil is given a descriptive name and a letter symbol which indicates its principal characteristics. Table 3–3 shows in outline form the method of identification employed by the *Unified Soils Classification System*. The characteristics of *U.S.C.S.* soil groups for road construction are given in Table 3–4.

Pavement Structure Design

The **pavement structure** is the surfacing material placed upon the roadbed of all-weather roads. On logging roads the pavement structure usually consists of a base course of coarse pit run rock or gravel, and a wearing course of crushed rock or fine gravel. Wheel load pressure is transmitted from the surface through the pavement structure to the subgrade in the form of a frustum of a cone, consequently the unit pressure on the subgrade decreases with increase in the thickness of the pavement structure.

The influence of a substantial pavement structure on a subgrade is to spread a wheel load over a larger area and thereby reduce the unit pressure. The unit pressure is not uniform throughout the base area of the cone frustum but ranges from a maximum in the area

TABLE 3–3. Unified Soil Classification*

Field Identification Procedures†				Group Symbols‡	Typical Names
Coarse Grained Soils (More than half of material is *larger* than No 200 sieve size)‖	Gravels (More than half of coarse fraction is larger than No 4 sieve size)#	Clean Gravels (Little or no fines)	Wide range in grain size and substantial amounts of all intermediate particle sizes	GW	Well graded gravels, gravel-sand mixtures, little or no fines
			Predominantly one size or a range of sizes with some intermediate sizes missing	GP	Poorly graded gravels, gravel-sand mixtures, little or no fines
		Gravels with Fines (Appreciable amount of fines)	Non-plastic fines (for identification procedures see CL below)	GM	Silty gravels, poorly graded gravel-sand-silt mixtures.
			Plastic fines (for identification procedures see CL below)	GC	Clayey gravels poorly graded gravel-sand-clay mixtures
	Sands (More than half of coarse fraction is smaller than No 4 sieve size)	Clean Sands (Little or no fines)	Wide range in grain sizes and substantial amounts of all intermediate particle sizes	SW	Well graded sands, gravelly sands, little or no fines
			Predominantly one size or a range of sizes with some intermediate sizes missing	SP	Poorly graded sands, gravelly sands, little or no fines
		Sands with Fines (Appreciable amount of fines)	Non-plastic fines (for identification procedures see ML below)	SM	Silty sands, poorly graded sand-silt mixtures.
			Plastic fines (for identification procedures see CL below)	SC	Clayey sands, poorly graded sand-clay mixtures

directly under the load to zero at an indefinite radial distance. While this condition is recognized, design procedures usually consider that the vertical pressure of a wheel load is distributed uniformly throughout the base of a cone frustum which has a base angle of 45 degrees.

Figure 3–7 indicates the relationship between a wheel load of 4050 pounds and the unit pressures on the subgrade for various pavement structure thickness. Based upon the foregoing considerations, it is possible to determine pressures on the subgrade for various wheel loads and pavement thickness. Following is the procedure to determine the unit pressure on a subgrade supporting a pavement structure of a certain thickness which in turn has contact with a tire supporting a certain wheel loading.

TABLE 3–3 (Continued)

		Dry Strength (Crushing Characteristics)	Dilatancy (Reaction to Shaking)	Toughness (Consistency Near Plastic Limit)	Group Symbols‡	Typical Names
Fine Grained Soils (More than half of material is *smaller* than No 200 sieve size)	Silts and Clays (Liquid limit less than 50)	None to slight	Quick to slow	None	ML	Inorganic silts and very fine sands, rock flour, silty or clayey fine sands with slight plasticity
		Medium to high	None to very slow	Medium	CL	Inorganic clays of low to medium plasticity, gravelly clays, sandy clays, silty clays, lean clays
		Slight to medium	Slow	Slight	OL	Organic silts and organic silt-clays of low plasticity
	Silts and Clays (Liquid limit greater than 50)	Slight to medium	Slow to none	Slight to medium	MH	Inorganic silts, micaceous or diatomaceous fine sandy or silty soils, elastic silts
		High to very high	None	High	CH	Inorganic clays of high plasticity, fat clays
		Medium to high	None to very slow	Slight to medium	OH	Organic clays medium to high plasticity
	Highly Organic Soils	Readily identified by color, odor, spongy feel and frequently by fibrous texture			Pt	Peat and other highly organic soils

IDENTIFICATION PROCEDURES ON FRACTION SMALLER THAN No. 40 SIEVE SIZE

* Courtesy: Bureau of Land Management, U.S. Department of the Interior.
† Excluding particles larger than 3 inches and basing fractions on estimated weights.
‡ *Boundary classifications*—Soils possessing characteristics of two groups are designated by combinations of group symbols. For example GW-GC, well graded gravel-sand mixture with clay binder. All sieve sizes on this chart are US standard.
‖ The No 200 sieve is about the smallest particle visible to the naked eye.
For visual classifications, the ¼″ size may be used as equivalent to the No 4 sieve size.

1. Determine the contact area of the tire, which is somewhat elliptical, and reduce this to the radius dimension (r) of a circle of comparable area.
2. Add the depth (d) of the pavement structure to the radius dimension (r).
3. Determine the area of a circle with a radius equal to $r + d$.
4. Divide the wheel load by the area in square inches and the result is the unit pressure on the subgrade.

TABLE 3–4. Characteristics of Soil Groups for Road Construction (Unified Soil Classification System)

U.S.C.S. Symbol		Value as Sub-Grade*	Field CBR Value	Drainage Characteristics	Erosion Index	Frost Action	Compressibility and Expansion	Compaction Characteristics	Dry Weight† lbs./cu. ft.
1		2	3	4	5	6	7	8	9
GW		Excellent	60-80	Excellent	100	None to very slight	Almost none	Good	125-140
GP		Good to excellent	25-60	Excellent	100	None to very slight	Almost none	Good	110-130
GM	d	Good to excellent	40-80	Fair to poor	60	Slight to medium	Very slight	Good	130-145
	u	Good	20-40	Poor to impervious	50	Slight to medium	Slight	Good	120-140
GC		Good	20-40	Poor to impervious	70	Slight to medium	Slight	Fair	120-140
SW		Good	20-40	Excellent	80-90	None to very slight	Almost none	Good	110-130
SP		Fair	10-25	Excellent	80-90	None to very slight	Almost none	Fair to good	100-120
SM	d	Good	20-40	Fair to poor	20	Slight to high	Very slight	Good	120-135
	u	Fair to good	10-20	Poor to impervious	10	Slight to high	Slight to medium	Good	105-130
SC		Fair to good	10-20	Poor to impervious	50	Slight to high	Slight to medium	Fair	105-130
ML		Fair to poor	5-15	Fair to poor	10-20	Medium to very high	Slight to medium	Good to poor	100-125
CL		Fair to poor	5-15	Practically impervious	40	Medium to high	Medium	Good to fair	100-125
OL		Poor	4-8	Poor	30-40	Medium to high	Medium to high	Fair to poor	90-105
MH		Poor	3-8	Fair to poor	30-40	Medium to very high	High	Poor to very poor	80-100
CH		Poor to very poor	3-5	Practically impervious	50-60	Medium	High	Fair to poor	90-110
OH		Poor to very poor	3-5	Practically impervious	50	Medium	High	Poor to very poor	80-105
PT		Unsuitable		Fair to poor		Slight	Very high	Fair to poor	

* Value as subgrade, foundation, or base course (except under bituminous) when not subject to frost action.
† Unit dry weight for compacted soil at optimum moisture content for modified AASHO compactive effort.
Courtesy: Bureau of Land Management, U.S. Department of the Interior.

Applying the procedure to a specific example involving a 4050-pound wheel load, a 10:00 × 20 tire, and a pavement structure thickness of 10 inches results in a unit pressure on the subgrade of 5.6 pounds per square inch. The result is obtained as follows: A 10:00 × 20 tire has a contact area of 81 square inches, which is

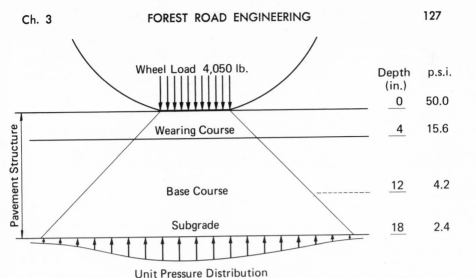

Fig. 3–7. Relationship between pavement structure thickness and unit pressure on the subgrade. (Courtesy of the Bureau of Land Management, U.S. Department of the Interior.)

equivalent to a circle with a radius (r) of 5.1 inches. The slope of the cone frustum is 45° with its base; therefore, to determine the radius of the base of a cone frustum the depth (d) of the pavement structure is added to the radius (r) of the tire contact area. In the example r is 5.1 inches and d is 10.0 inches which, when added together, is equivalent to the radius of the base of the cone frustum. The area of the base is 716 square inches, which divided into the wheel load of 4050 pounds is equivalent to an average unit pressure on the subgrade of 5.6 pounds per square inch. Average unit pressure may be determined for any wheel load and pavement structure depth d by the formula, where r is radius of a circle equal in area to tire contact area:

$$\text{psi} = \frac{\text{wheel load in pounds}}{\pi (r + d)^2}$$

Surfacing Bearing Strength

The design of the pavement structure starts with the determination of the **bearing strength** of the subgrade. The California Bearing Ratio (CBR) is recommended as a useful measure of the bearing strength of forest road subgrade soil. The CBR expresses the resistance of a compacted soil to penetration by a test piston as a percentage of the resistance to penetration of compacted crushed rock. The CBR may be determined by means of portable test apparatus for field use or in a soil mechanics laboratory. Interpretation of CBR test results to determine the CBR value is best done by a soil mechanics engineer. The services of consultants in this field are available. The required thickness of pavement structure for a range

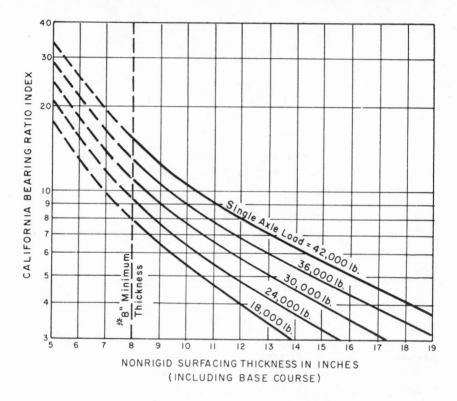

Fig. 3–8. Thickness chart: nonrigid surfacing based on California Bearing Ratio indices. (Courtesy of the Bureau of Land Management, U.S. Department of the Interior.)

of CBR values and single-axle loads is given in Fig. 3–8. The usual range of field CBR values for the various soil groups included in the Unified Soil Classification System is given in Table 3–4.

SUGGESTED SUPPLEMENTARY READING

1. BYRNE, JAMES J., ROGER J. NELSON, and PAUL H. GOOGINS, 1960. *Logging road handbook. The effect of road design on hauling costs.* 65 pp. U.S. Department of Agriculture Forest Service. Agricultural Handbook No. 183. Gives the factors affecting log truck travel time, and methods of calculating travel time and log-hauling cost.
2. HUGGARD, E. R. *Foresters' engineering handbook.* 1958. W. Heffer & Sons, Ltd., Cambridge, England. A handy source of useful information concerning the various aspects of forest engineering.
3. PEARCE, J. KENNETH. 1960. *Forest engineering handbook.* 220 pp. U.S. Department of the Interior, Bureau of Land Management. State Office, Portland, Oregon. Covers the subject of forest road engineering in detail.
4. *Roads handbook.* 1965. U.S. Department of the Interior, Bureau of Land Management, Washington, D.C., 20240. Release 9–20. August 1965. The manual for engineering roads on Bureau of Land Managements lands.

4

Forest Road Construction

Included in this chapter are those items which a forester or forest engineer should consider and examine if he is responsible for supervising or inspecting a forest road construction operation. The topic is approached from the standpoint of the supervisory duties usually performed by a forester or construction engineer. A timber sale forester, employed by a public timber managing organization, often is confronted with the task of inspecting a forest road construction project for compliance with design specifications as embodied in a timber sale contract. Attention is directed to those items which tend to cause the most trouble in regard to forest road construction. Suggested actions are presented which may be employed to detect discrepancies between contract requirements on forest road construction projects and what actually is being, or has been, accomplished on the job. The chapter also covers forest road construction, stressing those points which tend to produce a serviceable and safe forest road consistent with economy in both initial construction and subsequent maintenance costs.

The construction of forest roads involves a sequence of operations designed to convert a predetermined route through a timbered area to a condition which will permit log trucks to haul maximum payloads of logs or bolts from the woods to a mill, railhead, or log dump in a safe, efficient, and economical manner. Commercial timber grows on terrain which ranges from smooth and unbroken to extremely steep and rugged. Confronted with this range of topography, it is apparent that no one series or sequence of construction operations will conform to all conditions.

In general the road contruction procedure involves: removing the vegetation, trees, and brush, and certain stumps from a portion of the road right of way; placing **culverts** and constructing bridges; excavating or embanking to a certain grade elevation; forming and clearing the ditch; rock surfacing all-weather roads; and installing open-top culverts.

CLEARING AND GRUBBING

Clearing

The objective of clearing is to dispose of vegetative material, such as live merchantable trees, live unmerchantable trees, **snags,** and **windfalls,** from the road right-of-way.

The extent of clearing usually is designated by setting stakes or using **flagging** to establish a clearing line. The clearing line may coincide with the line marked by **slope stakes.** On moderate to steep slopes, the practice of establishing the clearing line a short distance, four or five feet, beyond the upper slope stake line tends to stabilize the upper portion of the cut slope, as shown in Fig. 4–1. Stakes may be lost during the clearing operation; therefore, the clearing line or slope stakes should be **referenced** to allow for re-establishment for use during the **grubbing** and earthwork operations.

Unmerchantable windfalls should be **bucked** into convenient lengths for removal from the right-of-way. Windfalls containing sufficient merchantable volume should be bucked into standard log lengths. Bucking windfalls during the initial phase of the clearing operation is easier and, above all, much safer than bucking them in conjunction with the green timber. The practice of skidding windfalls prior to the main felling and bucking in heavy stands of green timber tends to reduce breakage.

Snags next should be felled and bucked into standard log lengths if they contain sufficient sound volume to qualify as a merchantable log. However, if they are unmerchantable, the same procedure as used on unmerchantable windfalls would prevail. Snag felling requires constant vigilance to prevent committing an unsafe act. Dan-

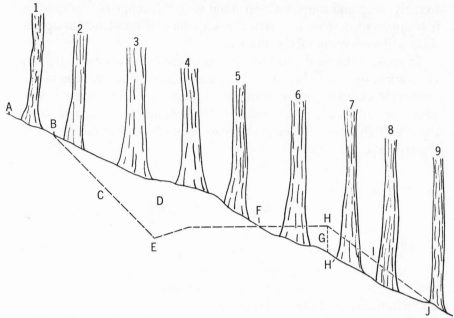

Fig. 4–1. Cross section of a forest road on moderate side slope showing: A, clearing line stake; B, slope stake; C, cut slope; D, section in cut or excavation; E, ditch; F, center line stake; G, section in fill or embankment; H-H', vertical projection of outer edge of subgrade; I, fill slope; and J, toe of fill.

gerous snags may be felled using dynamite. The practice of felling all snags within a distance equal to their height from the road is advocated by the more safety-minded.

Felling and bucking the standing green timber follows next in order. Figure 4–1 shows the extent to which standing timber should be cleared from the road right-of-way. Trees numbered 1 through 8 inclusive should be felled. Directional felling, reasonably parallel to the direction of the road, will facilitate more efficient log decking. Felling which placed the tops off the right-of-way reduces brush handling. The felled trees are then bucked into logs of standard length plus an adequate allowance for trim.

On steep to moderate side slopes, trees situated adjacent to the upper clearing line of the right of way which may have their roots undercut during the excavation operation should be felled. Removing these trees will prevent the possibility of having them fall onto the road in the future. Their early removal will also prevent the adjacent ditch from becoming clogged by soil disrupted when such trees fall. (Refer to Fig. 4–1: Tree number 1 is a case in point.)

The trees which have stumps of the size and type which will allow removal using a bulldozer or winch line should be felled to leave a stump three to four feet in height. A high stump greatly increases bulldozer or winch efficiency and production by providing added leverage.

The smaller trees and slash may have to be piled for burning in those instances where the terms of the road construction contract so specify and the degree of forest fire hazard permit using this method of disposal.

Pushing trees over by bulldozer action instead of the conventional felling operation generally is not permitted by public forest agencies. In situations where this practice is permitted, the bulldozer operator should be protected from falling branches and dead tops by a strong canopy on his bulldozer.

Grubbing

Basically the grubbing operation consists of preparing a portion of the road right-of-way for subsequent road construction activities.

The activities involved in removing all large, woody material from the right-of-way, such as logs, tops, chunks, windfalls, snags, and large pieces of stumps, are included under the general heading of chunking out.

Merchantable logs are moved to reasonably flat terrain located outside of the road-building activity zone, but still near enough to allow the logs to be loaded on trucks without additional handling. The logs are bunched or decked by bulldozer blade action or winch line yarding. Unmerchantable logs, chunks, and debris should be moved from the road-building zone and piled to prevent the possibility of being buried in a fill. Decomposition of this material will result in unequal fill settlement which, in turn, woud have a detrimental effect on the road by causing unevenness in the road surface. Pushing debris downslope, adjacent to the clearing line, is a recommended practice which tends to prevent this undesirable material from becoming incorporated in the fill.

Stump removal is accomplished by using dynamite, bulldozers, or a combination of both. It is somewhat general practice to remove stumps which are situated between the line marked by the upper slope stakes and the line designating the vertical projection of the outer edge of the road subgrade. Fig. 4–1 indicates that stumps from trees numbered 2 to 6 inclusive be removed in line with the aforementioned general practice. The figure also indicates the position of the vertical projection of the outer edge of the subgrade by the letters H and H'. Placing fill over stumps located within the subgrade width should not be allowed. Stumps buried in a fill will decompose and leave voids which eventually will cause the road surface to weaken and, in turn, will increase road maintenance costs.

Information regarding species, stump diameter, stump condition (green, old but solid, or partly rotted), root characteristics (tap, semi-tap, or lateral), and whether the soil is firm and dense or loose and dry, if ascertained prior to a stump removal operation, can be of considerable value to the **powder men.**

Dynamite develops power in the form of a gas which is generated during the explosion. Actually, dynamite is transformed from a solid state into a gaseous state in a minute fraction of a second. Confining the gas by proper charge placement, together with due consideration to the physical characteristics of the stump and its environs, should be considered to achieve maximum results. The *Blasters' Handbook* (DuPont, 1966) is recommended as a guide to the proper placement and size of charge for stump removal.

Strict compliance with federal, state, and local laws and regulations when transporting or storing dynamite and blasting caps is of utmost importance. Appendix B in the *Blasters' handbook* contains a list of precautions to be exercised when handling dynamite and blasting caps.

The information regarding stump characteristics and environs should also be considered when bulldozers are used to remove stumps. Here also the horsepower of the bulldozer can be utilized to the fullest extent when stump information is considered and applied. A very effective grubbing combination consists of a medium or large bulldozer, Fig. 4–2 (80 to 150 drawbar horsepower), which has the ability to remove most stumps, together with dynamite for those larger stumps which are beyond the dozer's capability.

Stump splitters offer a means whereby large or "stubborn" stumps may be halved or split, thereby facilitating their removal. Splitters are made of heavy gauge steel and resemble a large wedge. Customarily, splitters are mounted on the rear end of a bulldozer, which provides the power for the splitting operation. Fig. 4–3 shows a

Fig. 4–2. Bulldozer engaged in grubbing operations on a logging road location. (Courtesy of the Terex Earthmoving Division, General Motors Corporation.)

typical stump splitter mounted on the cross bar of a ripper attachment.

A road-building project, being conducted on other than flat ground and which requires construction activity at more than one locality, may require that a pioneer road be constructed. A pioneer road is a tractor-trail type of access to accommodate road building equipment and allows for the delivery of culverts and bridge materials.

Pioneer roads are often constructed in a somewhat haphazard manner, which makes subsequent construction to road design cross-section difficult and more costly. On side slopes, the best location for the pioneer road is considered to be along the top of the cut slope. On side hill fill locations, building the pioneer road along the toe or bottom edge of the fill will provide a bench for holding and compacting the fill. Yarding or bunching the logs from the right-of-way timber may also influence the location of the pioneer road. The practice of discussing pioneer road location with construction per-

Fig. 4–3. Stump splitter mounted on the tool beam of a ripper attachment. (Courtesy of the American Tractor Equipment Corporation.)

sonnel in advance of actual construction, to ensure that all problems involved have been considered, is advised.

Some degree of control should be exercised over the clearing and grubbing personnel to make sure that all debris, which ought to have been removed, actually is removed. This is of particular importance insofar as the **excavation** and **embankment** operations are concerned. An incomplete clearing and grubbing operation will tend to reduce the effectiveness of the subsequent phases of the road building process.

DRAINAGE STRUCTURES

Experienced forest and logging engineers are in agreement that no other single item of road building is as important as drainage.

Drainage structures associated with forest roads are: culverts, ditches, bridges, and open-top culverts.

Culverts

In order to function according to design specifications, proper installation of culverts is necessary. To ensure that culverts are installed at the proper locations, the practice of staking and referencing the points of inlet and outlet is advised, thereby assisting the installation crew in attaining proper alinement, elevation, and gradient. The effort involved in culvert size determination and field engineering is of little use if nullified by improper installation techniques.

Acceptable installation processes require that proper consideration be given to bedding, strutting, handling and laying, and backfilling and tamping.

Bedding. Adequate bedding provides stable, uniform bearing for the culvert throughout its supported length. The bed should be of firm, well-compacted granular soil free of rocks or boulders and be rounded on the culvert radius to a depth of approximately one-sixth the culvert diameter, thereby fitting the contour of the culvert. Rocky or soft unstable soil should be excavated to a depth of not less than 6 inches and a width of two diameters, and backfilled with sand or fine gravel.

Installing a culvert under a high fill on a uniformly unstable foundation may require cambering. Camber is achieved by raising the center portion of a culvert above its normal position to allow for subsidence after the fill is placed. Sufficient camber may be obtained by installing the upstream portion of the culvert on a flatter grade than the downstream portion. The amount of camber at a specific installation is a matter of judgment; however, the center of the culvert should not be raised to such a degree that flow will be impeded by the adverse gradient in the upstream portion of the culvert.

Strutting. A corrugated metal pipe (CMP) culvert deflects under a load. The deflection tends to shorten the vertical diameter and increase the horizontal diameter. The extent of such deflection is an indication of how well the backfill has been placed and compacted. The better the quality of the backfilling operation, the smaller is the deflection. Strutting involves elongating the vertical diameter of a corrugated metal pipe culvert and is done prior to backfilling, which allows additional side support to develop as a pipe resumes a full-round shape under the fill load.

The general practice indicates that CMP culverts larger than 48 inches in diameter be strutted. The use of shop-strutted pipe, which is strutted by means of heavy gauge wire, is recommended. Based upon the labor cost of field strutting, shop strutting pipe is generally competitive in installed cost.

Field strutting is done at the culvert site after the pipe has been laid but prior to the backfilling operation and consists of using timbers of appropriate size for the upper and lower sills and vertical struts. A soft wood compression cap must be used between the vertical struts and the top sills to permit a certain amount of deflection which develops as the fill material is being placed and compacted (National Corrugated Steel Pipe Association, 1965).

All struts must be removed after the fill has consolidated. However, in the event of danger from flood water, they may have to be removed prior to the time the fill has completely settled.

Handling and Laying. Care must be exercised in handling galvanized CMP to prevent denting or scraping the galvanized coating. Pipe may be rolled or hoisted but not dragged over the ground. It should not be dropped from a truck. When necessary to push the pipe with a bulldozer, protecting the pipe from blade action by a plank and from scraping along the ground by the use of skids or poles is advised. Strutted pipe must be laid with the longer diameter in a vertical position.

Backfilling and Tamping. Proper backfilling has an important effect upon the strength and load-bearing ability of any drainage structure. Corrugated metal drainage structures develop lateral support as they deflect under load. Consequently, to attain maximum load-bearing capacities and prevent washout and culvert settlement, it is essential that the backfill be made of suitable material properly placed and adequately tamped or compacted.

Backfill material with drainable characteristics is preferred; however, local fill material may be used, if it is placed and compacted with care. Backfill material devoid of large rocks and hard lumps of soil is most effective. The most effective material has a granular structure which contains enough silt or clay to provide a dense, stable fill (American Iron and Steel Institute, 1967).

The usual procedure is to place the backfill by hand or by means of a small bulldozer, in 6-inch layers. Each layer should be well tamped under the **haunches** of the pipe, and then filled and tamped up to the top of the culvert. Fill soil then is placed evenly in layers from both sides until the culvert is covered to a depth of at least 1 foot over pipe of diameters up to 36 inches and to a depth of $\frac{1}{3}$ the diameter in the case of larger sizes.

Heavy equipment travel over the culvert pipe should not be permitted until the backfill soil supporting the sides has been tamped and compacted, and the required cover thickness has been provided.

Ditches

A bulldozer with an **angle** and **tilt** type of blade is customarily used to form roadside ditches or, if soil conditions permit, a grader or motor patrol may be used. A very effective combination consists of having a bulldozer with an angle-tilt type of blade working in conjunction with a grader. The bulldozer forms a rough ditch and the grader smooths and finishes the ditch. Rock in the ditch area will require dynamite. Ditches usually are constructed during the earthwork phase of road construction. In no case should forming the ditch be allowed after the road has been surfaced. To ditch at this time results in spreading soil over the road surfacing, and subsequently may tend to muddy a road during wet weather and create a dust problem during dry weather.

In general, ditches ought to have the same gradient as the road and be deep enough to drain the subgrade. Abrupt changes in ditch slope should be avoided to prevent the possibility of pools forming in the ditch. Proper construction provides conditions whereby water is conveyed and not stored. Allowing water to pool in a ditch will have a detrimental effect on the stability of the adjacent subgrade (Fig. 4–4). Ditches must terminate at a stream or at one of the other types of drainage structures to effectively serve their purpose.

Care must be exercised in shaping the ditch side slopes. The condition and type of soil should be considered when any major deviation from the normal side slope ratios is contemplated. In ditch locations which have soils of an erosible nature steep ditch slopes tend to be unstable and sloughing takes place, which in turn retards the flow of ditch drainage. The situation is most critical when ditch slope ratios are beyond the ability of a grader or motor patrol to clean out or reshape ditches effectively. Circumstances such as these tend to increase maintenance costs.

Fig. 4-4. Example of ineffective ditching and cross-draining. Stationary water tends to saturate the subgrade and lessen the ability of the road to support adequate payloads. (Courtesy of the College of Forest Resources, University of Washington.)

Bridges

A waterway which is too large for a culvert or culvert-type structure will necessitate constructing a bridge. A cost comparison between a large culvert and a bridge may indicate that a bridge would be more economical. Although a large culvert may discharge a streamflow without any difficulty, the in-place cost of the structure may well be excessive when compared to the cost of constructing a bridge.

Constructing a bridge for a forest road may involve the expenditure of a considerable sum of money; consequently, it is imperative that controls and inspections be instituted to ensure compliance with the designed features of the structure. Usually bridge construction does not commence until road construction activities have progressed to a point which provides a means of access to a bridge site so that materials and equipment may be delivered to the location.

The bridge site should be identified clearly and the **abutment** or **pier** locations and center line of the bridge staked and referenced. Although it is somewhat general practice to prepare a rather complete **bridge site survey** for bridge design and construction projects, this is not always the case. Due consideration should be given to the effectiveness of a bridge not only to withstand extreme high-water conditions but also to the effect that drift, i.e., trees, logs, brush, heavy-rooted stumps, will have upon the structure during extreme high water. Many bridges have been washed away by drift building up against the piers, abutments, or the bridge stringers, beams, or decking to such a degree that the stream flow was impeded, developing a dam-like condition prior to washout (Fig. 4–5). Allowance for adequate clearance between the high-water elevation and the bottom of the bridge itself to permit free passage of drift is imperative. As high-water conditions subside, drift deposited against a bridge or abutments should be removed to reduce the risk of having fire destroy the structure and endanger adjacent timber.

Fundamentally, all structures are as sturdy as the foundation which supports them. Whether the foundation abutments or piers

Fig. 4–5. Drift accumulations against bridge abutments are a fire hazard and provide conditions which may result in a bridge being washed out. (Courtesy of the College of Forest Resources, University of Washington.)

for a bridge consist of log cribbing, concrete, or **pile bent** construction, the necessity of adequate foundation support is of paramount importance. The load-bearing ability of the soil or rock at the abutment sites should be determined by careful examination. Table 25 on page 18–45 in the *Forestry handbook* (Forbes and Meyer, 1955) presents the allowable bearing pressure for footings related to various soil and rock conditions. The combined use of the table together with field investigation at the site may resolve the decision whether to use a spread- or pile-type foundation. The information included in the table will also permit determination of foundation area in the case of concrete abutments. Table 26 on page 18–49 in the *Forestry handbook* (Forbes and Meyer, 1955) may be used to determine necessary pile penetration and number of piles in the case of pile bent abutments. Log crib abutments and piers are quite serviceable, particularly in the case of bridge installations of a semi-permanent nature (Fig. 4–6).

The action of water or scour, particularly during near peak flow conditions, may undermine bridge abutments and footings which in turn may seriously weaken the structures and jeopardize a bridge

Fig. 4–6. Log stringer bridge with log abutments. Note the heavy slab which serves as bulkhead planking and prevents road ballast from settling into the space between the stringers. (Courtesy of the College of Forest Resources, University of Washington.)

installation. Protection against scour generally is developed by constructing **wingwalls** to prevent water from eroding the fill in the vicinity of bridge abutments. Wingwalls should extend into the stream bank far enough to prevent water from getting behind the structure. Wingwalls may be constructed using logs backed with **riprap** or they may be made of concrete.

The trend to permanent bridges has led to a wider use of reinforced concrete construction in place of the somewhat short-lived log stringer type of bridge. The increased use of concrete necessitates a rather close control to ensure that proper proportions of the ingredients (sand, **aggregate,** cement, and water) are used to produce concrete which conforms to design strength. Proportions should be determined by the use of either weight or volume and not by guess or estimate. The consistency of concrete is usually measured by what is called the slump test. A sample of the concrete is placed in a sheet metal mold of a specific size having the shape of a **frustum of a cone.** The sample is tamped with a rod to ensure symmetrical distribution of the concrete in the mold. The mold is removed by lifting in a vertical direction and a measurement is made between the height of the mold and the average height of the top surface of the concrete after subsidence; if the difference amounts to 3 inches, the concrete is said to have a 3-inch slump. The slump of concrete for heavy mass construction is usually restricted to a maximum of 3 inches and a minimum of 1 inch, while the range for **reinforced** foundation walls and footings is a 5-inch maximum and a 2-inch minimum. Construction involving heavily reinforced members requires more fluid mixes than large members containing little reinforcing. The tendency to use **bank run** or local aggregate on forest construction projects involving concrete structures suggests the necessity of inspection to ensure that clean aggregate is used. Aggregate contaminated with silt, clay, or soil will not produce concrete of design strength. Adjusting the proportions of a concrete mix may be advisable when the cost of procuring or hauling aggregate of proper quality is excessive. The practice of increasing the proportion of cement and decreasing the proportion of aggregate may have definite cost advantages under these circumstances and may deserve consideration. The *Concrete Manual* (U.S.D.I., Bur. of Rec., 1956) is recommended as a source of valuable information for the forester who may be confronted with forest construction projects involving concrete.

When it is necessary to position bridge construction equipment on both sides of a stream at a bridge site, a "shoofly" may be required. A shoofly is a tractor-trail type of access which usually extends upstream from a bridge site to a crossing point. The actual crossing may be accomplished by fording the stream or by means of a temporary bridge. The shoofly would continue on to the far side of the bridge site, thereby permitting construction activities to proceed from both sides of the stream.

A bridge construction project usually involves various pieces of mechanical equipment, equipment operators, and materials. The necessity of maintaining a continuous and uninterrupted construction schedule is of basic importance, particularly when a concentration of machines and manpower is involved.

Construction planning refers to careful work planning to eliminate or minimize unnecessary delay, rehandling materials, and idle man power. It ensures completing the various phases of bridge construction in proper sequence and completing the structure within the time allotted. Construction planning generally requires:

1. A careful estimate of the material, equipment, and personnel required to construct a structure or complete a particular job.
2. A time schedule of specific construction operations embracing all work to be accomplished, equipment to be used, materials required, and labor to be employed.
3. A schedule of progress together with an uncomplicated system of periodic reports and progress records, thereby allowing a comparison between work completed and scheduled progress. The comparison tends to promote more effective supervision and job control.

A working drawing of the design which includes all pertinent information portrayed in a clear and understandable manner should be available to the construction foreman. A true working drawing includes all the data required to construct a structure to certain design specifications. Usually a bill of materials is included with a working drawing.

In the case of construction projects on a contractual basis, where there may be a possibility of disputes arising, it is desirable to document progress by means of photographs taken with a 10-second camera during various stages of construction.

Open-Top Culverts

Properly constructed roadside ditches collect runoff from both the road surface and adjacent terrain. The accumulation of water must be transferred to either a conventional-type culvert or an open-top culvert. In the case of the conventional-type culvert, the water is carried across and under the road proper; whereas, in the case of the open-top culvert, the water is carried across the road and through the upper part of the pavement structure. Actually, an open-top culvert is a continuation of a roadside ditch which is channeled through and across the surface of a road.

Pole culverts, box culverts, and rail culverts are the types of open-top culverts most commonly used. As the names imply, poles are used in constructing a pole culvert, sawn members for the box culvert, and railroad steel for the rail culvert. Specifications for all three types are presented in *Northeastern Loggers' Handbook* (Simmons, 1951).

Open-top culverts generally are installed after the road has been ballasted and the **surface course** has been placed. A bed is prepared by excavating a sufficient amount of road surfacing or ballast to allow the culvert to be placed at the finished grade elevation. A recommended practice consists of installing the culverts to provide for a fall of not less than $\frac{1}{2}$ inch to the foot; the culverts should be placed across the road at an angle (**skewed**) so that two wheels will not hit them at the same time. The skewing will also tend to develop a self-cleaning effect by the resulting increased gradient. The spacing between open-top culverts will depend on soil formation and gradient of the road.

EARTHWORK

Forming a road subgrade involves constructing a continuous bench of a certain specified width having a center line which coincides with a designed or designated center line for the finished road. Various types of earth-moving equipment may be used to accomplish the necessary excavation and embankment so that a road subgrade will be formed which conforms with specifications.

Inspection of Cleared Right-of-Way

Inspecting a cleared right-of-way prior to the start of earthwork operations may prove to be very worthwhile. Particular attention should be given to the construction zones, which generally include that portion of the right-of-way between the slope stakes and corresponding catch points. The practice of allowing accumulations of debris, resulting from felling and bucking right-of-way timber and grubbing operations, to remain on the construction zone should be discouraged. Incomplete grubbing may reduce the effectiveness of earth-moving equipment, which automatically increases costs. A practice which has definite advantages consists of restricting grading to those segments of a road where the grubbing operation has been completed.

No loose debris, which would slide down into the ditch at some future time, should exist on the area between the line marking the top of the cut slope and the line which marks the extent of clearing. All stumps, windfalls and other organic material also should have been removed to preclude being covered by embankment or fill earth. Debris, inadvertently or deliberately deposited in streams, must not be tolerated. Right-of-way logs to be loaded out at a later time should be piled in locations where they will not interfere with grading activities. On relatively flat ground, stumps sometimes are left too close to the shoulder of the road, thereby interfering with road grader maintenance operations. Slash burning, if required by contract or agreement, should have been carried out.

Staking and Restaking

Usually center line, clearing line, and slope stakes are obliterated during clearing operations. Stakes or high poles marking points of intersection may also be destroyed. The task of restaking to identify points, necessary to guide the construction personnel in building a road to design specifications, is greatly facilitated by having adequate reference points properly identified in the field and correctly recorded in the location survey notes.

The extent to which center line staking is done depends upon the standard of the road. A suggested minimum requires that center line stakes be set on sharp curves, switchbacks, fills, and at the ends of tangents. On higher standard roads, center line stakes are set at intervals of 50 feet at each **station** and +50 point, and at the beginning and end of each curve.

Slope stakes are of considerable assistance to the operators of road building equipment such as bulldozers or power shovels. The stakes give the operators the exact point to start cutting and, thereby, ensure grading to the correct **cut slope ratio.** While the experiences of certain equipment operators are adequate to form cut slopes to a certain ratio, many are lacking in this ability and, consequently, stakes are of great value to the latter group. If a bulldozer operator starts his cut too high on a slope, he would be wasting or **sidecasting** too much soil and would also be removing more land area from production than is necessary. On the other hand, if he started his cut too low on a slope, he would have to violate the designed cut slope ratio, resulting in a steeper and usually less stable cut slope and, if the correct cut slope ratio is maintained, the result would be a too narrow road width in a **bench** condition. Steep cut slopes, in other than rock, tend to slough off and impair the effectiveness of roadside ditches.

The use of "Swede levels" for the guidance of the earthmoving machine operator is recommended. A Swede level is a high pole or stake with a cross bar nailed to it at a height above the finished grade elevation equal to the eye height of the bulldozer or power shovel operator above the bottom of the tracks of the machine. Swede levels are usually set on or below the lower shoulder line. When the operator has excavated or cut down to where the cross-bars are vertically aligned, he is on grade in accordance with the prescribed specifications. When his line of sight is above the cross-bars, he has a good indication of how much he has yet to cut. Paint pellet pistols have been effective in spotting grade at the operator's eye level on trees which are located on the downhill side of a road location. This method of marking grade is reported to have acceptable accuracy up to 75 feet, depending upon the diameter of the tree being marked. Poles marked with flagging also may be used to identify finished grade in relation to the eye level of the equipment operator.

In those instances where the operator is required to do the restaking, assistance should be offered. If the forester or engineer is able to interest members of the construction crew in assisting with setting construction stakes, they will know how to replace stakes in his absence and much time and wasted effort would be saved. The construction crew should be encouraged to obtain and learn to use a **hand level** together with a **rod** and a 50-foot tape. A serviceable level rod may be made quite easily by marking foot and half-foot points on a narrow stick of lumber or light weight pole. The rod, level, and tape may be used to set or reset construction stakes at times when the forester or engineer are engrossed in other duties.

Take-off from Existing Roads

The take-off of new construction from an existing road involves certain specific problems which are rather difficult to resolve by office design. This is particularly true in the case of a road design from a *P*-line survey, and usually involves adjustment in the field. The take-off center line should be staked and levelled and the profile plotted together with a segment of the existing road to ensure adequate grade separation. The grade on the new road must coincide with the grade on the existing road until the center lines of the two roads are separated by the sum of the half-widths of the two road beds. Disregarding the relationship between the new road and the existing road insofar as grade separation is concerned may entail additional expense to bring the new road to grade.

In the case where a road system, consisting of a **main** and **spur** roads, is being developed, the practice of constructing the spur road take-offs at the time when the main road is being constructed has definite advantages. This practice will obviate many difficulties later on, particularly on hillside take-offs which present a special problem of soil removal and soil placement insofar as the main road and nearby drainage structures are concerned. Grading spur take-offs a short distance beyond the grade separation point allow these areas to serve as turnouts until such time as the spur roads are constructed.

Excavation and Embankment

Excavation, or cut as it is more commonly identified, includes two specific types—through and sidehill. Through cuts are made through a small hill or over the top of a large hill to cut the ground to the designed grade of the road. Sidehill cuts are made along the sides of hills or canyons, a bench being cut into the hill for the roadway. In sidehill cuts, part of the road may be on fill and part on cut. The proportion of the road on fill is dependent upon the ability of the side slope below the road to retain, or hold, the fill.

Embankment, or fill, refers to building up those areas adjacent to the center line of a road to coincide with the designed or desired gradient.

Ordinarily, road locations on easy or relatively flat terrain do not require extensive excavation or embankment. Oftentimes, scratch grading such as removing the humus layer is sufficient to prepare a satisfactory subgrade or base for the placement of surface material.

Bulldozers or angledozers with straight or "U" type blades (Fig. 4–7), are usually the best equipment for excavation when the ma-

Fig. 4–7. Large size bulldozer with "U" blade in heavy excavation. (Courtesy of the Caterpillar Tractor Company.)

terial is to be moved a short distance. The recognized economic haul for a bulldozer with a straight blade is 150 feet, while the same unit with a "U" blade would have an economic haul of 450 feet (U.S.D.A., Forest Service, Region 6, 1958). There are two types of bulldozer blade control—hydraulic and cable. The hydraulic type provides a positive downward pressure to the blade which allows the operator to exercise a high degree of operational control when making a pass. The hydraulic control is of particular importance when operating in soft muddy ground. Should the machine become mired down, it is usually possible to raise the front end of the tracks by exerting downward blade pressure on poles or small logs placed under the blade; thereby allowing poles to be placed under the tracks to provide flotation. However, the cable control type reacts faster, which is advantageous when the operator drops the blade at the start of a pass or chooses to raise it at the completion of a pass.

Fig. 4–8. Self-powered motor scraper. (Courtesy of Allis-Chalmers.)

Tractor-drawn or self-powered scrapers or carryalls (Fig. 4–8), are excellent for making cuts but have somewhat limited use in forest road construction. A tractor-drawn scraper unit with a capacity of 10 to 12 cubic yards would require a space with a width of approximately 21 feet to negotiate a minimum turn. A self-powered scraper having the same capacity would require a space with a width of approximately 35 feet for making a minimum turn. This requirement would therefore limit the use of scrapers to those construction projects involving roads of higher standard where the **subgrade** would provide the necessary width for a turn-around maneuver.

Roads of lower standard, where the subgrade does not provide the minimum turn-around space, ordinarily are not suited to scraper operation. Constructing turn-around sites in steep mountainous terrain to be served by lower standard roads is considered too costly and therefore generally precludes using scrapers; however, under conditions where it is necessary to make a heavy through cut or build a large fill, the cost of preparing turn-around sites in steep terrain may be justified. The advantage of the scraper lies in its ability to haul large volumes long distances. The economic haul for a scraper is approximately 1500 feet (U.S.D.A., Forest Service, Region 6, 1958).

Power shovels are used for making side hill cuts when they can cast the soil over the side of the embankment, or for making through cuts when trucks are available to haul the material away. The economic haul for a power shovel and dump truck combination is approximately 1 mile (U.S.D.A., Forest Service, Region 6, 1958).

Under certain conditions, such as in low-lying swampy terrain or easy, relatively flat terrain, power shovels may be used as the basic road-building machine (Fig. 4–9). A shovel can form the road

Fig. 4–9. Power shovel, positioned on "pads," excavating on a logging road construction project. (Courtesy of the Harnischfeger Corporation.)

bed and shape the ditches very effectively without churning or miring the base soil. A bulldozer operating under similar conditions, with its repetitive forward and reverse travel, would disturb the base soil to such a degree that the machine would not be able to produce up to its capacity.

Power shovels are very effective in building roads through swampy or relatively flat areas by the turnpike method of construction, where the shovel excavates the soil from both ditches and deposits it on the road area thereby providing a road bed adequately elevated to allow drainage to take place.

Graders or motor patrols rarely are used for heavy cutting. However, when other equipment is not available, light sidehill and through cuts may be made with towed or motorized graders. On relatively flat ground such as exists in sections of the ponderosa pine and southern pine regions, a grader may be used to form a road subgrade. When using a grader for this purpose, it is imperative that all stumps be removed during the grubbing operations.

Where possible, fills or embankments should be started at the lower slope stakes. This practice will allow the fill to be built up with well-compacted horizontal layers of soil not to exceed 12 inches per lift. The earthmoving equipment should be routed to cover the entire width of the fill layer, thereby taking advantage of the compaction effect generated by the equipment.

Obtaining maximum **compaction** necessitates keeping the fill soil at near **optimum moisture content.** Fill material which does not contain sufficient moisture may be dampened at the excavation area or a layer at a time on the fill. Soil which is too wet, as after a rain, should be allowed to dry before compaction. Manipulation of an overly wet layer by some means of **scarification** will aid evaporation. Large fills may require specialized compacting equipment to ensure adequate compaction and stabilization.

Three types of rollers are used for compaction purposes: **sheepsfoot,** rubber-tired, and grid or mesh. Care should be exercised to ensure that the proper type of roller is used to provide the correct compaction method. Fills of plastic or cohesive soils must be built up in uniformly compacted layers, preferably 6 inches thick, at a controlled moisture content. The layers should be rolled with a sheepsfoot roller or a heavy rubber-tired roller.

If fill material is composed of clean sand or sandy gravel, moisture control is not too important. These soils may be compacted effectively even when very wet or very dry. Rubber-tired equipment may be used in compacting these soils. Very effective compaction may also be obtained from the combination of saturation and the vibratory effects of crawler tractors, particularly when tractors are operated at fairly high speeds so that vibration is increased. Sands and gravels with silt or clay binders require effective control of moisture for adequate compaction. Rubber-tired rollers are best for compacting these soils, although sheepsfoot rollers may be used effectively.

If the fill material contains large rocks, they ordinarily are placed at the bottom of a fill, covered with soil, and compacted in layers as explained previously. Care should be exercised in filling and compacting soil between rock fragments to avoid subsidence resulting from compacted soil bridging across the voids between rocks. Compaction of this type is difficult and is generally done by vibration from the passage of track-type equipment over the fill area.

The mesh type roller may be used to compact sands and gravels. Combinations of these two materials are difficult to compact by mechanical means because they are readily displaced when disturbed by compacting equipment. The mesh roller is said to compact these materials efficiently because of the confining action of the mesh upon the material while it is being rolled. Mesh rollers may also be used to crush and compact pit run material on haul roads (Fig. 4–10).

Rock Excavation

Rippers of the singletooth type mounted on bulldozer blades and multitooth rippers mounted on the rear of bulldozers (Fig. 4–11) are gaining in popularity. These units very effectively break up or rip **hardpan** and certain classes and formations of rock, thereby facilitating excavation and transportation of waste or fill materials.

A successful ripping operation hinges upon obtaining information regarding the extent and characteristics of the material below the surface of the ground. A portable refraction seismograph was developed to enable engineers and contractors to examine subsurface material in an indirect manner. This instrument incorporates the principle of seismic analysis which is based on the fact that shock

Fig. 4–10. Grid (mesh) roller crushing and compacting pit run rock. (Courtesy of the Hyster Company.)

waves travel through each kind of soil (rock, gravel, sand, clay, etc.) at varying velocities. The speed of shock waves in a hard, tight rock is fast; while in loose soil, it is slow.

The seismic or shock wave is produced by striking a steel plate with a sledge hammer. The steel plate is positioned at various distances, in increments of 10 feet, from a geophone receiver. Depending upon the model being used, the time interval in milliseconds between the hammer blow and the receipt of the seismic wave at the geophone may be read directly on the instrument panel of a geophone receiver or by viewing the seismic wave on a cathode-ray tube.

The time in milliseconds and distance in feet are recorded in tabular form. A graph is plotted with distance on the x axis and time on the y axis which allows the determination of velocities for the various layers of material and their respective thicknesses. For example: top soil has a seismic wave velocity of between 1,000 and 2,000 feet per second in contrast to shale, which ranges between

Fig. 4–11. Crawler tractor bulldozer with multitooth ripper. (Courtesy of the International Harvester Company.)

2,000 and 12,000 feet per second, depending on the degree to which weathering has taken place or the physical characteristics of the formation.

Knowing or measuring the shock-wave speed in a particular location would indicate the kind or type of material involved. The depths of layers of successively harder material may also be computed. The refractive seismograph has been used by more-progressive road builders to evaluate rock hardness and the extent and depth of rock formations. The results of a test in a locality would indicate whether it would be possible to rip the rock or whether drilling and blasting is required (Fig. 4–12).

The comparative costs of ripping, where rippers can be used on a production basis, and of drilling and blasting reveal a decided advantage to ripping. On nine operations, widely dispersed geographically, the ripping cost amounted to a trifle over one-third the cost of drilling and blasting (Caterpillar Tractor Co., no date).

A ripper could be used to advantage on material which exhibits

Fig. 4–12. Seismic subsurface exploration on a log haul road location. (Courtesy of the College of Forest Resources, University of Washington.)

certain physical characteristics such as fractures, faults, or planes of weakness of any kind; weathering resulting from temperature and moisture changes; brittleness and crystalline nature; high degree of stratification or lamination; large grain size; moisture permeating clay, shale, and rock formations; and low compressive strength. Ripping will be difficult if the rock formation is massive and homogeneous; non-crystalline and, therefore, not brittle; without planes of weakness; fine-grained with a solid cementing agent; and containing moisture which sometimes consolidates the formation, but once exposed to air, disintegrates and is easily handled by a bulldozer or scraper. The *Handbook of ripping* (Caterpillar Tractor Co., 1966) is suggested as a reference for additional information on the subject.

Drilling and Blasting

While ripping rock, under the proper circumstances, is effective and economical, dynamite continues to play a major role in forest road construction activities. Heavy rockwork usually requires drilling and blasting using the proper type and amount of dynamite.

Stripping the soil and organic material to expose seams, crevices, and rock formations should be done by machines where practicable. The actual locations for drilling should be located by experienced powder men and adequately identified to assist the drilling crew.

Larger rock formations should be cross-sectioned. This practice serves to produce more effective blasting by providing data for determining proper charge size and most effective charge placement.

There are various grades of dynamite available. Certain conditions require a specific grade to satisfy the particular objective for making the shot. Rock which is to be **wasted** would require a certain grade of dynamite; whereas, rock which is to be used as **road ballast** would require a different grade. The choice of type depends upon the desired effect. The *Blasters' Handbook* (DuPont, 1966) is recommended as a valuable guide in this connection.

The selection of the proper grade of dynamite, size of charge, and charge placement is a highly technical operation; therefore, only adequately trained and competent personnel should be entrusted with this responsibility. Complex blasting problems and the removal of large rock formations beyond the technical ability of the blasting personnel on a project may require outside assistance. All of the larger powder distributors have experts available to assist in solving the more difficult blasting problems.

The wagon drill or crawler drill in conjunction with an air compressor, either tractor-mounted or separate-unit, is customarily used to drill holes in rock (Fig. 4–13). A tractor-mounted hydraulically controlled rock drill is also used and is very effective. The air compressor which powers the drill also is mounted on the tractor.

A self-contained rock drill, designed in Sweden, for use in remote areas, has been used by the U.S. Forest Service for several years and has proved to be a useful tool. It is powered by a gasoline engine and is completely self-contained. The drill weighs 53 pounds and can be back-packed over rough terrain without difficulty. The unit has particular merit for drilling test holes and for isolated rock blasting where compressed air is not available.

The hazards involved in rock blasting require strict adherence to accepted safety practices for the handling, transportation, and storage of explosives.

Fig. 4–13. Track-mounted rock drill with hydraulic control. (Courtesy of the Chicago Pneumatic Tool Company.)

PAVEMENT STRUCTURE

The subgrade of a road consists of natural soil prepared, and occasionally compacted, to support the **pavement structure** and the traffic loads. The pavement structure, also termed surfacing, topping, or ballast, is customarily used to designate the situation when material such as bank run rock with soil binder is used throughout the pavement structure.

The function of the base course is to distribute surface loads to a unit pressure capable of being supported by the subgrade, provide drainage, and minimize frost action. The function of the surface course is to provide a smooth, durable running surface of low **rolling resistance** for rubber-tired vehicles. Furthermore, the surface course serves as a cover or roof by protecting the subgrade from rainwater.

Prior to the start of the surfacing operations and also during the actual placement of surfacing, several points need to be checked for compliance with contract specifications. The practice of keeping complete field notes on all items of noncompliance is advised. Establishing flagged stakes at locations which need correction identifies them to the construction foreman, and corrective measures can be marked on the flagged stakes.

Checking Subgrade Prior to Surfacing Operations

Ditches. A convenient ditch template, for checking ditch depth and width, may be made of light crossbar and a staff. The crossbar is attached to the staff at the proper ditch depth distance from one end of the staff and also at the proper horizontal distance which conforms to the design. If the ditch is made correctly, the longer segment of the crossbar will touch the shoulder of the roadway and the shorter segment will touch the bank.

Cut Slopes. Cut slope ratios should be checked with an Abney level. If the slopes appear to be deficient, checking may be done from either slope stake points or slope stake reference points. Note whether any overhanging debris is present which would slide down into the ditch.

Subgrade Width. Direct measurement of roadbed width may be accomplished by the use of either a tape or a staff graduated in feet and half-feet.

Grade. There should be reasonable compliance with the profile on designed roads, or maximum grade specifications on other roads. Grade checking is most easily accomplished with an Abney level.

Curvature. There should be reasonable compliance with the plan on designed roads or maximum **degree of curvature** on other roads. The most rapid way of measuring degree of curve is to lay off a 62-foot **chord** between two points on a curve. The measurement in inches of the middle ordinate, which is the perpendicular distance between the 31-foot mark on the chord and the curve, equals the degree of curve. Convenient equipment for this method of checking the degree of curvature is a 100-foot tape and a flexible carpenter's rule for measuring the middle ordinate in inches.

Subgrade Finish. The subgrade should be smoothed to design **crown slope** specifications with a grader just as though the subgrade were to be the finished road (Fig. 4–14). If the road contractor has a grader which will allow the blade to be raised, cut slopes may also be smoothed.

Fig. 4–14. Grader smoothing out the subgrade prior to the ballasting operation. (Courtesy of the Caterpillar Tractor Company.)

Turnouts. **Turnouts** are vital to the safety of everyone who will travel over the road, yet they are often found to be deficient, both in number and dimensions. Experience has shown that turnouts require special attention from the forest engineer or forester who is charged with the responsibility of inspection on a forest road construction project. When the exact location of turnouts is not specified, the intervisibility and maximum spacing ought to be checked on the ground. If more turnouts are needed, they should be built prior to the start of the surfacing operation. In the case where the turnouts are specified on the plan and profile of the road, the locating forest engineer may not have visualized a sufficient number of required turnouts. Under these circumstances, the forest engineer should persuade the operator or road contractor to build the additional turnouts needed to provide safe working conditions for his operations.

Checking During Surfacing Operations

Thickness and Width of Pavement Structure. In those localities where surfacing material is scarce or costly, the operator may be inclined to skimp on the thickness of the rock surface layer. The forest engineer or compliance engineer should be present at least during the initial stages of surfacing operations to guard against this practice. Under circumstances where the forest engineer cannot be present, the practice of setting short stakes on the road shoulders with the tops of the stakes at surface elevation will help to obtain the designed surfacing thickness. The stakes may be set with a hand level and rod. Such stakes are termed "blue tops" in highway engineering practice.

Surfacing width should also be checked for compliance with specifications. The placement of surfacing material should commence on that part of the road nearest to the rock source, so that the dump trucks will always be traveling on surfacing and not on the subgrade.

Quality of Surfacing Material. Usually material for the pavement structure is obtained from a gravel pit which contains rock of acceptable size and quality. The rock, depending on its condition, may be ready for loading or it may have to be prepared for loading by ripping or dynamite.

Volcanic cinders, mine tailings, coal cinders, and slag are other materials which may be available in certain localities. These materials have qualities which tend to produce a serviceable pavement structure.

Placement of Surfacing Material. A front-end loader or power shovel generally is used to load dump trucks which transport the rock to locations on the road subgrade. Dumping the rock to allow a certain degree of spreading is the usual practice. A small- or medium-size bulldozer is very effective to assist with the spreading operations. Scrapers also may be used to haul and spread surfacing material providing it is somewhat homogeneous in size and does not contain large rocks which would impair loading and spreading activities.

The economics involved in hauling and spreading pit run rock and blading off the large pieces should be investigated. Under certain circumstances this practice may have definite cost advantages.

Regardless of which type of material is used in the pavement structure and what method of spreading is employed, the haul road should be bladed smooth during the hauling operation to permit the dump trucks to perform at an acceptable rate of production. A haul road with a rough, uneven running surface increases truck travel time and breakdown time, thereby decreasing production and increasing cost.

Unstable Areas in the Subgrade. Colloidal clay and slippery soils tend to lubricate pavement structure fragments and facilitate their shifting under traffic loads. A sand "blanket" or layer of filter sand, 4 to 6 inches thick, extending the width of the subgrade, is commonly used. It serves as a lateral drain for **capillary water** and for water infiltrating the surface. The sand layer will also prevent silt or clay from "pumping" up into the base (Hennes and Eske, 1955).

A suggested practice for areas where heavy plastic clay is a problem involves grading to coarse material and back to fine surface material. Placing road ballast directly on clay of this type normally is ineffective; however, grading to sand, gravel, rock and back to gravel and finally to surface course material tends to provide adequate subgrade load-bearing capacity.

Crown Slope. The crown slope of the road surface should also be checked for compliance with specifications or accepted practice, which is one-half inch per foot.

Compaction. Dump truck drivers should be instructed to refrain from driving the same wheel track when traveling to and from the rock pit. Traveling the same set of wheel ruts limits any compactive effect to a relatively small percentage of the pavement structure.

Surfacing Turnouts. The practice of postponing the surfacing of turnouts until after the road is surfaced should be discouraged. This practice usually results in uncompacted turnout surfacing, in lack of bonding effect between turnouts and roadway surface, and often in insufficient length and width of surfaced turnout. Staking the extent of turnouts as a guide for the surfacing operation and having them surfaced concomitant with the adjacent roadway precludes the construction of turnouts which are inadequate both in size and serviceability. Stakes should be set where the turnout approaches and leaves the roadway and at the beginning and end of the full width of the turnout.

MAINTENANCE

Road maintenance operations normally consist of all activities required to maintain the road subgrade and pavement structure at the standard to which the road originally was built, with due allowance made for wear and tear.

Included under this section are those measures designed to keep the road surface, shoulders, ditches, slopes, and drainage structures in a serviceable condition. Preventive maintenance of surfaces, shoulders and drainage systems is an important phase of proper maintenance activity. Prompt and adequate repairs at the first indication of failure and efficient planning for adequate repair measures are essential in assuring a reasonably rapid, unhindered flow of traffic.

The importance of preventive maintenance and the necessity for prompt maintenance cannot be overemphasized. Neglect and delay permit traffic and weather to develop minor defects into major problems. Progressive failure of roads is a serious matter, and the speed with which a failure becomes worse increases greatly with the seriousness of that failure.

Limitations on Use of Roads

The practice of prohibiting hauling on forest roads during spring periods when there is evidence of frost leaving the ground tends to reduce road maintenance cost. Truck traffic should be barred until such time as road use will not damage the road. The same restrictive measures ought to be used immediately following heavy or extended periods of rainfall when it is obvious that the road will not support the loads without damage to the pavement structure or the subgrade.

Surface Maintenance

Maintaining a smooth surface on earth or gravel roads requires frequent smoothing operations. The type of surface and amount of traffic will determine how often the process needs to be repeated. The smoothing process is best accomplished by using a motor patrol or truck-towed grader; however, a road drag may also be used to advantage. Surface grading ought to be done in such a manner that the dimensions of the original surface cross-section may be restored. The smoothing operation should not waste any surfacing material over the shoulder line. In general, operations are directed toward moving surface material which has been displaced from the traveled portion of the road, such as in the vicinity of the shoulders, back to its original placement location. Truck travel may cause low spots to develop in a road surface. Suitable material should be used to bring the low spots up to grade. Compacting, other than by normal truck travel, also may be advisable to develop the same degree of compaction as exists in the adjacent road surface.

Generally, surface grading on earth roads is done after a rain, when the soil pulverizes readily, not when the soil is too sticky or when the soil is completely dry. Grading in the spring as soon as the frost is out of the ground is beneficial. The practice of smoothing the surface prior to freeze-up is advised as it allows the surface to freeze smoothly and remain in this condition during most of the winter.

Gravel roads may be graded when in a wet condition because gravel smooths more effectively when wet. A gravel road surface in a saturated condition cannot be graded effectively.

Washboarding or corrugation of a road surface is caused by the tangential force transmitted to a road surface by the wheel of a vehicle. The effective traffic capacity of a road which has a washboard condition is lowered because of reduction in speed, payload, and tire life.

Washboarding of sand and gravel roads develops where loose surface materials are in contact with a somewhat resistant base material. Should the base material become softened, which would allow the loose particles to become embedded instead of displaced, washboarding is not noticed. Washboarding develops most frequently under conditions where the loose particles are one-half inch or more in diameter and becomes less noticeable with a decrease in particle size.

Blading or scarifying may alleviate the washboard condition temporarily; however, stabilization of the road base and removal of the loose surface material, which is tossed back by tire action causing the washboard-like undulations, is suggested as a remedial measure for more permanent relief.

Reshaping a roadbed to design specifications requires a skilled grader or motor patrol operator who should, during this operation, correct inslopes, outslopes, **superelevation** of curves, fill settlement, and smooth the roadbed.

Mudholes in a road are a problem which cannot be eliminated merely by dumping rock in the holes. The suggested practice consists of draining the hole, removing the mud, and filling the hole with high-quality material. Usually mudholes are caused by poor drainage; therefore, correcting deficiencies in the drainage system oftentimes eliminates the problem.

Fill Protection

Repairing road shoulders and preserving or maintaining original fill width may require hauling in material or drifting available material on the site. Fills on side hills can be protected from wash by diverting water above them. Water should not be allowed to flow down the sides of fill slopes if it will cause gullying. The practice of building a raised shoulder or berm along the edge of fill sections and diverting road surface water into **shoulder spillway culverts** or berm outlets at suitable intervals tends to protect a fill.

Snow Fences and Snow Removal

Logging operations in areas where the problem of keeping roads open to traffic during the winter months usually involves the placement of snow fences and the operation of snow plows.

Reconnaissance to determine where snow fences will be required to control the drifting of snow should be made before winter begins. Drifts form when windborne snow picked up in open spaces loses velocity and is deposited in sheltered places. Danger spots, accordingly, are roads at ground level or in cuts adjacent to large open areas. High snowbanks left by snowplows close to a road furnish both the conditions and snow for extensive drifting. Snow fences normally are not required in the vicinity of high fills, wooded or brushy areas. Commercial snow fences of the wooden lath type are effective and easily erected. Fences of a more temporary nature utilizing evergreen brush suspended on and woven into wire strands also are effective.

A variety of equipment is used for snow removal. Snowplows of either V or one-way type mounted on trucks are designed to discharge snow rapidly. Truck-mounted plows may be used to handle recently drifted snow (not packed or crusted) up to a depth of 36 inches. Displacement plows of the V type are used to discharge snow to the side beyond the shoulder line. Moving heavy masses of snow with a displacement plow requires that it be tractor mounted. Rotary plows or snow blowers are used when conditions have exceeded the capacity of the V-type plows. Road graders or motor patrols are satisfactory for light snow. Publications 13 and 15 in the Supplementary Reading list are suggested for reference and further reading on the subjects of snow fences and snow removal.

Ditches

Drainage ditches should be kept clear of weeds, brush, sediment, and other accumulation of debris that tends to obstruct the flow of water. Cleaning out ditches or re-ditching is usually done by means of a road grader, and consists of obtaining the proper slope to the ditch and re-establishing the back slope of the ditch. Care should be exercised that the cut slopes are not undercut during the re-ditching operation. Material removed from ditches should not be drifted across the pavement structure as this will leave undesired material in the road. A rubber-tired front-end loader is an ideal piece of equipment for removing the material resulting from a ditch maintenance operation.

Culverts

Evidence that a culvert requires cleaning may point to the possibility that it was installed incorrectly. Culverts must be kept clear of debris and sediment, and water prevented from cutting around or undermining. Frequent inspections should be made to determine whether culverts are functioning properly. Clearing is usually necessary after heavy rains and is normally done by hand. Inspection of catch basin approaches and outlets for clogging debris should be of a continuing nature. Particularly thorough inspections ought to be made late in fall in preparation for the winter season and again in the spring to ensure minimum spring breakup difficulties.

Bridges and Abutments

The frequency of inspection and maintenance of bridges varies with the amount and intensity of rainfall, adequacy of construction, and general condition of the structure. A minimum program consists of a complete inspection made in the fall in preparation for winter and another in the spring to determine the extent of repairs required. In addition, all bridges and abutments should be inspected during and after floods, and after any other possibly damaging occurrence. Following each inspection, any of the following operations may be necessary: tighten structural members, bolts, rods, wedges, floor boards, stringers, wire rope wrappings, wheel guards, and handrails; seal cracks in members; re-aline truss members; remove dirt and debris; correct scour, undermining, and settlement of **bents**, piers or abutments; and remove any accumulation of drift in contact with or adjacent to any part of the bridge structure. Inserting shims between a cap and stringers should be done where subsidence of pile bents has caused the adjacent bridge deck to tilt or settle below the design elevation.

There must be a smooth transition between the road surface and the bridge deck. Any settlement in the bridge approaches should be corrected as soon as possible. This condition, if neglected, subjects the bridge to excessive impact stresses and tends to weaken the structure.

Replacement of floor boards or run plank should be accomplished prior to the time that failure is imminent. Nails or spikes used to secure floor boards to bridge beams tend to work loose and need to be redriven to preclude damaging tires.

Right-of-Way

Overhanging brush, as well as brush obstructing visibility on curves, should be removed. Roadside brush may also retard the drying out of the roadbed after storms. Chemical brush control is now used widely and is quite effective. New developments in this field should be investigated as they occur.

Surface Loss

During dry weather, dust caused by vehicle traffic on forest roads is not only a nuisance, but dangerous and wasteful. Dust is also detrimental to motor vehicles, a factor which has an important bearing on vehicle maintenance cost. Flying dust represents a loss of binder which in turn leads to potholing and raveling of the road surface. Treatments or processes used to control dust are termed palliatives.

Treating the surface of a road with water, road oils, spent sulphite pulping liquor, and chemical salts, such as calcium chloride, all tend to bond the surface and prevent displacement of fine particles.

Treatment with palliatives usually is restricted to those segments of a road system when the volume of traffic is heavy and accident hazard is high. Where overall treatments of roads is not necessary, dust palliatives may be applied with good effect at sharp curves, abrupt vertical curves, road intersections, and at other critical locations.

Estimates by the logging industry of the amount of surfacing lost range from one-half to two inches per year (Nixon, 1958). This material must be replaced if the road is to be kept in the best possible condition for hauling logs and maintain its load-carrying capacity. In areas where good quality hardrock is scarce and expensive to obtain, the value of this material lost each year is considerable.

PRODUCTION COST DATA

The forest engineer or forester who may periodically inspect a road construction project has an opportunity to collect accurate production data from which cost data may be compiled. Such data are of considerable value in estimating road construction costs for subsequent projects. Keeping a detailed record of construction progress, somewhat comparable to a case history, involving equipment, number of men working, and the amount of work accomplished during the periods of time between inspections or visits, has considerable merit. Such records provide a sound basis for sharpening one's judgement regarding personnel, equipment, and production as related to cost. All pertinent information which is closely associated with production, such as condition of forest cover, topography, soil, rock outcrops, and weather, should be noted.

SUGGESTED SUPPLEMENTARY READING

1. ADAMOVICH, J. P., J. P. TESSIER, and F. M. KNAPP. 1964. Cost and production analysis of drilling and blasting rock on forest truck road construction. Research Paper No. 59. University of British Columbia, Vancouver, Canada. This paper covers in considerable detail the various cost and production elements as related to rockwork.
2. Caterpillar Tractor Co., Peoria, Illinois. *Fundamentals of earthmoving*. This booklet, as the title implies, covers the fundamental principles of earthmoving and equipment application. Good presentation of power requirements and power limitations.

3. DAHMS, W. G. and G. A. JAMES. 1955. Brush control on forest lands. Pacific Northwest Forest and Range Experiment Station, U.S. Forest Service, U.S. Department of Agriculture, Portland, Oregon 97208. Research Paper No. 13. Presents a complete treatment of several methods of brush control. While emphasis is on methods for the Pacific Northwest, the booklet does contain a good bibliography which should be very useful.

4. FORBES, R. D. and A. B. MEYER (eds.). 1955. *Forestry handbook.* The Ronald Press Co., New York. Section 18 contains good source material on forest truck road construction and related operations.

5. HAUSSMAN, RICHARD F. 1960. Permanent logging roads for better wood lot management. Division of State and Private Forestry, U.S. Forest Service, U.S. Department of Agriculture, Upper Darby, Pa. A booklet prepared to meet the need for assistance in the construction and maintenance of a road system in a wood lot.

6. HUGGARD, E. R. and T. H. OWEN. 1959. *Forest machinery.* Adam and Charles Black, London, England. Presents a comprehensive survey of the field of forest machinery. Chapter VI, Road Construction Plant, is particularly appropriate.

7. Hyster Company, Tractor Equipment Operations, Portland, Oregon. 1966. *Compaction handbook.* 4th ed. This handbook provides a working knowledge of soil and rock mechanics together with the application of mechanical tools in artificially densifying the soil.

8. International Harvester Company, Chicago, Illinois. *Basic estimating.* 1st ed. While slanted toward International Harvester equipment the booklet contains much useful information on equipment selection, production estimating, and cost estimating.

9. NICHOLS, HERBERT L., JR. 1954. *How to operate excavation equipment.* D. Van Nostrand Company, Inc., Princeton, New Jersey 08540. Supplies detailed non-technical information about how to operate the important types of excavating, hauling, and grading machinery.

10. NICHOLS, HERBERT L., JR. 1962. *Moving the earth,* 2nd ed. North Castle Books, Greenwich, Conn. This text is excellent for construction men to read. Virtually all earth-moving problems are covered along with much descriptive data on earth-moving equipment and operating techniques.

11. NICHOLS, HERBERT L., JR. 1956. *Modern techniques of excavation.* North Castle Books, Greenwich, Conn. Provides complete information relative to the planning and execution of excavation and grading projects of all kinds.

12. PECKWORTH, HOWARD F. (Managing director). American Concrete Pipe Assn. 1958. *Concrete pipe handbook.* American Concrete Pipe Association. While concrete pipe does not have wide use as drainage structures on forest roads, conditions may require that it be used. The Handbook contains much valuable information regarding hydraulics, runoff, etc.

13. SILVERSIDES, C. R. and A. KOROLEFF. 1949. *Construction and maintenance of forest truck roads.* Pulp and Paper Institute of Canada, Montreal, Canada. A rather complete treatise on forest roads, including location, construction, maintenance, equipment and production data. Bridges are discussed quite thoroughly.

14. STAMM, E. P. 1950. Logging road construction. *The timberman.* Vol. III, No. 2, 64–78. A very good article by an expert in this field. All phases of construction as practiced in the Pacific Northwest are covered in a thought-provoking manner.

15. U.S. Department of the Army. 1957. Technical Manual 5–250 *Roads and airfields.* Headquarters, Department of the Army, Washington, D.C. This manual contains much valuable information on all phases of road construction. As is to be expected the military organization plays a very important part; however, the fact remains that it is well-written and quite easily understood.

5

Timber Production Management

Management in the broad sense refers to the manner in which a business is directed, conducted, or regulated so that a desired result will be achieved. Timber production management embraces a rather broad spectrum of both direct and indirect non-operational considerations together with those operational activities associated with providing the many mills and plants of the forest products industry with the basic raw material—wood. Regardless of whether timber harvesting operations involve a large complex organization, employing many men, or a small undertaking, with comparatively few employees, timber production management is being practiced to a greater or lesser degree, depending upon circumstances.

TYPES OF OPERATIONS

Nationwide, the timber production facet of the forest products industry is comprised of a multitude of various kinds and types of timber harvesting operations engaged in fulfilling the demand for raw material. From the management or control standpoint these operations may be categorized into two distinct groups, company operations and contract operations. A company operation refers to a situation where a particular company or corporation provides the personnel and equipment necessary to harvest timber from its own land or from land in public ownership. In a company operation, direct control over the disposal of the raw material produced (logs, pulpwood, chips, etc.) is exercised by the company. In contrast, a contract operation refers to a negotiated agreement between either a timberland owner or a successful bidder on a public timber sale and a contract operator. A contract operator, depending upon the size of his organization and corresponding equipment availability, may agree to undertake a certain part or portion of the timber harvesting process or, should his organization have the capability, perform the complete operation. Generally, in a contract operation the contractor has no control over the disposal of the logs or other forest products which are produced. In terms of number, there are many more contract-type operations than company operations; while sizewise, in terms of personnel and equipment, a company operation almost always is larger.

Company Versus Contract Operations

Each type of operation may exhibit certain advantages when related to a particular set of circumstances. A good case may be made for both types of operations, and any comparison tends to be rather controversial.

Often, operational flexibility has certain advantages. In this respect changes in operational procedure may provide profit and/or production benefits. In a company operation, changes may be made in a much more direct and hence quicker manner; a new method may be instituted by means of either a memorandum or verbal instructions, depending upon the size of the organization. However, in a contract operation, where activities usually are carried on according to normally rigid contractual provisions, operational changes may require a new contract. In this respect, contract operations do not have the degree of flexibility generally associated with company operations.

Often contractors take a contract job at a price equal to a company's cost, and still make a profit. Situations such as these generally are the result of sound planning and effective leadership on the part of the contractor, coupled with a capable and hardworking crew.

When a company is engaged in timber harvesting as a means of supplying its mills with wood, labor unions tend to be most influential and the union precept of division of labor is reflected in operational procedures and practices which are the result of negotiations between labor and management. Each job, both in the woods and the mills, carries a written description of the specific task the worker is to perform and the hourly wage he will receive. A worker hired for a particular job is restricted to performing those tasks directly associated with that job and normally is prohibited from assisting on jobs other than the one for which he was hired. A relatively minor incident, such as a truck driver assisting with the loading operation, could lead to controversy between labor and management.

Generally, a crew on a contract operation consists of the best men a contractor can hire in a locality. Often crew members may be a contractor's friends or relatives. A contractor usually is engaged in a much smaller size of operation than a company operation and has more choice in hiring and discharging crew members. His main objective is to provide himself with men who are dependable, hardworking, and versatile. A successful contractor has the ability to imbue a certain *esprit de corps* in his crew which generally is reflected in a better than average level of production. In order to hire and retain good workers a contractor may pay higher wages than comparable union scale; however, this higher rate can only be justified if a correspondingly high rate of production is maintained. Many times a contractor is the working boss on the job and, as such, is in a position to give somewhat closer supervision to the operation than that generally exercised on a company operation.

Observations of both types of operations indicates that in contract operations, crew members will help each other and do whatever is necessary to maintain a level of production consistent with the contractor's objectives. Along with this mutual assistance, the practice of alternating jobs during the day, particularly in those cases where certain jobs are more fatiguing than others, tends to bolster morale.

Company Operation. The activities associated with timber production management within a company may be portrayed by means of an organization chart. By means of this type of chart one may trace the lines of responsibility from top management down to the various levels of operational activity. Generally, a logging company has a staff of specialists who have the experience and ability to plan, organize, and conduct a complete timber harvesting operation, which also could include a wood procurement program. Examples of representative timber or woodlands departments or divisions are presented. Figure 5–1 represents a timber company or department typical to the Douglas-fir region, Fig. 5–2 shows the organization of a woodlands department or division in the Northeast, and similarly Fig. 5–3 represents such an organizational arrangement in the South. In all cases, these organizational arrangements include numerous key personnel together with highly skilled individuals. However, some company operations, depending upon circumstances, such as size, location, and operational capability, may exhibit a radical departure from those which have been presented. It is well to note that in a true company operation complete control over timber harvesting and attendant activities rests entirely with the organization.

Companies often engage contract operators to provide a portion of their wood usage. When operating conditions are reasonably similar, this arrangement provides an opportunity whereby the efficiency and costs associated with company operations may be examined and compared with contract operator performance. Thus, a company has provided itself with a yardstick which may be of considerable value in effecting improvements in both production and costs on future company operations.

Company operations which function within the framework of large lumber or pulp and paper concerns often enhance their effectiveness by seeking assistance from specialists in other areas within the parent organization. This is of particular significance in regard to tax, legal, and accounting matters. The benefits of a company computer center in data processing and linear programming as related to decision making are of particular significance and importance. A facility of this type provides a means whereby alternate operational proposals may be tested and comparative results predicted quickly and accurately.

In general, company operations have a higher degree of stability than that which is normally exhibited by contract operations. This condition is directly attributable to the comparatively large capital expenditure required when a permanent mill or plant is constructed. Generally, a mill or plant signifies permanency, and this image tends to identify practically all activities associated with the various ele-

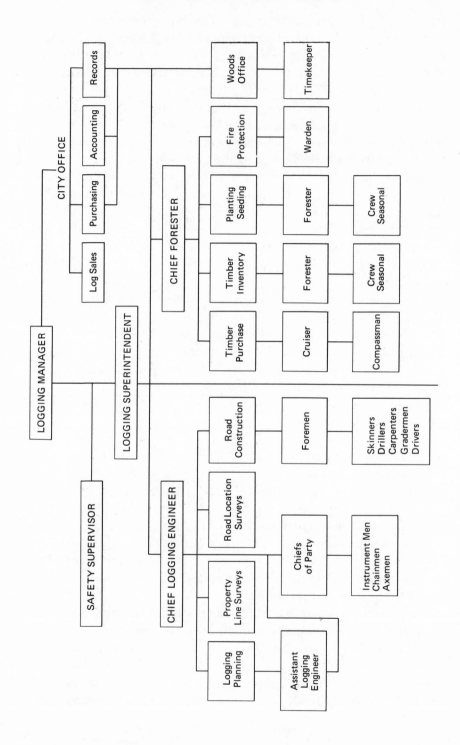

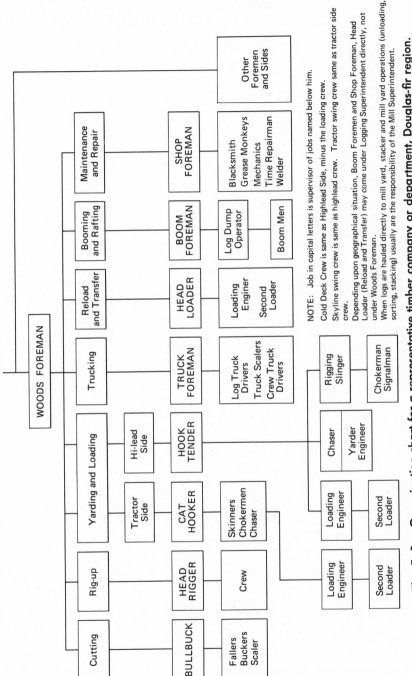

Fig. 5–1. Organization chart for a representative timber company or department, Douglas-fir region.

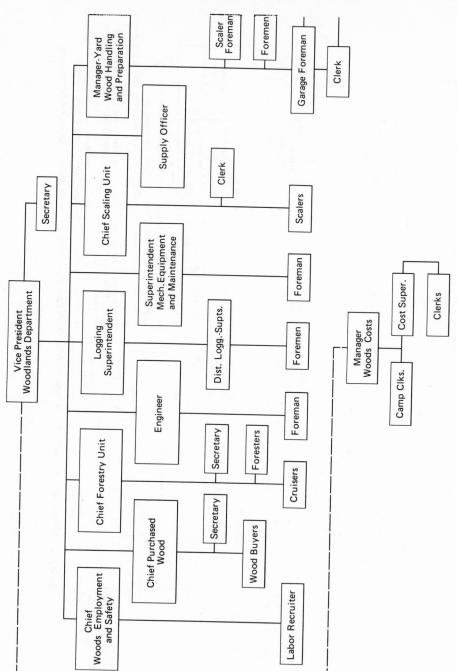

Fig. 5–2. Organization chart for a representative woodlands department, northeastern United States.

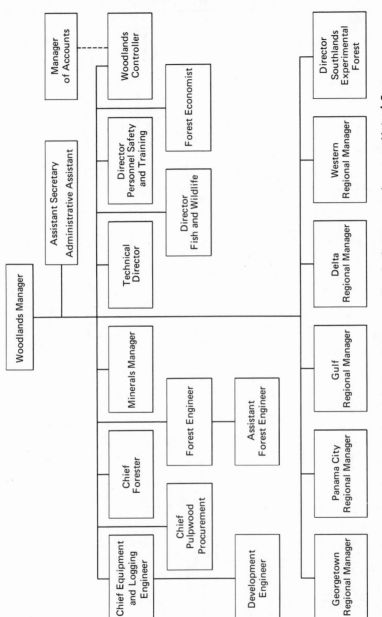

Fig. 5–3A. Organization chart, representative woodlands division, southeastern United States.

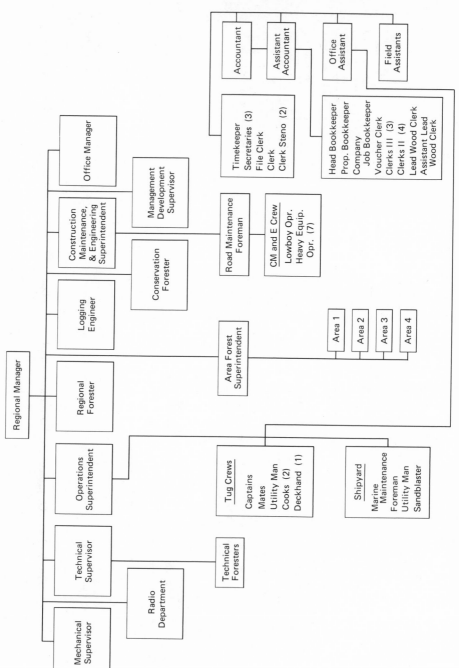

Fig. 5-3B. Region organizational chart.

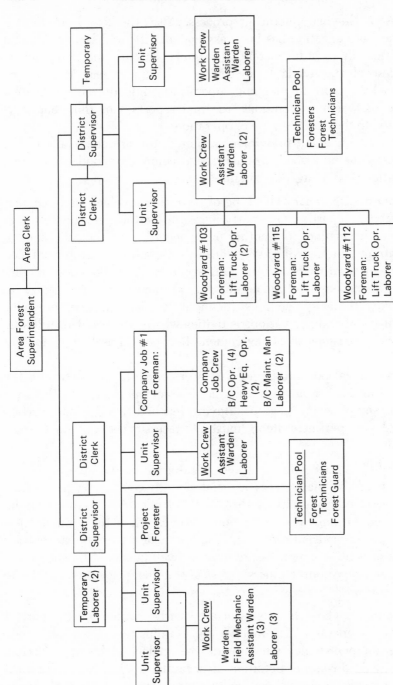

Fig. 5–3C. Area organizational chart, woodlands division, southeastern United States.

ments of the manufacturing process. Also, the length of time a business organization has been associated with, and has maintained its stature in an industry tends to signify stability and permanence. In view of the available specialists a company may have at its disposal to assist in planning, operating, and controlling its own timber harvesting operations, the major portion of this chapter is devoted to the contract type of operation.

The management principles presented also are applicable to the independent pulpwood producer, who commonly conducts a small operation employing less than nine men.

Contract Operation. This type of operation involves an agreement between two parties. One party agrees to perform certain activities in a certain manner for another party according to the provisions of a contract which has legal significance. A contract operator is faced with somewhat the same array of problems as are common to company operations, but generally the problems are not nearly as complex. Basically, he must be able to plan, finance, organize, operate, and manage various activities which will result in fulfilling specific provisions of an agreement between himself and another party.

The logical point to begin would of course be the desire on the part of an individual to become a contract operator, and this may be for any of several reasons. Having reached this decision, certain logical and pertinent items must of necessity be examined thoroughly:

Ability. A successful contract operator should be able to do any and all of the many tasks associated with contract requirements. This level of capability is the stock-in-trade of a contractor who will tend to maintain production at times when the normal operational routine is interrupted. The mere desire to be a contract operator is not enough. Desire must be well bolstered by such traits as ingenuity, resourcefulness, physical stamina, pride, and ability, to name a few.

Planning. The activities associated with planning on the part of a contract operator may involve the examination of available men and equipment which will provide the nucleus of his organization. He may do this either before or after he has a job in sight. He must have enough operating capital to support his organization during the period when he will be moving in equipment and setting up the operation and also until he receives payment for work accomplished.

He must be able to visualize how the flow of logs will move and what men and equipment will be needed to bring this about. Should he have a particular location in mind which may involve his services as a contract operator, he can be more specific in his deliberations. In this case he should become as familiar as possible with the ground, timber, road construction, improvements, and the general desires of the parent organization. It is mandatory that he do this so that he may be in the best possible position to bargain when a contract is forthcoming.

The more experienced contract operator will tend to relate production and attendant costs on previous contracts to the one at hand. In this manner he is able to determine whether a certain contract price is a reasonable offer which will allow him to pay his expenses and provide him with a profit. The matter of records will be examined more thoroughly in another part of this chapter; however, at this point a well-organized set of cost and production figures experienced on previous contracts can be of great value.

Once the contract is signed, the contract operator may find that he needs a certain amount of financial assistance. Of necessity, he will be expected to provide the lending agency with certain information about himself so that his loan application may be processed.

Financing. In the broad sense, all activities involve a somewhat direct relationship with various financial considerations. While initially this relationship may not be apparent, a more meaningful impression is achieved when cost is viewed as a means of acquiring the necessary goods and services generally associated with a particular undertaking. The degree to which goods and services may be acquired is usually reflected by a person's financial position, which normally is determined by examining one's assets and liabilities so that a net worth may be ascertained. In general, net worth is the extent by which the dollar value of a person's assets (what is owned) exceeds the dollar value of his liabilities (what is owed).

Credit plays a most important role in practically all business ventures and is of particular importance in timber harvesting. Mechanization of woods operations necessitates large capital investments in various pieces of equipment, such as power saws, skidding units, loaders, trucks, and road construction machinery. From the standpoint of capital investment, credit is a very important part of any financial endeavor. The extent to which credit may be granted is closely associated with net worth dollar value, which reflects the combined value of cash, property, and equipment. Reputation and business acumen, while somewhat intangible, often have a marked bearing upon transactions which involve credit. The tangible ingredients of net worth—cash, property, equipment—when combined with the intangibles provide a means whereby a credit rating may be established. A prudent individual would make every attempt to retain at least an acceptable credit rating and, if possible, improve it. Credit is a very important consideration and has a direct bearing upon actual or potential financial stability.

Depending upon circumstances, certain phases of contract operations may be financed by manufacturers and sellers of equipment, banks, equipment finance companies, and business associates or friends. All these sources of financial assistance have one point in common. They are vitally interested in the risk factor involved when extending credit or loaning money. Arrangements for financing will vary widely among the sources listed; however, bank financing is somewhat uniform throughout banking circles and, consequently, best lends itself to this discussion.

Customers have every right to expect service, regardless of whether they are dealing with a storekeeper, equipment dealer, or banker. Service insofar as a bank is concerned refers to providing the customer with money and credit when he needs it, and in necessary amounts. Of vital importance to the lender is a thorough understanding of what is involved in a timber harvesting operation so that he may better comprehend the effects of weather, shutdown, mechanical breakdown, competition, and price fluctuations. The uninitiated has no way of knowing that these problems even exist, much less the adverse effect they may have upon certain phases of a timber harvesting operation. While it is the lender's responsibility to be knowledgeable about conditions and circumstances in connection with a loan, the borrower should make a sincere effort to apprise the lender of contingencies associated with the operations to be financed. A mutual assistance or cooperative attitude should be developed between banker and borrower so that effective communication may be established and maintained.

A banker will require certain information from a customer upon which to base his decision regarding credit. Generally, the information relates to what the credit man refers to as the "C's" of credit:

character, capital, capacity, collateral, circumstances, and coverage (insurance). Information regarding a borrower when presented within the framework of the "C's" of credit, and coupled with an intimate knowledge of a business, provides a sound basis for customer evaluation.

A brief description of the six "C's" of credit will tend to emphasize their importance to both banker and borrower:

CHARACTER. One of the most valuable assets a borrower can possess is moral character. Generally, this suggests that the borrower is a man of his word and that he will make every effort to conserve his assets and thereby provide assurance that his indebtedness will be repaid.

CAPITAL. A borrower who has invested in his own venture tends to show that he has faith in its future. In contrast, one who expects others to assume his financial burdens on a permanent basis, by reason of a weak financial position or by lack of a sense of business responsibility, had better reflect upon the advisability of becoming an entrepreneur in the forest products industry.

CAPACITY. The combination of operational and managerial skill demonstrated by a borrower is reflected in how well he has made use of his own investment and the degree to which he has added to it. Past experience may be of somewhat limited value when a person embarks on a business career. For example, a man may be a very good skidding-tractor operator or truck driver, but this alone does not qualify him to manage and direct his own operation.

COLLATERAL. A good credit record as a borrower may often allow the lending agency to loan money or extend credit under conditions which are more favorable than could be tolerated in the case of a less impressive credit record. The status of a person's credit rating has a direct bearing upon the degree to which a lending agency will require collateral to secure a loan. Equity in home ownership, improved real estate, marketable securities, and cash surrender value of life insurance are assets and may serve as collateral for securing a loan or establishing credit. Serious consideration should be given to the consequences of having to withdraw from a venture in the face of financial obligations associated with an undertaking. Premature withdrawal, in such instances, usually involves a loss which may have an adverse effect upon both the party involved and his family.

CIRCUMSTANCES. Numerous factors are involved when a lending agency makes a loan or credit is extended. While many "circumstance items" are included in those "C's" of credit already discussed, certain additional items are pertinent. The degree to which a venture is seasonal in nature is an important factor when a loan is being considered. Unfortunately, timber harvesting is one business which is seasonal, particularly in those regions where winter conditions or spring break-up tend to either shut down or restrict operations. The detrimental effect upon continuous production caused by the suspension of timber harvesting activities, by either voluntary or agency action, during periods of extreme fire hazard must be recognized.

COVERAGE. This area refers to the various types of insurance protection against basic risks. A borrower who has taken adequate precautions against certain hazards or risks generally is looked upon as a more knowledgeable customer than one who has not provided himself with such protection. Contract operators along with other small business organizations are subject to various kinds of business losses. Losses in this sense may result from death of the owner or partner, operational interruptions caused by fire or other violent causes, losses from theft or other acts of dishonesty by employees or others, and public liability not included under workmen's compensation insurance. The need for adequate insurance coverage should be a major consideration of anyone associated with a business venture. In many cases the lender may make adequate insurance coverage a loan provision.

Equipment. Generally timber harvesting equipment on contract operations includes various pieces of mechanical equipment which exhibit varying degrees of newness. Many times the operational effectiveness of the equipment may be related to the length of time it has been in use; however, operator abuse and lack of maintenance may cloud this generality. Sooner or later a decision must be made regarding replacement of mechanical equipment. In the final analysis, production together with relevant costs will tend to focus attention on the necessity of making such a decision. In this regard, certain circumstances, such as unscheduled downtime, obsolescence, overhaul, maintenance and turn-in value will often precipitate this action. The effect of crew frustrations, as a result of work stoppages caused by breakdown, while somewhat intangible, may be reflected in a loss of production and therefore have a direct bearing upon a decision to replace.

While the decision to replace undoubtedly has been made many times in what appeared to be a somewhat arbitrary or off-hand manner, the move probably was the result of a dollar-oriented opinion. This procedure may provide satisfactory results in isolated cases, but the astute contract operator usually bases his decision to replace upon more factual knowledge. Equipment operators have the responsibility of handling a piece of mechanical equipment in a manner which will allow it to function within the limits of its capability. Extending this responsibility to include minimal record keeping on the equipment is a sound policy which has been adopted by the more progressive contract operators. Information relative to the performance and behavior of a piece of mechanical equipment should be recorded and periodically tabulated so that a proper cost and production evaluation may be made. An uncomplicated system of keeping records should be established which will provide the necessary information when decisions regarding equipment replacement are required. Record keeping, to be effective, must be mandatory and be an integral part of the job.

Certain equipment dealers and manufacturers supply their customers with forms and booklets which facilitate record keeping. This material provides a convenient means whereby equipment operators can enter cost and production data as related to machine performance on a particular job. Generally these records are of two basic types: one involves a daily record of work and cost and the other provides a means whereby the daily record may be tabulated by month and year. The daily record may be in booklet form, of shirt pocket size, with sufficient pages to allow daily information to be recorded over a period of a month. Costs of fuel, lubrication oil, grease, filters, tires, repair parts, and labor may be entered daily, along with job description and work accomplished. Tabulating daily costs into a record book allows monthly totals to be determined for all of the various cost items incurred along with a grand total of all costs per month or year. When service meter readings are recorded or hours worked determined, hourly operating cost may be calculated quite readily.

Equipment may be acquired by either direct purchase or by means of a leasing arrangement. The advantages and disadvantages of both possibilities should be examined and evaluated according to the financial position or structure of the operator who contemplates acquiring equipment.

PURCHASE. Acquiring equipment through purchase offers certain advantages which often are seemingly contradictions of those advantages associated with leasing equipment. A thorough understanding of the effect that either method of acquisition will have on a certain financial position is mandatory so that the most advantageous course of action may be followed.

Advocates of purchasing equipment suggest the following benefits:

1. Interest rates are higher for leased equipment.
2. Purchasing provides more attractive tax benefits.
3. Residual value in equipment.
4. No fixed obligation.
5. Ownership provides prestige and status.
6. Can dispose of obsolete equipment.

A decision to purchase equipment should not be made until the relevant income tax laws have been examined thoroughly. This may be done by the purchaser or by securing the services of a tax expert. The objective is to purchase in a manner and at a time which will provide the maximum financial benefit to the purchaser. The relevant income tax considerations may be classified into three main categories:

1. Incentives as a result of investment credit.
2. Reduction of income tax through depreciation of equipment.
3. Tax treatment of gain or loss when equipment is sold.

Investment Credit. Initially, investment credit was intended to provide an incentive to purchase new equipment; however, it also applies to purchases of used equipment.

The investment credit provides a credit against the federal tax liability equal to a maximum of 7 percent of an investment in both new and used machinery and equipment in the year the property is acquired. Actually, the credit has the same effect as a cash rebate and therefore is more beneficial than deductions from income. A credit reduces tax liability dollar for dollar, while the tax benefit of a deduction ranges between 22 and 48 percent for businesses.

According to directives of the Internal Revenue Service (U.S. Treasury Department, 1966) an investment credit is granted when new or used property is acquired. To qualify, the property must: (1) be depreciable, (2) have a useful life of at least four years, (3) be tangible property used as an integral part of manufacturing, production, or extraction, and (4) be placed in service during the year it is acquired. While in general all tangible property used in a business qualifies, machinery and equipment are the principal types of property which qualify.

The useful life of qualifying property, at the time it is placed in service, determines what portion of the acquisition cost qualifies for investment credit. There are three useful-life groups, each of which provides a different level of cost qualification for investment credit: (1) only one-third of the amount of money invested in qualifying property with a useful life of at least four years but less than six years to the owner is subject to credit; (2) in the case of property with a useful life of at least six but less than eight years, two-thirds of the amount invested is subject to credit; and (3) the entire amount invested in property with a useful life of eight years or more is subject to credit. The investment credit allowed is 7 percent of the amount of investment subject to credit.

An example will tend to clarify the foregoing. Assume that an operator purchased the following equipment: (1) new equipment with a useful life of three years for $5,000; (2) used equipment with a useful life of four years for $10,000; (3) used equipment with a useful life of five years for $50,000; (4) new equipment with a useful life of six years for $60,000; and (5) new equipment with a useful life of ten years for $70,000. The amount of investment subject to credit is calculated as shown in Table 5–1.

TABLE 5–1. Investment Credit Base

Property	Cost	Estimated Useful Life	Part To Be Counted	Amount Subject to Credit
New equipment	$ 5,000	3*	0	0
Used equipment†	50,000	5	1/3	$ 16,667
New equipment	60,000	6	2/3	40,000
New equipment	70,000	10	all	70,000
				$126,667

* Equipment or machinery with a useful life of less than four years does not qualify.
† Used equipment purchases are limited to $50,000 in any one year.

The investment credit allowed under these conditions would amount to 7 percent of $126,667 or $8,867. This amount would be deducted from the total tax liability on the operator's tax return.

LEASE. The alternative to purchasing timber harvesting equipment is to enter into a lease agreement with an organization which provides this service. Under a lease arrangement, record-keeping for income tax purposes is simplified; for example, no thought need be given to the item of depreciation expense. Many large well-financed companies lease equipment, particularly automobiles, for use on company business. Major items of construction and logging equipment, including log trucks, may be leased.

The decision to lease instead of buying often is difficult to make. There are pros and cons which should be scrutinized closely to determine whether leasing or buying is the better move in light of existing circumstances. While some of the benefits of leasing logging equipment are rather straightforward, others are somewhat obscure and difficult to assess. A thorough examination of a proposed leasing program by an accountant, tax expert, or other knowledgeable individual may be advisable to determine whether certain anticipated benefits will accrue.

The case for leasing includes benefits such as the following:

1. Capital is conserved or freed for more productive uses and not tied up in a purchase price.
2. Total instead of partial financing.
3. Generally, leases are not shown as a liability on a balance sheet.
4. One lease payment which can be readily substantiated instead of separate and varied costs.
5. Not necessary to speculate in the used equipment market or locate buyers for equipment no longer needed.
6. Provides a hedge against **obsolescence.**
7. Equipment may be procured immediately for either planned or unanticipated situations.
8. Generally, modern top quality equipment is available at all times.
9. Normal credit sources not affected.
10. Deducting lease payments from gross income (before taxes) and not out of net profits (after taxes) as is the case when equipment is purchased.

Depending upon circumstances, additional benefits may develop and due consideration should be given to the effect they may have upon a particular venture.

Planned Equipment Replacement. After what period of time is it advisable to replace a machine with a similar, but newer, model? On the surface this may be assumed to be a simple question and should have a simple and uncomplicated answer. However, upon examination one will conclude that the answer to this "simple" question is far from uncomplicated.

The practice of retaining a machine beyond its useful life or disposing of it too soon may result in costs which are unnecessarily high and profits correspondingly low. Decisions to replace will be more meaningful and have greater validity when they are based upon appropriate cost and performance data. Equipment replacement decisions may be influenced by circumstances such as necessary engine overhaul, extra available capital or starting on a new job or contract. While these reasons may tend to justify machine replacement, they do not provide a sound basis for a planned replacement program.

Generally, all mechanical timber harvesting and construction equipment has a somewhat typical life cycle. Initially, a unit is able to perform at a high level of production when in a new, near new, or recently rebuilt condition. As in-use hours accumulate on a unit and improved equipment is placed on the market, some of the key machines on an operation are either traded in or relegated to work which is less demanding. In the case of a large crawler tractor being used both for skidding and road construction, a typical cycle might include: ripping, dozing, skidding, clearing, standby, and ultimately, junk or spare parts.

The necessity of adequate record keeping is emphasized. A complete record of maintenance and repair costs together with downtime must be available and related to the machine involved. This information, coupled with a thorough examination of pertinent costs together with a consideration of price increases and changes in dollar purchasing power, provides a sound basis for machine replacement decisions as part of a planned program. Various cost items, such as depreciation costs, investment costs, replacement costs, maintenance and repair costs, downtime costs, and costs of obsolescence, have a direct bearing upon replacement decisions. When related to a particular machine, the cumulative cost of the listed items divided by cumulative machine hours provides a cumulative cost per hour. The most economical time for replacement may be determined by noting whether the cumulative cost per hour for a particular machine becomes increasingly higher or lower with added machine hours.

The expense of owning and operating construction equipment used on a highway project generally is reckoned on a 2,000-hour year, while timber harvesting and forest road construction equipment usually is limited to a 1,600-hour or 200-day operating season, depending upon geographical location. Usually winter weather or the rainy season limits the effectiveness of mechanical equipment involved with timber harvesting and forest road construction activities. Generally, operations are either curtailed or suspended when weather conditions are such that acceptable levels of production cannot be maintained. In order to compensate for this situation, the timber harvesting industry uses a shorter operating season when costing equipment. A period of shutdown, generally during the winter season, allows machinery to be overhauled and readied for the start of the next operating season. This arrangement tends to provide a higher machine availability than when machine overhaul activities are conducted during the actual operating season, as is generally the case with highway construction machinery.

The relative effects of the various cost items pertinent to replacement decisions will be shown by means of an example. A hypothetical sitaution involving a piece of mechanical timber harvesting or road construction equipment with an initial price of $20,000 will be used. It is assumed that the machine will operate for a period of 1,600 hours during a 200-day season.

DEPRECIATION COSTS. For purposes of planned equipment replacement, depreciation costs identify the decline in actual value as a result of usage, obsolescence, and general overall condition. Machine make and model may also influence value decline. Depreciation costs in this instance refer to machine owner costs as reflected in the dollar difference between purchase price and resale or trade-in value. In contrast, depreciation expense refers to that portion of the purchase price the Internal Revenue Service will allow the owner to charge off as an expense against income. Depreciation expense is determined by means of the usual straight-line, sum-of-the-digits or declining-balance methods.

Generally, a greater proportion of the total decline in actual value occurs within the first few years a new machine is in use and is the result of more frequent model changes and innovations, such as power shift transmission along with an increased rate of product improvement.

The procedure for determining depreciation cost per cumulative hour for a particular machine requires the determination of trade-in value on a yearly basis throughout the effective life of the unit. Subtracting trade-in value from initial price determines yearly depreciation cost. Cumulative depreciation cost is determined by adding a succeeding year's depreciation cost to the previous year's cost. Cumulative job hours are determined in the same manner. Dividing the cumulative depreciation cost by the appropriate number of cumulative job hours provides a depreciation cost per cumulative hour. Depreciation costs for the hypothetical machine are presented in Table 5–2.

TABLE 5–2. Depreciation Costs

Initial Price—$20,000
Working Hours per Year—1,600

Item	Year					
	1	2	3	4	5	6
Trade-in Value (% of delivered price)	75%	60%	50%	40%	35%	30%
Trade-in value	$15,000	$12,000	$10,000	$ 8,000	$ 7,000	$ 6,000
Yearly depreciation	$ 5,000	$ 3,000	$ 2,000	$ 2,000	$ 1,000	$ 1,000
Cumulative depreciation	$ 5,000	$ 8,000	$10,000	$12,000	$13,000	$14,000
Cumulative working hours	1,600	3,200	4,800	6,400	8,00C	9,600
Depreciation cost per hour	$ 3.10	$ 2.50	$ 2.10	$ 1.90	$ 1.60	$ 1.50

Examination of depreciation cost-per-hour values within a six-year period reveals a steady drop in depreciation cost. If this were the only consideration, the data strongly favor retaining the machine.

INVESTMENT COSTS. Regardless of whether a machine is acquired by outright purchase on a cash basis, installment contract, rental-purchase, or lease, items such as finance charges, interest, insurance, and taxes must be considered so that true costs may be determined. When a unit is rented or purchased by means of an installment contract, practically all of the interest is the result of charges for financing the transaction. In the case of a cash purchase, the interest charged against a machine is equivalent to that which would be earned had the same amount of money been invested.

In actual practice, an owner would, of course, use the type of charges applicable to his particular financial circumstances. However, in order to provide a reasonably uncomplicated approach, cash purchases will be assumed in the example. In this case, it is generally accepted that the combined cost of interest, insurance, and taxes is equivalent to 10 percent of the average yearly investment for a given year. Investment costs for the hypothetical machine are presented in Table 5–3.

TABLE 5–3. Investment Costs

Item	Year					
	1	2	3	4	5	6
Investment (beginning of year)	$20,000	$15,000	$12,000	$10,000	$8,000	$7,000
Less depreciation	$ 5,000	$ 3,000	$ 2,000	$ 2,000	$1,000	$1,000
Investment (end of year)	$15,000	$12,000	$10,000	$ 8,000	$7,000	$6,000
Average yearly investment	$17,500	$13,500	$11,000	$ 9,000	$7,500	$6,500
Investment cost (10) %	$ 1,750	$ 1,350	$ 1,100	$ 900	$ 750	$ 650
Cumulative investment cost	$ 1,750	$ 3,100	$ 4,200	$ 5,100	$5,850	$6,500
Cumulative working hours	1,600	3,200	4,800	6,400	8,000	9,600
Investment cost per cumulative hour	$ 1.10	$ 0.97	$ 0.88	$ 0.80	$ 0.73	$ 0.68

Investment costs on a per cumulative hour basis decrease as a machine passes through its period of useful life. This trend indicates that these costs moderately favor retaining a machine.

REPLACEMENT COSTS. These costs take into account the price trend of construction machinery and equipment together with changes in the purchasing power of the dollar. Based upon the past behavior of the wholesale price index for construction machinery and equipment, the projected wholesale price index is likely to increase 5 percent per year for the next several years. The purchasing power of the dollar has decreased approximately 2 percent per year during the past two decades. Assuming that this trend will continue during the next several years, one may estimate the purchasing power of future dollars. Replacement costs for the hypothetical machine are presented in Table 5–4.

TABLE 5–4. Replacement Costs

Dollar Valuation and Price Index Correction
Trading at End of Year

Item	Year					
	1	2	3	4	5	6
Price of replacement of equipment	$21,000*	$22,000‡	$23,000	$24,000	$25,000	$26,000
Price adjusted to 1968 dollar	$20,580†	$21,120§	$21,620	$22,080	$22,500	$22,880
Less original cost	$20,000	$20,000	$20,000	$20,000	$20,000	$20,000
Loss on replacement	$ 580	$ 1,120	$ 1,620	$ 2,080	$ 2,500	$ 2,880
Cumulative work hours	1,600	3,200	4,800	6,400	8,000	9,600
Loss of replacement cost per cumulative hour	$ 0.36	$ 0.35	$ 0.34	$ 0.33	$ 0.31	$ 0.30

* $20,000 × 1.05 = $21,000 (increase of 5 percent per year)
† $21,000 × .98 = $20,580 (decrease of 2 percent per year)
‡ $20,000 × 1.10 = $22,000
§ $22,000 × .96 = $21,120

Here again, there is an indication that the machine should be retained; however, the magnitude and trend of the yearly losses do not suggest anything more than a slight advantage if the machine were to be retained.

MAINTENANCE AND REPAIR COSTS. Generally, the cost of maintenance and repair is the largest single item of operating expense for timber harvesting and construction equipment. While it is possible to spend an amount equal to the initial purchase price of a machine for maintenance and repair items within the life cycle of a particular machine, severe operating conditions may bring this equalization at an earlier date.

In actual practice, maintenance and repair costs for an individual machine would exhibit a very irregular pattern. This condition is the result of major overhauls or expensive replacements, such as tracks, tires, etc., which would show abnormally high costs in the year they occur. This situation stresses the importance to machine owners of maintaining accurate cost records on an individual machine basis. Despite these yearly variations in maintenance and repair costs, there is a certain degree of correlation between these costs and downtime. Using available data based on averages developed by a leading tractor manufacturer for its track-type tractors, downtime and maintenance will be prorated accordingly. However, when computing actual hourly costs equipment owners should, of course, use their own maintenance and repair cost data as related to a specific machine.

If conditions were such that maintenance and repair costs were the only factors to consider, one could conclude from Table 5–5,

TABLE 5–5. Maintenance and Repair Costs

Item	Year					
	1	2	3	4	5	6
Availability	95%	93%	90%	86%	82%	80%
Maintenance and repair cost	$1,400	$1,800	$2,800	$3,800	$ 4,800	$ 5,400
Cumulative repair cost	$1,400	$3,200	$6,000	$9,800	$14,600	$20,000
Cumulative work hours	1,600	3,200	4,800	6,400	8,000	9,600
Cumulative repair cost per hour	$ 0.88	$ 1.00	$ 1.25	$ 1.53	$ 1.82	$ 2.08

which presents maintenance and repair costs for the hypothetical machine, that yearly trading is advised in order to avoid the financial burden associated with increasing costs and resulting downtime.

The marked yearly increase of maintenance and repair costs together with the resulting unavoidable downtime heavily favor trading a machine yearly.

DOWNTIME COSTS. Downtime refers to the time during a work period, i.e., day, week, or month, when a machine is unavailable to perform its usual functions in an acceptable manner. Percent availability is the relationship between a certain period of work time and the corresponding time a machine is available within this work-time period. When a unit is unavailable, the time lost must be regained if production schedules are to be maintained. In this respect, downtime is a cost and should be considered as an operating expense. As a machine becomes older, the effect of normal wear will be reflected by an increase in downtime.

Downtime may vary considerably from one machine to another depending upon make, age, and the quality of preventive maintenance provided. While availability percentages have been included in the example to indicate the effect of this factor, downtime expense should be computed from individual machine records. The effect of machine unavailability may extend beyond the individual machine involved. This is particularly important on timber harvesting operations and road construction projects where production achieved by one machine usually is dependent upon other machines in the operational cycle. The downtime charge of $6.00 per hour used in the example is the approximate cost of owning and operating a $20,000 machine to replace the one which is unavailable, thereby maintaining production. Downtime costs for the hypothetical machine are presented in Table 5–6.

TABLE 5–6. Downtime Costs

Item	Year					
	1	2	3	4	5	6
Availability	95%	93%	90%	86%	82%	80%
Hours available (out of 1,600 hours)	1,520	1,488	1,440	1,376	1,312	1,280
Hours to be regained	80	112	160	224	288	320
Cost at $6 per hour	$ 480	$ 672	$ 960	$1,344	$1,728	$1,920
Cumulative cost of downtime	$ 480	$1,152	$2,112	$3,456	$5,184	$7,104
Cumulative work hours	1,600	3,200	4,800	6,400	8,000	9,600
Cumulative downtime cost per hour	$ 0.30	$ 0.36	$ 0.44	$ 0.54	$ 0.64	$ 0.74

The magnitude and trend of downtime costs significantly favor trading a machine yearly.

COST OF OBSOLESCENCE. Machinery manufacturers are spending ever-increasing sums of money on research and engineering in an effort to provide industry with machines which can surpass the performance of older models. While most equipment manufacturers have no established pattern or time table when new models are introduced, past history indicates most companies introduce new models every two or three years. Within recent years, **powershift** transmissions, **torque converters,** hydraulic winches, and **turbochargers** have increased significantly a machine's ability to do work. Consequently, most new units are considered to have a productive potential in excess of 20 percent over their predecessors. **Gas turbine** developments suggest that still higher production and performance levels will be possible.

In order to incorporate the effect of increased production potential and new model availability into the framework of the example, a 20 percent increase in production potential and a new model period of three years will be assumed. Once a new model is available, the older model must operate extra hours to compensate for its inability to match the production capability of the new model on an hour-to-hour basis.

Obsolescence cost for the hypothetical machine is presented in Table 5–7.

TABLE 5–7. Cost of Obsolescence

Item	Year					
	1	2	3	4	5	6
Obsolescence factor				20%	20%	20%
Hours required to match production of current unit (20% of 1,600 hours)				320	320	320
Cost at $6.00 per hour				$1,920	$1,920	$1,920
Cumulative cost				$1,920	$3,840	$5,760
Cumulative work hours	1,600	3,200	4,800	6,400	8,000	9,600
Obsolescence cost per cumulative hour				$ 0.30	$ 0.48	$ 0.60

As new models are placed on the market, obsolescence costs decidedly favor trading a machine for the new model at that time.

SUMMATION OF CUMULATIVE PER-HOUR COSTS. Several cost factors have been examined and related to a $20,000 machine in an effort to estimate both their individual and collective effect upon the economics of replacing the unit. A summarization of the cost

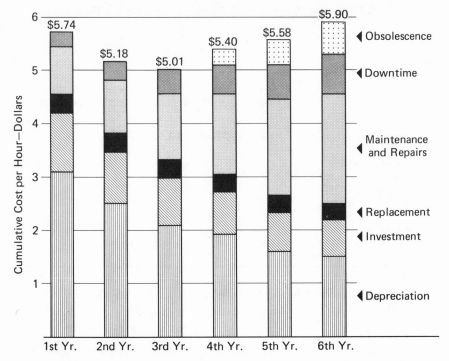

Fig. 5–4. Summary of cumulative costs per hour for a $20,000 machine, 1,600-hour work year.

factors is presented in Table 5–8, while Fig. 5–4 portrays the identical data in graphical form.

Examination reveals that certain factors, when considered on an individual basis, favor retaining the unit indefinitely while others suggest early replacement. An indivisible relationship exists among all of the cost factors examined and their collective effect provides a sound basis for making replacement decisions. The data developed for the hypothetical machine indicates that it should be replaced after it has been in use for a period of three years. This does not mean that every owner should replace every machine every three years. The main point to consider is that in every equipment user's operation, there is one ideal time when equipment should be replaced, regardless whether it was acquired in a new or used condition. Under certain circumstances a two-year replacement cycle would be appropriate and in others a longer period would provide increased economies. It is well to note that the totals shown in Table 5–8 and Fig. 5–4 do not include certain additional cost items;

TABLE 5–8. Summary of Cumulative Costs per Hour

| | Years and Hours | | | | | |
Factor	1 1,600	2 3,200	3 4,800	4 6,400	5 8,000	6 9,600
Depreciation costs	$3.10	$2.50	$2.10	$1.90	$1.60	$1.50
Investment costs	1.10	0.97	0.88	0.80	0.73	0.68
Replacement costs	0.36	0.35	0.34	0.33	0.31	0.30
Maintenance and repair costs	0.88	1.00	1.25	1.53	1.82	2.08
Downtime costs	0.30	0.36	0.44	0.54	0.64	0.74
Obsolescence costs	—	—	—	0.30	0.48	0.60
Total cumulative cost per hour	$5.74	$5.18	$5.01	$5.40	$5.58	$5.90

such as fuel, grease, lubrication oil, and, depending upon the type of machine, wire rope. These costs when combined with operator's wages would increase each total approximately $4 to $5 per hour.

IMPROPER REPLACEMENT PRACTICES. Up to this point, cost comparisons have suggested that replacement has dollar advantages if made at a certain time after a machine has been purchased. Another important consideration involves the loss associated with replacing a machine either too soon after purchase or retaining it beyond the most economical year. Table 5–9 shows the loss in this respect for the hypothetical machine. Furthermore, a comparison between the last two columns in Table 5–9 (loss by not replacing in most economical year and investment credit) shows that the size of the losses more than offsets any possible gain which might accrue by retaining the equipment long enough to obtain investment credit.

The loss on the late side is much more costly for the reason that the owner's loss per cumulative hour is multiplied by an increasingly greater number of hours. Mistakes in judgment in selecting the replacement year are less costly if replacement is made too soon than if made too late.

The necessary prerequisites for implementing an effective equipment replacement program are performance and cost records. Such records provide the basis for valid and meaningful replacement decisions.

TABLE 5–9. Effect of Improper Replacement Practices

Replace-ment Year	Timing	Cumulative Hours	Loss per Cumulative Hour	Loss by not Replacing in Most Economical Year	Invest-ment* Credit
1st	2 Years too soon	1,600	$0.73†	$1,168	—
2nd	1 Year too soon	3,200	$0.17	$ 544	—
3rd	Most economical year to replace	4,800	—	—	—
4th	1 Year too late	6,400	$0.39	$2,496	$467‡
5th	2 Years too late	8,000	$0.57	$4,560	$467
6th	3 Years too late	9,600	$0.89	$8,544	$933§

* Refer to Page 186.
† From Table 5–7. Difference between cumulative cost per hour: $5.74 for 1st year and $5.01 for 3rd year.
‡ $20,000 $\times$ 1/3 $\times$ .07 = $467
§ $20,000 $\times$ 2/3 $\times$ .07 = $933

Having considered the effect of certain direct and indirect cost items, as related to a hypothetical machine, a sound basis for making dollar-savings time-of-replacement decisions has been presented. In contrast, such decisions when made in a haphazard manner may be costly.

This method was developed by the Caterpillar Tractor Company in 1965.

Records. Typically, most contract operators seem to have an aversion to paperwork. They would much rather harvest timber. While record keeping may not by itself show a profit, the usefulness of this endeavor should not be underestimated, particularly in light of the benefits which may be derived. The ability to know exactly how much cost is involved in performing a particular function provides a management yardstick which is of considerable value.

Records are effective when they show the degree of success being achieved by a business enterprise. They also serve to support statements made on tax forms. Furthermore, records generally provide the preliminary basis for making sound decisions. Records should provide needed facts at the proper time and they should be collected and entered in an uncomplicated manner.

There are many different record keeping systems in use within the timber-harvesting industry and, generally, they reflect the particular needs and desires of the user. Basically, records serve to document the when, where, and how much, in relation to various situations. While it may be possible for the responsible person to have much of this information in his "cranial filing cabinet," the possibility of overloading the "miscellaneous niche" or "misfiling" an important item may lead to cost or operational conditions which are less than optimum or, in extreme situations, have disastrous financial consequences.

Generally, all records which are associated with harvesting forest products, involve both people and money in either a direct or indirect manner. Using this generalization, records involving inventory, purchasing, sales, production, cash, credit, employees, equipment, insurance, and taxes, to name some of the most obvious, provide a contractor with factual information relative to his cash position and tend to provide him with some measure of protection against unreasonable demands from individuals, organizations, or taxing agencies.

While the criticism relative to the ever-increasing amount of forms, reports, and data demanded by governmental agencies may or may not be justified, the fact remains that generally an employer must comply or accept the consequences.

Perhaps the most important records for which an employer is responsible involve various taxes which are paid to the federal and state governments. Included are: federal and state income taxes, social security taxes, and federal and state unemployment taxes. While the provisions of the Fair Labor Standards Act and state Workmen's Compensation laws are not of a tax nature, they do require an employer to maintain satisfactory records so that his financial posture will not be jeopardized. Often the counsel of a tax expert or accountant may provide a valuable service in establishing an adequate record-keeping system. Tax investigations resulting from improper record-keeping procedures are time consuming and frequently embarrassing.

Fortunately, certain of the more progressive logging organizations, both large and small, have developed forms which facilitate record keeping and have reported their methods through trade journal articles or by means of association releases. Standard forms available from commercial sources may provide the necessary framework for an adequate record-keeping system.

Production and Payment Control. As is the case in all business enterprises, some measure of production must be applied to activities so that costs may be ascertained and production control may be exercised. Production in timber harvesting may be expressed in several ways: board foot, cubic foot, linear foot, cord, **cunit,** piece, and weight. Basically, scaling is considered to be the measurement of logs to determine their usable contents in board feet. In its broadest interpretation, scaling is the process which expresses the net usable contents of forest products in an acceptable unit of measure.

Possibly the most controversial unit of measure is the board foot. It refers to a piece of lumber 1 inch thick, 12 inches wide and 1 foot long. When logs are to be scaled for board foot content, the scaler uses a particular log rule as directed by law or contract agreement. Since 1846, over forty-five log rules have been used at one time or another; however, only ten or twelve of them are in general use today. As is to be expected, there are differences and inconsistencies between and among various log rules; however, price adjustments on a per unit of volume basis can be used to offset such disparities. Certain states and Canadian provinces have adopted an "official" or "legal" log rule.

The control of an entire timber harvesting operation is very closely involved with scaling, which is necessary for the following reasons:

1. It provides a measure of forest products bought or sold.
2. It provides the means whereby work accomplished may be measured at various stages of the entire operation.
3. The accuracy of a timber cruise may be determined.
4. It provides inventory data.
5. Marking logs during the scaling operation may assist in log sorting by species, routing logs to proper destinations and establishing ownership.

Depending upon the organization of a timber-harvesting operation, certain information regarding production is necessary. For example, where felling and bucking is a separate operation from yarding or skidding, a measure of work accomplished at various stages of the operation must be determined so that proper payment may be made.

Within recent years, considerable emphasis has been given to weight scaling, particularly by the pulp and paper industry. This method in effect converts loads of pulpwood, logs, or tree-length sticks to board-foot volume, cords, or other acceptable units. The advantages of using weight as a means of conversion to volume are most apparent in those applications where grade or quality is not a major consideration. The pulp and paper industry, where the raw material, pulpwood, has a rather low value per unit of volume, did much of the early investigative work in this direction. Excessive void space in a truckload of pulpwood or logs may go undetected by a scaler whereas the dial on a scale is impartial. Obviously, weight scaling will not provide an exact measure reflecting the volume of pulpwood being delivered to a yard or weighing station on any one truck since logs vary in shape or form. This variability brings about a corresponding variation in void space when logs are hauled by truck. However, when put to continuous use the method will provide an accurate total measure when a great number of loads are used to arrive at an average.

Converting factors are used when a forest product which has been measured in one unit is to be expressed in some other unit of measure. As an example, cords are converted to board feet. Generally, the main reason for converting from one unit to another is to more effectively compare or estimate costs and production. Cost accounting procedures or timber purchase payments may necessitate the use of conversion factors.

COMPLIANCE INSPECTION

Many foresters, in both public and private employ, have the responsibility of inspecting timber harvesting activities on both company and contract operations for what is termed compliance. This type of inspection provides the means whereby a comparison may be made between what is actually being done in contrast to what had been agreed upon previously.

The guide lines for making a compliance inspection of necessity must be associated closely with the type of agreement in effect.

Depending upon circumstances, the agreement may be either verbal or spelled out in the form of a contract. For all practical purposes, one may eliminate verbal agreements as generally unsatisfactory in this day and age. Prior to the time a compliance inspection is made, the inspector must be thoroughly familiar with the contents of the timber sale contract so that he may make an effective and meaningful inspection. This type of inspection should be made soon after activities commence on a timber sale so that the seller may determine: (1) whether the buyer is aware of his responsibility insofar as compliance is concerned and (2) to what degree the activities of the buyer or operator satisfy the desires of the seller. Compliance inspection is of particular importance in those cases which involve operators who have not yet had the opportunity to prove that they are competent. An example might be those situations where operators move into new locations. Under these circumstances the seller would be unfamiliar with the operator's willingness to recognize or comply with the provisions as embodied in the timber sale contract.

Inspection report forms are many and varied, depending upon the organization responsible for making the inspection. While a verbal interchange of comments relative to a compliance inspection may be effective in certain instances, a written report tends to formalize any misunderstanding or misinterpretation of agreements. Inspection reports, of course, should be distributed to the principals involved so that corrective action can be taken, should it be required.

Some of the more troublesome areas which seem to appear on inspection reports often involve:

Utilization Standards

Removal of all merchantable material. Proper dimensions for: stump height, stump diameter, minimum top diameter, log lengths and trim allowance. Removal of damaged trees.

Progress

Of particular importance in those localities where weather and season tend to have a strong influence on subsequent timber harvesting activities.

Carelessness

Excessive breakage as a result of poor workmanship in the log-making process. Excessive damage to "leave" trees or residual stand during the skidding operation.

Cutting Boundary or Tree Marking

Deliberate or inadvertent trespass by cutting crew. Cutting unmarked trees.

Orderly Log Removal

Failing to follow an orderly progression of log removal, such as leaving logs in difficult or remote locations.

Landing Cleanup

Failing to observe "good housekeeping" rules, such as removing empty containers and unsightly reminders of timber harvesting activity from landings or other areas where equipment has been serviced.

Compliance inspection should be made in a reasonable manner with due respect for those circumstances which may be unavoidable. While strict compliance may be a worthy goal, the fact remains that in practice something less than this level is usually acceptable. A competent compliance inspector should recognize when corrective action is necessary. He also should be cognizant that certain aspects of the timber harvesting processes at times are unavoidable and not generally considered to warrant corrective action.

IMPORTANT CONSIDERATIONS

Because public opinion is such a potent factor insofar as the forest products industry is concerned, one must examine certain items which tend to have strong influence upon both the administration and operation of timber harvesting activities. Naturally, all facets of the industry are involved, including the various public agencies that offer timber for sale, private timberland owners, and the operators who harvest the timber.

While the more conservatively minded may object to the controls and regulations already in existence, indications suggest that the industry will have to accept more regulatory measures in the future. The areas of prime importance are those which have a direct effect upon people such as recreation, soil disturbance as related to water quality, and air pollution. Usually regulations and controls are the result of legislative enactment; consequently, politics cannot be ignored.

Recreation

The demands for outdoor recreation are very closely associated with the much heralded population explosion—now a fact of life! While the present attitude of the forest industry is much more cooperative toward recreational demands than in the past, it is quite obvious with increased population pressures that greater demands will be forthcoming. Such demands may well develop into some form of governmental control in the event that a reasonable and sincere effort by the industry is not made to satisfy these demands.

Recreation is rapidly developing into the major use of this nation's forests, and this includes both public and private ownership. Furthermore, all other uses of the forest resources, such as timber, grazing, water, and wildlife, may well be subordinate to the demands of recreational use. Should this relationship develop, timber harvesting activities would be affected more than any other use.

The reason for the increased demand for recreational use of forest land stems from people with money and leisure time. Our society has provided us with a situation which places money in people's pockets so that they may acquire mobility and use their leisure time as they see fit.

Outdoor recreation means many things to many people, and this in itself tends to create a problem. In an effort to show the interrelationship of desire on the part of the recreationists and what is available to satisfy this desire, one may consider air, water, and soil as the basic elements of outdoor recreation. Various combinations of these elements provide the tangibles of recreation which include trees, streams, wildlife, rivers, lakes, beaches, and mountains. A blending of these tangibles tends to identify various recreational activities. Hiking, boating, trail riding, camping, hunting, fishing, photography, beach combing, rock hounding, and mountain climbing are some of them. The link between people and these recreational activities is provided by a network of roads together with vehicles.

Some of the most effective means of advertising and promotion are being used to entice the "recreation dollar" to all parts of the country, and records indicate that a high degree of success has been achieved. Projections of past trends into the future suggest a near fantastic level of expenditure on the part of the outdoor recreationist. It would appear that the forest products industry must decide not only what it can do but also what it must do to satisfy, at least in part, those recreational demands which surely will be forthcoming.

There is a rather direct relationship between timber harvesting activities and the aforementioned elements and tangibles of recreation. In this respect, timber harvesting operations which enhance recreational potential will be accepted by the recreational fraternity— generally without fanfare. Conversely, criticism in respect to operational procedures which tend to destroy or jeopardize recreational values generally will receive prompt criticism and censure from both individuals and organizations. To many individuals, the forest products industry of today still operates in the shadow of the early so-called "timber barons," and follows the "cut out and get out" policy. Critics of the industry tend to strengthen their effectiveness by joining or supporting organizations which advocate restrictions in regard to timber-harvesting activities. It may well be that time has run out insofar as the industry's ability to do what it has to do in respect to recreation facilities; however, in case some time may be bought, it behooves the forest products industry to inundate its critics with factual information in regard to what has been done, along with future plans for further recreational development. It should be noted that many of the companies within the forest products industry have provided the recreationist with many and varied recreational facilities such as camping and boating areas, hunting and fishing areas, picnic and scenic areas, boat launching ramps, and bathing areas, to name some of the more important.

The various federal and state organizations responsible for managing the public timberlands are very heavily committed to recreational demands. These organizations prepare and administer timber sales according to the provisions of a timber sale agreement acceptable to the purchaser. Within recent years, much more emphasis has been placed upon those provisions of the agreement which enhance or protect recreational values. In this respect the successful bidder is committed to perform various timber-harvesting activities in a manner which follows the recreational policy of the timber-selling agency. Obviously, a clear understanding should exist between the organization harvesting the timber and the selling agency as to just what is to be done and how it is to be accomplished. In those instances where the timber-selling agencies direct operators to develop recreational facilities or values, which require additional or special effort, proper cost allowances are both necessary and justified.

Aesthetics

A program which involves the correlation of aesthetics with timber harvesting has recently been made mandatory in timber management and operations on United States Forest Service timber areas in Region 6. The successful bidder for a Forest Service timber sale in this region will be required to perform various timber harvesting operations in a manner which will leave the logged area in a more attractive condition than heretofore. Timbered areas will be zoned according to their location insofar as the effect the "cut-over picture" will have upon the recreationist and the public in general. In essence, the objective of this program involves a means which, hopefully, will lead to a greater public acceptance of timber management activities.

All timbered areas within national forests in Region 6 will be classed as either timber key value areas or landscape management areas under this program. The majority of timber sales will be located in timber key value areas where the aesthetic and recreational objective will be to develop and maintain the attractive appearance of a thrifty, healthy, productive, used and well-managed forest accessible and usable for various forms of the dispersed types of forest recreation including hunting, fishing, and berry picking. Generally, these lands will provide the remote type of scenery when viewed from the more traveled road systems. There are two major subdivisions of the landscape management area, the immediate foreground and primary foreground. Included in the immediate foreground portion of a landscape management area are roadside zones, trailside zones, waterfront zones, buffer zones around recreation occupancy areas, and other areas of forest that are viewed at close range by many people.

In contrast, the primary portion is behind the immediate portion and is viewed as a forest canopy. Relating the foregoing categories of land to a road or highway one would have this arrangement:

Timber Key Value Area

Primary Foreground   Landscape Management Area

Immediate Foreground   "E" Road

The forest land included in a landscape management area will be managed according to certain objectives and restrictions depending upon its recreational and aesthetic potential. Obviously, more intensive treatment will be administered to the land which is classed as immediate foreground where the protection and maintenance of scenery is the guiding principle.

While this example refers to a specific region in the U.S. Forest Service, all regions operate according to a Multiple Use Management Guide. These regional guides prescribe necessary and desirable coordination of the various national forest system resource uses and activities. Obviously, much emphasis is placed upon the basic renewable resources: air, soil, water, timber, range, recreation and aesthetics, and wildlife and fish. Certain of these resources are closely associated with timber production management, while others may appear to be far removed. However, the fact is that timber harvesting activities on national forest land will be carried on according to the local Multiple Use Management Guide, which places strong emphasis on scenic natural beauty and recreational values.

Labor and Labor Relations

Today, the logging-camp type of operation with its generally migratory labor force is being replaced rapidly by timber harvesting operations where the workers are of the commuter or home-guard type. Suitable access between woods operations and adequate living centers has provided the forest industry with a generally more

dependable type of woods worker. While in the past the various logging activities required a high degree of skill of a special type, the
situation in timber harvesting, now, and in the future, involves
proficiency in a different set of skills. Competition from substitute
materials has caused the forest products industry to examine criti-
cally production and cost relationships. Mechanization of timber
harvesting activities had to be undertaken in order to remain competitive in cost with wood substitutes. As a result of mechanization,
woods workers, with the ability to learn and use appropriate skills,
had to be trained and guided. While much has been accomplished
in this respect, there remains a continuing need for improvement in
both cost and production.

The necessity of providing woods workers with year-around employment at a wage which is competitive with alternate types of
employment has been recognized by the more progressive timber-
harvesting organizations. The forest products industry gains nothing
by training an equipment operator who, after becoming proficient,
chooses to leave the industry and work where he can be employed
the year around, and thereby better himself financially. This is not
to say that workers should be subsidized to remain in the industry.
Every effort must be made to utilize men and machines in such a
manner that acceptable production levels will be achieved. Of parallel importance is the enhancement of worker satisfaction and the
minimization of the desire on the part of labor for a change in employment.

In order to provide year-round employment for workers involved
with timber harvesting activities, much advance planning is necessary. This is of particular importance in those regions and localities
where weather conditions tend to curtail operations. While it is
true that some progress has been achieved in this respect, particularly by the more progressive organizations, greater emphasis must
be given to this important consideration. The traditional operating
season period of employment must give way to year-round employment if the forest products industry is to compete effectively in the
labor market.

Radio Communication

Since the forest industry is directly involved with a rather wide
spectrum of activities, often somewhat far removed one from the

other, communication by means of radio is essential. Furthermore, an increased use of radio communication within the industry is to be expected in the future, as additional activities are introduced. As the forest industry is not unique in this respect, future competition from other industries for both existing and additional channels may be brisk.

The Federal Communications Commission (F.C.C.) is the agency which is ultimately responsible for the allocation of channels to both public and private agencies requesting a radio communication system. The Forest Industries Radio Commission (F.I.R.C.) located in Eugene, Oregon, is the agency which coordinates channels, or air space, for use within the forest industry. Radio users who encounter interference on channels assigned to them can obtain assistance from the engineer-in-charge of the F.C.C. in their local district.

Radio communication activities are administered according to certain rules and regulations; consequently, violators are subject to citations which in turn may lead to fines. Incomplete logbooks, irregular operating procedures and improper equipment maintenance are some of the more troublesome areas of concern.

Presently most of the major uses of radio communication within the forest industry may be listed under three headings: (1) people, (2) weather and fire, and (3) men and equipment.

People. There is an ever-increasing number of people seeking some form of outdoor recreation in forested areas; consequently, an increasing number of problems is to be expected. Radio serves to provide the means whereby timely assistance may be rendered. This is of particular importance in regard to accidents or emergencies.

Weather and Fire. Sudden and abrupt changes in weather conditions may have serious consequences for both men and equipment. Alerting personnel to impending hazardous wind and weather oftentimes is done by means of radio communication. Timely interchange of weather information prior, during, and subsequent to periods of forest fire danger is one of the most important uses of radio. Present-day forest fire detection techniques, along with pre-suppression and suppression activities, are involved directly with radio communication.

Men and Equipment. Radio communication allows management to direct the movement of men and equipment from one location to another. Haul trucks may be directed to landings which have logs

ready for loading instead of having trucks wait at landings which
are empty. Radio controls are an integral part of certain skyline carriages. Electronic signaling devices usually are used as a means of
controlling certain activities associated with cable yarding systems.
By means of such a device directions from the person controlling
the operation are relayed to the yarding engineer who activates the
system accordingly. An example of such a device is the Talkie-tooter
which relays directions in either voice or whistle tones. Video apparatus of the television type also is being considered as a means of
providing the yarding engineer with more direct control over cable
yarding activities.

SUGGESTED SUPPLEMENTARY READING

1. Falk, Harry W., 1958. *Timber and forest products law.* Howell-North, Berkeley,
   California. This book covers the entire field of timber and forest products law.
   Everyone in forestry or the forest industries will find understandable, non-
   technical answers to his legal questions in Falk's book.
2. Johnson, Peter. 1959. Cost accounting protects logger. The 1959 Timberman
   Logging Handbook. Miller Freeman Publications, Portland, Oregon. Describes
   a custom-made cost accounting system used by a progressive contractor. The
   presentation starts with the final summary form and works back to the source
   of various costs.
3. Luiten, Irvin H. 1966. No escape from politics. *Proceedings,* Society of American
   Foresters. 1966 National Meeting, Seattle, Washington. Discusses the relationship between the politician and professional forester within the framework of
   communication. The author emphasizes the necessity of political involvement
   at all levels of government.
4. Small Business Administration, Washington, D.C. The S.B.A. has made available many publications which deal with all aspects of running a business. Upon
   request the S.B.A. will provide interested parties with a list of material currently
   available much of which is free of charge.
5. Symposium on Financial Management. 1967. *Journal of forestry.* Vol. 65, No. 7.
   While all of the papers included in the symposium are interesting and pertinent,
   Part V by G. O. Baker, Jr., entitled "Forest Management Records, Reports, and
   Controls" is of particular significance insofar as this chapter is concerned.
6. U.S. Department of the Interior, Federal Water Pollution Control Administration, 1970. *Industrial waste guide on logging practices.* Portland, Oregon. This
   publication is dedicated to the logging practices which must be adopted by the
   logging industry if water quality is to be protected in the streams of the Pacific
   Northwest. Logging roads, logging, and watershed restorations are examined
   and recommended practices are spelled out.
7. Zwick, Jack. 1965. *A handbook of small business finance.* Small Business Administration, Washington, D.C. Points out the major areas of financial management and describes a few of the many techniques that can be utilized to
   understand the results of past decisions and apply this understanding in making
   decisions in the future.

6

Felling and Bucking

The primary process in logging is the conversion of trees to logs, which includes felling, limbing and bucking. Simple hand tools, the axe and saw, were used for this purpose until two decades ago when the power chain saw was introduced. Since then the mechanical saw has made steady in-roads on the axe and saw to the extent that it has practically eliminated hand tools from the felling and bucking scene. The axe and saw are still used in those localities which have rugged topography and small-sized trees. And the axe still is used as a limbing tool, especially in timber which has small-diameter limbs.

FELLING

Felling is the process whereby a standing tree is severed from its stump so that subsequent logging operations may be undertaken (Fig. 6–1). The severing or felling cut is made at a point on the trunk above the root collar. This activity is identified as felling and is carried on by a faller; if two men are required for this task, a set of fallers. The ease with which the felling operation can be accomplished depends upon several factors. It should be noted that these factors also are the main reasons why logging methods vary so markedly throughout North America and the world.

Fig. 6–1. Faller making the back cut using a chain saw. (Courtesy of Homelite, a division of Textron Inc.)

Factors Affecting the Felling Operation

Type of Cutting. The demands imposed by the cutting method may vary from partial cutting to clear-cutting. When selected trees are felled there is a much greater chance that the trees may lodge against others instead of falling to the ground—a condition termed hang-up —than when all trees are felled as in clearcutting.

Climate. Weather conditions may interfere with the efficiency of men and machines, particularly in those areas which have subzero temperatures, heavy snowfall, or an extended rainy season. Access to a logging chance may be limited during periods of adverse weather.

Timber Size. Generally the larger trees have the higher value. Large trees, especially those with a decided lean, require that a far greater degree of control be exercised by the faller so that breakage is minimized and proper positioning on the ground is achieved.

Small trees may be made to fall in a desired direction and generally present no great problem. Production in terms of unit volume per faller-day is much higher when felling large trees than when small trees are felled.

Timber Density. The number of stems per unit of area has a bearing upon the number and occurrence of hangups. Stumps provide a situation which increases the possibility of excessive breakage.

Subsequent Operation. Greater directional control must be exercised when tree length or long logs are to be skidded using mobile units, whereas a short-log **show** does not require the same degree of adherence to directional control during the felling operation.

Topography. Generally, steep topography provides conditions which tend to produce excessive breakage particularly when timber is felled downhill instead of along the contour or uphill. The walking time increases, as do the circumstances for accident and injury.

Ground Conditions. Rocky surface conditions tend to increase the possibility of excessive breakage; whereas swampy conditions tend to impede the efficiency of men and machines.

Underbrush. Dense underbrush requires greater effort to swamp out a place to work and a safe get-away route.

Experience and Type of Labor. In areas where timber production is an important facet of the economy, generally there is a source of experienced labor available for logging operations. The trend is toward replacing the logging camp by the "home guard" type of woods worker who daily travels to the job from his home. The commuting woods workers are considered to be more dependable and safety conscious than the somewhat transient logging camp type of woods workers.

Type of Payment. Generally one of two possibilities is used: (1) payment based upon volume produced in terms of board feet, cords, cunits, or other expressions of volume, and (2) payment based upon time, such as hour, day, or week.

Usually a higher rate of production is realized when the volume method of payment is used. However, in timber which is of high quality such as **veneer** or **peeler grade** where greater care must be exercised in felling the trees to minimize breakage, time payment is preferred.

Felling Procedures

Whether hand or power chain saw methods are employed, the prime objective is to fell the tree at the least cost and with a minimum amount of breakage.

Prior to starting the actual felling operation the faller should select a safe get-away route, inspect the tree for overhead hazards, and determine the direction in which the tree is to be felled. Generally small trees can be cut and pushed over in any desired direction. The choice of felling direction usually is based on having the felled tree so positioned that the subsequent operations of limbing and bucking can be accomplished most effectively. Felling trees over stumps and rocks will increase the possibility of loss through breakage which, however, may be minimized by selecting the correct felling direction and using proper felling procedures. Consideration should also be given to the direction of the lead—the direction in which the logs will be moving during the skidding or yarding operation.

Control of the felling direction is accomplished by some form of undercut which is made at the desired stump height and on the side of the tree selected as the felling direction. Three types of undercuts are shown in Fig. 6–2. The conventional type may be

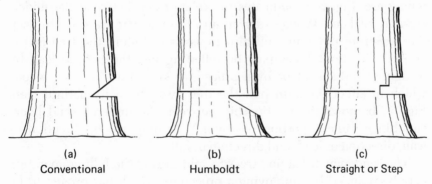

      (a)   (b)   (c)

Conventional   Humboldt   Straight or Step

Fig. 6–2. Types of undercuts.

made solely with an axe or a combination of axe and saw. The horizontal cut is made first and then the sloping cut is made to clean out the undercut. The Humboldt type, which originated in the redwood region, is nothing more than making the undercut in the stump portion of the tree instead of in the **butt-log** portion. This type can be cut easily by a power saw and has the advantage in that the undercut is cleared more easily in contrast to the conventional type, which usually requires additional effort to clear the undercut with a **Pulaski** or axe. The straight or step type generally is used when felling large-size timber and is therefore restricted to the Pacific Coast regions, primarily in the Douglas-fir and redwood regions. The horizontal cuts are made with a power saw and the undercut section is removed with the aid of a Pulaski tool. The step type may be made either in the butt-log portion or in the stump portion of a tree.

Estimates indicate that up to 7 percent of the best lumber in the butt log in large timber (40 inch D.B.H. and larger) can be saved by making the undercut in the stump portion of a tree, i.e., using either the Humboldt or step-type undercut.

The depth to which the undercut is made depends upon the amount of lean a tree has in the direction of the undercut. Deep undercutting tends to reduce the possibility of splitting during the felling operation.

Following the completion of the undercut, the backcut is made. The backcut should be sawed so that it will be slightly above the flat part of the undercut to prevent the stem from sliding off the stump in a backward direction as the stem moves through the felling arc. Fig. 6–3a indicates the relationship between the undercut and backcut. Wedges driven into the backcut are used to keep the saw from pinching and also to assist in felling the tree in the desired direction. Trees may be felled against the natural lean by wedging the backcut or by a procedure known as "holding wood" which in effect tends to pull the tree away from its natural lean. Fig. 6–3b shows the relationship between the undercut and backcut, hinge (uncut portion of stump), holding wood, direction of lean, direction of lead, and direction of fall.

The tendency of leaning trees to split during the felling operation may be reduced by employing a procedure which has proven to be effective (Stenzel, 1957). In this procedure, the undercut is made in the conventional manner on the "leaning face" of the tree to be felled. Next, a shallow cut, from one or more inches in depth, depending upon the size of the tree, is made at the point where the backcut would normally be started. This is followed by making a "boring cut" from the side, at a point which will leave an adequate

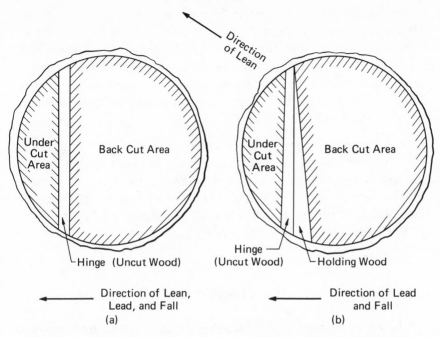

Fig. 6–3. Cross-section of stump areas. (a) Tree leaning in the direction of the lead; and (b) tree leaning away from the direction of the lead.

hinge of uncut wood. After the "boring cut" has been completed the faller guides the saw in such a fashion that the remaining uncut wood (other than the hinge) is severed. Felling leaning trees using conventional backcut procedure tends to greatly increase the possibility of loss in merchantable volume generally associated with such mechanical defects as **"barberchairing,"** splitting, or **stump pull.**

When large trees of high quality are to be felled, it may be advantageous to employ some mechanical means which will assist positioning the felled tree most effectively and thereby tend to keep breakage, both in the tree being felled and adjacent uncut trees, to a minimum.

Introducing hydraulic wedges or jacks into the backcut area of a tree during the felling operation, and directionally pulling trees at the proper time, also during the felling operation, are methods whereby a force, other than the conventional wedge and sledge procedure, may be applied. In effect, a hydraulic wedge or jack serves the same purpose as the conventional wedge except that the lifting force is provided by a means of a manually operated pump instead of driving a wedge with a sledge hammer.

The other method, directionally pulling trees, involves a tractor fitted with a winch together with certain wire rope accessories. From a cost standpoint this method is limited to those steep side-hill locations in the Douglas-fir and redwood regions which support large high-quality fir and redwood trees. This method involves so-called uphill felling where the tractor provides the force to tip the severed stem uphill (Stenzel, 1953).

Tree Shears. A recent development in felling involves a hydrauli-cally activated **shearing unit,** which may be mounted on tractors and also on certain rubber tire carriers. Felling by means of shearing action offers several advantages which, according to the manufac-turer, enhance production. Shear felling is discussed in greater de-tail in Chapter 11 on pulpwood production.

LIMBING

Sawing or chopping the branches from a felled tree refers to limbing, and is done prior to the bucking and skidding operation. The method of skidding governs to a considerable degree who does the limbing and also during which particular stage of the logging it is accomplished.

Usually limbing is done at the location where a tree is felled. If tree-length logging is being practiced, the faller or felling crew usu-ally does the limbing. If logs are to be skidded and a two-man felling and bucking crew is utilized, one man will do the limbing and the other man will measure log lengths and mark the bucking points.

In either type of operation, tree length or log length, oftentimes it will be necessary to complete the limbing operation at the landing, particularly on those logging shows which have timber of larger diameters. The larger trees do not allow the knot bumpers or limbers to effectively limb the side of the felled tree which is in contact with the ground; hence, it is done at the landing. Many loggers do not attempt to limb the lower side of a down tree for the reason that the bucking operation is more effectively accomplished when the down tree is supported by the limbs on the underside. These underside limbs or stubs are not removed until the logs reach the landing.

When trees have relatively large limbs and limbing may be a separate and distinct operation, such as in the ponderosa pine region, a power chain saw with a short bar or a mechanical bow saw normally is used.

LOG MARKING

Considerable saving of wood and a higher cash return is possible with proper log marking procedures which conform to local practice or timber sale contract specifications. Log marking is of particular importance when high-quality timber is being logged, and some operators employ a log marker to designate the bucking points. A log marker can assure that the maximum return will be obtained from high-grade timber.

Some of the more important marking rules which will produce a greater value return are:

1. Wherever possible logs should be marked so that surface defects are concentrated in one log, thereby placing the adjacent log in a higher log grade.
2. Prior to marking a tree for bucking, the entire length should be studied to ensure the best return whether it be grade or volume.
3. Trimming allowance in keeping with regional practice must be maintained so that maximum scale will be achieved.

    Adequate trimming allowances are necessary to compensate for:

    a. Large undercuts of the conventional type made on large trees to avoid splitting the butt log.
    b. The impossibility of cutting all logs square-ended.
    c. Damage to log ends caused by loading hooks.
    d. End brooming, when logs are river driven or rafted.

    The extra length for trimming allows the trim saws at the mill to even off any irregularity in the ends thereby leaving square-end boards of the full specified length. On public land timber sales, logs without this allowance may be scaled in the next lower permissible length, while logs with excessive allowance may be penalty scaled for unnecessary wastage.

BUCKING

Bucking is the process whereby a felled tree is cut into logs in preparation for the skidding or yarding phase of logging. Most bucking is done using power chain saws (Fig. 6–4), and in most cases

Fig. 6–4. Bucker or sawyer cutting a felled tree into logs by means of a chain saw. (Courtesy of the McCulloch Corporation.)

the same saw used in the felling operation is also used in bucking. The increasing popularity of skidding tree length and long logs has led to more bucking at the landing and correspondingly less in the woods.

Factors Affecting the Bucking Operation

Timber Size. Large trees should be bucked into lengths which are practical and economical to handle from the standpoint of transportation (skidding and hauling).

Market Demands. Custom has established the rule that boards be trimmed to even-foot lengths in softwood species and to either odd- or even-foot lengths for certain species of hardwoods; therefore bucking practice coincides with these requirements.

Equipment Limitations. Yarding or skidding equipment may not function effectively if logs beyond the optimum size are produced. Sawmills also are limited in the length of logs they can saw by carriage length, feed, setworks, and log-handling facilities.

Transportation Restrictions. Local highway regulations restrict the G.V.W. (gross vehicle weight), length, width, and height of loaded log trucks traveling on public roads. While special permits may be acquired for transporting over-length loads, the cost of the permit and providing pilot vehicles when required precludes hauling under permit as general operating procedure. As a general rule the maximum log length which may be transported by log truck and trailer combination and log cars is 40 feet; however, longer logs, poles and piling require special consideration which is reflected in higher hauling cost.

Removal of Cull and Unmerchantable Portions of Felled Trees. Included are areas of defect, breakage, and the top portion of the bole beyond the limit of merchantability. Within certain bounds, usually specified by management or by provisions set forth in a timber sale contract, cull or defect is allowable up to a certain percentage of the gross volume of a log. Removing portions of the bole containing excessive cull or defect may result in fitting logs in lengths somewhat less than optimum. Breakage as a result of felling usually is handled in the same manner as defect because the part of the bole containing a break will not produce lumber of acceptable quality. The small, heavily branched tops are severed from felled trees at a point identified as the limit of merchantability and designated by the upper or top diameter limit.

Log Grades. Trees should be bucked so that as much of their surface clear lengths is in clear logs having dimensions of diameter and length as specified in the log grading rules germane to the region.

Bucking Procedures

The prime objective in bucking is to sever a felled tree at bucking points in such a manner to produce logs with a minimum of splitting or otherwise damaging the logs when making the cut.

Prior to starting the actual bucking operation the bucker should **swamp** out a safe place to work. Bucking felled trees on steep side-hills is a hazardous occupation; therefore a bucker must know how to cut logs in all types of dangerous and disagreeable situations without injuring himself or wasting merchantable wood.

The points covered under Log Marking are important to the bucker who does his own marking because he is responsible for getting the "most out of the timber." A measuring tape or pole will eliminate guesswork when marking logs and is of particular importance in timber of high quality.

Few bucking problems are encountered with felled trees lying on reasonably flat terrain. Furthermore, felled trees which are supported on the underside by uncut limbs or stubs tend to ease the bucker's task. More difficult situations such as a tree swinging in a bind, underbind, sidebind, and tree traps require special bucking techniques.

Tree Swinging in a Bind. A felled tree lying across a depression or on other logs with only the butt and top making contact is swinging in a bind. During the bucking process such a tree would quickly bind in the **kerf** resulting from the top cut, while cutting from below would result in opening the kerf with subsequent likelihood of splitting. The objective is to buck in such a manner as will prevent the saw bar from becoming bound in the upper kerf and avoid giving the log a chance to split and be damaged underneath. Figure 6–5 shows the various cuts required to buck a tree swinging in a bind. The numbers refer to the sequence in which the cuts should be made.

The procedure is to make the first cut, in the area of 1 in Fig. 6–5, to a depth at which a slight bind is noticeable. Cut number 2 is then made so that there will be an overlapping of the kerf, area *a*. Cuts 3 and 4 are made next and a condition termed "breaking it down the middle" is developed. Side cutting in area 3 and 4 is continued until the tree begins to settle, which indicates that the center area of uncut wood is separating. Usually, a small amount of undercutting in the area designated by 5 will be sufficient to break the logs apart. This procedure is suggested for trees which are too large to be propped up from below.

Underbind. A tree supported at the midsection, such as lying over a ridge or over another felled tree, is in an underbind condition. The log to be bucked may have its full length extended in the air with nothing to hold it up. Underbind is the reverse of the situation

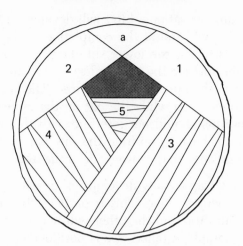

Fig. 6–5. Sequence of cuts when bucking a tree swinging in a bind. Numerals indicate sequence, letter (a) refers to overlapping kerf; and the shaded areas in the center, depending upon conditions, may either be sawed or allowed to break.

described under a tree swinging in a bind. Figure 6–6 shows the various cuts in numerical sequence required to buck a tree at a point which has an underbind condition.

Sidebind. A felled tree lying parallel to the contour on a steep hillside and held there by a stump or tree is in a sidebind condition.

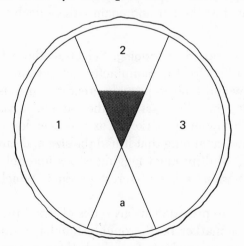

Fig. 6–6. Sequence of cuts when bucking a tree in an underbind condition. Numerals refer to sequence and letter (a) to overlapping kerf. Shaded area indicates uncut wood which breaks off.

During the bucking operation a bind will develop on the lower side and usually the log will roll at the completion of the cut. The procedure involved is to cut the lower side first and continue the cut far enough under so that the remainder of the cut may be made with the bucker on the upper side of the tree. The kerf on the lower side will tend to close up while the cut on the upper side is being made. This condition should not interfere with the operation since the bucker will not have to position the saw bar in such a manner to get into the tight spot on the lower side of the tree.

Log Traps. A log trap usually develops in the case where a bucker makes a cut at a wrong slope. This is particularly true in the case where one log will drop and the other will not move at completion of the cut. For example, a felled tree is positioned so that it is tight against its stump and is supported by another felled tree short of the first bucking mark. The slope of the cut should be such that the bottom of the cut will be closer to the butt log than the top of the cut. This procedure will allow the upper portion of the tree—the part above the butt log in this case—to drop free and not hold or trap the bar and chain. The part of the tree above the butt-log would not drop free if the angle of cut was sloped in the opposite direction, i.e., the bottom of the cut was farther from the butt-log than the top of the cut. In this case, the saw would be held fast in the completed cut by the weight of the upper portion of the felled tree settling and being lodged against the bar and chain (McCulloch Corporation, no date).

Bucking for Multiproduct Logging. Selecting those tree species or log sizes best suited for the manufacture of certain wood products, which in turn would tend to provide greater return, is integrated or multiproduct logging. For example, the value of a yellow birch log would be much higher if marketed as a veneer log than if it were sold as pulpwood, assuming that it had the size and quality to qualify as a veneer log. Multiproduct logging is possible only if there is an outlet for the various types of forest products which may be produced.

Bucking trees to provide various types of wood products requires that a bucker or marker be thoroughly familiar with the specifications for the various products for which there is a market and also the condition of the respective markets.

The trend toward tree length and long-log skidding, where the bucking is done at a landing or mill yard, has popularized multi-product utilization. Bucking at the landing allows the bucker to make a more thorough examination prior to actual bucking than would be possible if the bucking were done in the woods. He is not encumbered by brush, **jackpots,** or **lays** which might result in selecting a bucking point in the interest of safety or expediency instead of one based upon product specifications.

SUGGESTED SUPPLEMENTARY READING

1. CONWAY, STEVE. 1968. Timber cutting practices. Miller Freeman Publications, 500 Howard Street, San Francisco, California 94105. This publication was prepared by a practical man who discusses the topic from the standpoint of maximum profit. All facets of felling and bucking are examined and discussed in a clear and concise manner.
2. Crown Zellerbach Canada Ltd. Safety regulations for fallers and buckers. *Forest Industries,* Vol. 91, No. 4, April 1964. Pp. 84, 85. While these regulations were established on a large company operation in British Columbia, Canada, they may be applied almost anywhere.
3. Idaho State Board for Vocational Education. 1958. *A manual of instruction for log scaling and the measurement of timber products.* A very complete coverage of log scaling and log grading, particularly as practiced in the Intermountain Region.
4. McCulloch Corporation. How to use a power chain saw. Los Angeles, California 90045. Mr. Charles Stovall, an expert in felling and bucking techniques, prepared much of the material. A very complete coverage of log making.
5. Power Saw Manufacturers Association. 1966. *Power saw safety manual,* third edition. Power Saw Manufacturers Association, 2217 Tribune Tower, Chicago, Illinois 60611. Realizing that training and safe practices are as important to the forest products industry as they are in every other industry, the Association prepared the Manual to assist various safety groups in their training programs. The Manual includes sections on: Training Methods, Safety Rules, Preventive Maintenance and Fire Precautions.
6. SIMMONS, FRED C. 1951. *Northeastern loggers handbook.* Agricultural Handbook No. 6. Northeastern Forest Experiment Station, United States Forest Service, Upper Darby, Pennsylvania 09082. A comprehensive coverage of logging methods and procedures as practiced in the Northeast. A wealth of information presented by a man who is an authority on logging in the region.
7. SIMMONS, FRED C. 1955. Logging farm wood crops. Farmer's Bulletin No. 2090, U.S. Department of Agriculture, Washington, D.C. 20250. Explains all phases of the logging operation as related to farm woodlots, from the stump to the log truck.
8. STENZEL, GEORGE. 1953. More wood per tree by careful felling and bucking. Bulletin No. 14, Institute of Forest Products, University of Washington, Seattle, Washington 98105. 17 pp. A rather complete summarization of felling and bucking; while the material was prepared for Pacific Northwest logging conditions much is applicable to any area where trees are felled and bucked.

7

Tractive Skidding

Following the initial log-making process, logs are transported by the most convenient and economically acceptable means to a **landing, yard** or gathering place for further transportation by some other means. The initial log movement, also identified as minor transportation, usually involves some form of tractive skidding or cable yarding, depending on conditions and custom within a region.

Tractive skidding is the process whereby power, sufficient to transport the logs, is applied by means of draft animals or mechanical equipment. The logs are attached to the power source and dragged along the ground to the delivery point. The effectiveness of the power source may be increased by certain attachments which either lift one end of the load clear of the ground or the load may be wheel, track, or runner supported.

While mechanization has replaced the use of animals for skidding operations to a very marked degree, many mules and horses still are used for skidding and hauling logs. Mules are used over a rather large part of the southern and southeastern sections of the United States. Horses work more efficiently in a cooler climate and consequently are more effective than mules in those more northern areas which receive considerable snowfall together with sub-zero temperatures, such as in New England, the Lake States, and eastern Canada.

ANIMAL SKIDDING

Available animal power when used in ground skidding may be limited by: slope, surface conditions, timber or log size, skid distance and skid road spacing.

Slope

Moderate slopes with favorable grade in the direction of skid are ideal. Adverse grades of 3 percent or more for extended distances produce a severe strain on animals, particularly horses.

Surface Conditions

Rocky ground tends to produce hangups which in turn may cause injury to the animal. Heavy brush may have to be swamped if its presence prevents the animal from functioning in an acceptable manner. Swampy conditions which might mire and frighten an animal may also cause injury. Mules do not frighten as easily as horses do when mired.

Timber or Log Size

Normally, animals are not capable of skidding large-diameter logs or tree-length sticks at a per unit cost which is competitive with mechanical skidding equipment. In those regions where the timber is in the small- to medium-size class, such as in the southern pine, New England, Lake States, and eastern Canada areas, animal skidding still continues to be an effective skidding method.

Skidding Distance

In general, the maximum skidding distance under favorable conditions is considered to be in the vicinity of 500 feet.

Skid Road Spacing

Road spacing greater than the maximum skidding distance of 500 feet may necessitate additional log handling for delivery to roadside or landing, which in turn would increase skidding cost.

The foregoing limitations to the use of animals for skidding logs suggest maximum efficiency under a combination of operating conditions where: slopes are moderately steep in favor of the load, ground surface is fairly smooth, timber is small, and relatively short skidding distances may be tolerated. Additional considerations, such as restricting animal skidding to small operating units and operations which can only support a small initial investment, tend to keep animal skidding competitive with machine methods even during this era of mechanization.

Woods Horses

Skidding requires more intelligence on the part of the horse than does routine farm work. The average farm horse does not have the temperament to cope with woods conditions, even though many are relegated to the dual role in keeping with seasonal farm and woods operations.

The best horse for woods work is one between five and ten years old and weighing between 1,300 and 1,800 pounds (Simmons, 1951). Legs and neck should be short and stocky with well developed muscles in the shoulder and hip areas (Fig. 7–1). A somewhat calm and unexcitable temperament is most effective. Usually a horse which is nervous, excitable, or skittish will not work as effectively when skidding logs as will one which pulls quietly and evenly.

Attachments and Skidding Devices

Small logs are attached to the harness of horses or mules by means of **skidding chains** for ground skidding. Heavier logs are attached by a combination of chain and **skidding tongs** or **crotch grabs** (Simmons, 1951).

While of limited importance in many parts of the United States and Canada, sled-type vehicles are used in New England, the Lake States and eastern Canada. Normally sleds are associated with snow; however, they also are used during the other seasons as a means of transporting logs. Certain wood-using industries, including wood turners, boxboard manufacturers and many sawmills in the northeast and Lake States insist that logs be delivered to them in a "clean

Fig. 7–1. Woods horse skidding or twitching a log to a yard. (Courtesy of the Brown Company.)

condition" and free from embedded dirt. This is accomplished by loading logs onto these sled-type vehicles at the location where the trees are felled and bucked. The loading operation consists of manually loading the logs onto the bunks or crossbeams of the vehicles and involves a combination of muscles, **skid poles,** and dexterity with a **peavey.**

There are several different types of sled-type vehicles in general use. Some support the logs so that they are free of any contact with the ground, while others just support one end of the logs leaving the other end to drag on snow or ground, depending upon the season. These vehicles may be either horse or tractor drawn; however, each method of power requires a certain type of sled construction so that the available pulling ability of either animal or machine is best utilized. Names such as **go-devil, scoot, sloop,** yarding sled, and **dray** "identify" some of the more common types of sled-type vehicles used for initial log transport (Simmons, 1962).

TRACTORS

The introduction of tractors to logging and subsequent improvments to the basic tractor unit, together with the development of various attachments and accessories, has brought about a most noteworthy change in the initial transportation of logs. No longer is the logger limited to the relatively short skidding distance and comparatively small log and turn size generally associated with animal skidding.

Since the introduction of the first practical tractor in 1880, which was steam-powered, had large rear driving wheels, and was employed primarily for agriculture, a series of improvements have been developed. Many of the improvements are a credit to the inventiveness and ingenuity of men associated with the logging industry.

The crawler-type tractor was developed in 1904 and was steam powered. This type of unit out-performed the older wheel-mounted model even though it was powered with an engine which generated less horsepower. In 1905 the gasoline engine replaced the large and somewhat cumbersome steam engine. The first practical diesel tractor, designed specifically for woods use, was developed in 1931 and rapidly gained favor by providing considerably more power at a lower fuel cost. Larger-sized tractors are all diesel-powered and with a few exceptions the same is also true for the smaller sizes. The diesel tractor is universally accepted by the logging industry as the prime log mover.

Logging with Tractors

Tractor logging is generally associated with some form of selection or partial cutting. This is in contrast to moving cable yarding, which is generally associated with clear-cutting. The mobility of the tractor, either wheel or track mounted, allows the unit to maneuver into position among the residual trees with comparative ease.

While tractors, particularly those with **winches,** may be used to skid logs up adverse slopes the practice usually results in decreased production and increased skidding cost. The most efficient use of any tractive skidding method is achieved when logs are skidded down favorable slopes. The initial planning and layout of a cutting unit should give due consideration to the effect that the topography will have upon the skidding operation. Skid trails and tractor roads on slopes greater than 30 percent should be laid out so that they approximate a direction which is at right angles to the contours. Tractor roads laid out in this manner provide operating conditions which tend to maximize production and minimize accidents. Tractors skidding logs down steep slopes at right angles to the contours have a greater tendency to remain upright than when they are traveling in the same general direction of the contours. This situation is of particular importance in areas where sideslopes are steeper than 30 percent.

Presently the trend in the industry is toward greater use of rubber-tired wheel units for skidding logs while the crawler tractor remains the undisputed favorite for forest road construction and associated activities. Each type has certain distinct advantages, depending on operating conditions.

Crawler Tractors

This type is manufactured in a rather wide range of sizes and specifications. The range of **drawbar horsepower** extends from less than 20 for the smallest unit to over 230 for the largest, while the weight range extends from two tons to thirty tons (Fig. 7–2).

In an effort to make comparisons among the various sizes and models marketed by the manufacturers, a standard test was developed by the Department of Agricultural Engineering of the University of Nebraska during the early 1920's to provide information regarding drawbar horsepower, **belt horsepower,** fuel and oil consumption, along with the manufacturer's specifications and claims for his machine. Since then, practically all tractor manufacturers use the testing facilities available at the University of Nebraska to evaluate their machines. In some instances engine horsepower is used as a performance criterion, particularly in the case of the larger machines, those weighing over nine tons.

Tractor Accessories

The tractor, when first introduced to logging, merely replaced the animals as the source of power. As with animal skidding, the

Fig. 7–2. Examples of large and small crawler tractors ground skidding. (Courtesy of Caterpillar Tractor Company and John Deere and Company.)

logs were skidded by attaching them directly to the drawbar by means of **skidding chains, slip hooks,** crotch grabs, or skidding tongs, and dragging them along the ground. While tractor ground skidding is still practiced, the introduction of various devices to more fully and effectively utilize the available drawbar pull has done much to increase production and reduce tractor maintenance.

Tractors, as they are delivered from the factory, require certain special equipment to adapt them for logging. The special equipment includes a steel guard to protect the underside of the crankcase and oil pan, a heavy duty radiator guard and protection for the lower tracks and wheel bearings. A tractor should also be fitted with a heavy front bumper plate or a bulldozer blade. Operators who are obliged to build tractor logging roads consider that a bulldozer blade is standard equipment and also that it allows the tractor to make its way into areas ordinarily inaccessible. The blade-equipped tractor has the ability to reposition felled trees and logs which, in turn, results in more efficient skidding. The blade may also provide better tractor balance during skidding by producing a more even load distribution over the track area. Where damage to seedlings and saplings and unnecessary gouging of the soil is probable, removing the blade from a tractor may be in order.

The tractor operator should be protected from overhead hazards such as falling limbs and cable breaks. Certain states have a law which requires the installation of tractor guards, similar to those shown in Fig. 7–2, to protect the operator from above and behind.

The winch, while a separate part of a skidding tractor, is so commonly associated with the basic tractor that it might well be considered an integral part. It is necessary for sulky and arch skidding and is highly desirable when ground skidding; therefore, a winch is considered to be a most important item for efficient tractor skidding. It consists of a set of gears, drum, housing, and brake.

The winch together with the wire rope on its drum has many uses in skidding. It allows the tractor to remain on relatively stable terrain while the winch line is carried or dragged to the log. The log is **choked** and the winch is used to pull it out. This is of particular importance in swampy locations, down steep pitches, or in rocky or heavy brush areas. Another situation which emphasizes the importance of the winch refers to a condition when a tractor may be bogged down in an area, such as a swamp or similar situations, where little if any traction may be available. The winch line is paid out, attached to a tree or stump, and the tractor actually pulls itself out and positions itself on areas which have better traction. Oftentimes a tractor may stall when skidding up an adverse pitch or on muddy or rocky ground. Under these conditions the brake on the winch may be released and the skidding unit moved ahead to better ground. The turn of logs remains stationary as the winch line is paid out until ground which offers better traction is reached. The winch gears are then engaged and the turn is moved toward the skidding unit by reeling in the winch line. This procedure allows the skidding unit to apply 50 to 80 percent more pulling power than is available at the drawbar (Simmons, 1951).

Hydraulic and power-controlled winches are becoming more popular. In contrast to the gear-driven type, the hydraulic winch has complete freedom from the conventional power transmission-gear linkage. Flexible pressure hose connects the winch to an oil pump power unit. This feature allows a hydraulic winch to be mounted in a location where it will be most effective. For example, it may be mounted on a **sulky** or **arch** to serve as a supplementary winch to the main tractor winch and thereby increase the skidding capacity of the unit. The power controlled winch used in conjunction with a **power shift** tractor allows the unit to winch a load without stopping the tractor, and minimum operator effort is required.

Log choking devices, such as chain **chokers,** wire rope chokers, **choker hooks, butt hooks,** and **drawbar hooks,** are some of the accessories which are necessary so that effective use of a tractor's pulling ability may be applied to a turn of logs during the skidding operation. A choker is a short length of flexible wire rope or chain and is used to attach logs to the winch line or, in some instances, directly to the tractor.

Chain-type chokers are more effective when skidding smaller diameter logs and are particularly effective when skidding smooth-bark species in regions which receive considerable snow along with sub-freezing temperature. The links tend to develop a better hold and, under the conditions mentioned, are more effective than wire rope chokers. The holding qualities of chain chokers may be improved by electrically welding teeth to both sides of the links. These protruding teeth tend to put a better bite on a log and hold fast during the skid. Usually chain chokers have a ring on one end and a flat choker hook on the other end. The ring end is passed around a log and the choker hook end is threaded through the ring. The choker hook is then hung on the winch line along with the other hooks making up a turn.

Wire rope or cable chokers are in use over a wide range of logging conditions. The first chokers were nothing more than a length of wire rope with an **eye splice** on both ends. The loop was made by threading one eye through the other. While this method was reasonably effective, the chokers were somewhat difficult to position on logs and wore out rather quickly.

Improvements and developments have resulted in providing the logging industry with a highly effective wire rope choker. The wire rope choker most commonly used consists of a length of wire rope with either an eye splice or eye socket on one end and a steel **ferrule** or nob on the other end. Under certain conditions there will be a ferrule on both ends. A Bardon choker hook, which is a short metal sleeve made of hardened steel, runs freely on the wire rope between the eye splice and ferrule (Fig. 7–3). On one side of the choker

Fig. 7–3. Wire rope choker, showing Bardon choker hook and ferrule. (Courtesy of the Esco Corporation.)

hook there is a tapered socket which fits the ferrule. A log is choked by passing the ferrule end of the choker around the log and inserting the ferrule into the tapered socket of the choker hook.

A somewhat similar device, the drawbar hook, allows wire rope chokers, which are double-end ferrule fitted, to be attached to the tractor drawbar. Drawbar hooks are made to accept from two to twenty-four chokers.

Attachments

Various attachments were developed to reduce the friction between the log and the ground over which it was being skidded in order to prevent the log or logs from "digging in" during skidding and reduce the occurrence of "hangups" on stumps and down material. Attachments such as **skidding pans,** arches, sulkies, scoots, and sleds became popular by delivering logs which were not encrusted with mud and gravel to the same degree as when they were ground skidded.

A skidding pan is a flat piece of heavy gauge steel, boiler plate or other hard metal with a turned-up front end. A chain is securely fastened to the middle of the front edge of the pan and to the tractor drawbar. The lead end of the logs making up a turn rests upon the pan and the individual logs are attached to the drawbar or drawbar hook by means of conventional wire rope chokers as shown in Fig. 7–4.

The tractor arch is a track-mounted framework of heavy steel construction consisting of a yoke, arched axle, reach, and crawler tracks. The reach is supported by the arched axle and yoke which in turn are supported by the crawler tracks. One end of the reach is attached to the tractor drawbar, usually by means of a universal-type hitch. The other end of the reach extends beyond the arched axle and supports the fairlead, which is an arrangement of horizontally and vertically mounted rollers. The rollers guide the winch line so that it may be drummed evenly. Figure 7–5 shows a tractor arch, tractor, and winch.

In a tractor-arch operation, chokers are set and attached to the winch line. The logs then are winched to the arch and hoisted, thereby positioning one end of the turn of logs below the fairlead and clear of the ground, leaving the other end of the turn in contact with the ground. The tractor arch was developed on the Pacific Coast for skidding large logs and has proven to be a very effective

Fig. 7–4. Ground skidding using a pan. Drawbar hook mounted below winch. Three-way arch hook attached to winch line. (Courtesy of the Esco Corporation.)

Fig. 7–5. Tractor-arch yarding unit. (Courtesy of the Hyster Company.)

piece of equipment for logging all kinds and sizes of timber over a wide range of operating conditions and on practically all types of terrain.

Certain tractor manufacturers have developed an integral arch (Fig. 7–6) which is attached directly to the tractor and winch

Fig. 7–6. Integral arch-tractor skidding unit. (Courtesy of John Deere and Company.)

housing, thereby eliminating the trailing feature of the conventional arch. The prime advantage of a tractor with an integral arch lies in its greater maneuverability, which is of particular importance in heavily forested areas. An integral-type arch arrangement where the short reach is attached directly to the winch and the combined arch-winch is mounted on a tractor is also available.

Rigging for Tractor Arch Skidding

Several possible types of rigging may be employed, depending on log size and tractor-arch capability. The least complicated and probably most frequently used method consists of attaching the chokers to a single arch or tractor hook as shown in Fig. 7–2. Double arch-hook rigging is somewhat the same except that it is possible to accommodate more chokers on the double hook than when using the single arch hook.

Another type is the double-tagline rigging and is often employed when a large number of logs make up a turn. It provides more spread to the load instead of bunching the logs under the fairlead and thereby reduces fantailing of the free ends. This method allows the logs to drag freely and tends to reduce the number of hangups.

When logs are small and a large number are required to make up a turn, triple-tagline rigging is particularly effective (Fig. 7–7).

Fig. 7–7. Triple-tagline rigging. (Courtesy of the Esco Corporation.)

Using taglines of varied lengths and chokers all of the same length tends to spread the turn under the fairlead, thereby increasing the capacity of the arch.

The tractor sulky (Fig. 7–8) is similar to the tractor arch and often is identified as a wheeled arch. The main difference is that the sulky is wheel-mounted in contrast to the arch, which is track-mounted. Structurally the two are identical. One manufacturer has one model with an interchangeable axle feature which allows either wheels or crawler tracks to be used. Generally the sulky is used when skidding small- and medium-sized logs; however, large sulkies are available for use in combination with the largest crawler tractors.

In an effort to increase the flotation of sulkies certain operators

Fig. 7–8. Tractor-sulky skidding unit. (Courtesy of John Deere and Company.)

mounted their sulkies on large airplane-type tires which proved very effective. However, tire damage, particularly in the sidewall area, caused by snagging and cutting by sharp rocks and limbs and followed by eventual penetration, led to the development of a tire designed exclusively for logging arch operations. The arch tire has increased rib height, a thick layer of cut-resistant rubber added to the sidewalls, and a one-half-inch-thick shield of steel-reinforced tread built-in under the tread and around the entire sidewall area. The steel-reinforced tread consists of several thick layers of tread rubber containing a great number of short lengths of hardened steel filaments which form a practically impenetrable mat that protects the cord body of the tire.

Track-mounted arches are 30 percent heavier in weight in the smaller sizes and 50 percent heavier in the largest sizes than tractor sulkies of comparable capacity. This feature, a distinct advantage in favor of the tractor sulky, allows a tractor-sulky skidding unit to climb steeper grades when in an unloaded condition, as on the return trip from a landing. This ability to climb steeper grades allows a tractor-sulky skidding unit to return to a log site via a shorter and therefore correspondingly quicker route than would be possible if a tractor-track mounted arch skidding unit was used.

Tracks Versus Wheels

Although a definite trend has developed toward rubber-tired wheel skidders, the crawler tractor still has many advantages. First and foremost it is a very versatile piece of equipment. It may be used for logging, road building and as a basic hauling unit for a wide variety of trailing units. It also serves as a mounting for various attachments and is a rugged mobile power unit for winches and air compressors.

The crawler tractor provides superior flotation and traction on nearly all surfaces. Its compact design, with all of the weight on the driving tracks, provides the ability to pivot-turn and enables the unit to go practically anywhere—up, down, or through, regardless of weather conditions.

The disadvantages of the crawler tractor include relatively slow travel speed, and that tracks tend to damage roots of standing trees, compact the soil, and churn the soil during rainy periods. The churning action tends to cause siltation of streams, particularly when the units are operated on slopes supporting erosible soil types.

Rubber-tired wheel skidders have the advantage of greater speed and mobility, particularly when operating on bulldozed skid roads, and is the main reason they are used as roading units.

The disadvantages of rubber-tired wheel skidders are the limitations in traction and flotation. These limitations have been overcome to a certain degree by the introduction of large-diameter, low-pressure tires. Wheel skidders do not have the versatility of the crawler tractor.

Rubber-Tired Wheel Skidders

These units very definitely have entered the logging scene to the extent that they now are serious contenders to track-mounted tractors over a rather wide range of operating conditions.

The first wheel tractors to be used in logging were the agricultural and industrial models and, as was the case with the early track-mounted tractors, were not overly efficient in this role. Within recent years much development work has been done to produce rubber-tired wheel skidders designed specifically for logging. The ever-present search for logging methods which will decrease logging costs and increase production seems to favor the wheeled tractor, particularly under certain operating conditions.

Wheel skidders are available in somewhat the same range of sizes as the crawler type. **Brake horsepower** ranges from 30 to 275 while the weight range is from three to twenty-eight tons. The various accessories, such as protective equipment, chokers, hooks, and winches, discussed in the crawler tractor section are also necessary to provide protection and increased production.

The smaller two-wheel-drive tractors, with few exceptions, are of the industrial or agricultural type, and although they are rather widely used in farm woodland logging, they have limited utility when used for logging on a full-time basis.

Certain manufacturers have developed a very effective and efficient skidding unit in the small-size class (6,000 to 10,000 pounds in weight, and developing up to 50 horsepower), which has gained rather wide acceptance in logging circles, particularly in pulpwood harvesting. Basically the unit consists of a compact, **articulated frame**, rubber-tired skidder. Four-wheel drive and a short turning radius enables it to maneuver around stumps and obstacles and climb moderately steep slopes with little trouble. A light bulldozer-type blade, while listed as optional equipment by certain producers, is useful to clear obstacles from skid roads and move logs at the landing area. Instead of a trailing-sulky or -arch attachment, the lifting effect is provided by means of an integral arch. Individual manufacturers have certain features which identify their particular machine. For example, the Garrett Tree Farmer, as do certain other makes, accommodates itself to the terrain by an oscillating action in the front wheels. The front axle is cradled so that either wheel is free to ride approximately three feet higher than the other, which tends to keep all four wheels in even contact with the ground, thereby providing better traction and stability. Additional information relative to small-size skidders is presented in Chapter 12, Pulpwood Production.

The medium-size wheel skidder class (11,000 to 27,000 pounds in weight and developing 50 to 160 horsepower) includes several different makes and models. Both four- and two-wheel drive units are available. Practically all makes have power or some form of power-assisted steering which tends to provide greater maneuverability and lessen operator fatigue (Fig. 7–9). The integral arch

Fig. 7–9. Medium-size, diesel-powered wheel tractor. (Courtesy of the Pettibone Mulliken Corporation.)

feature is common to all makes. Wheel skidders in the medium-size class are used for conventional stump to landing skidding and also serve as efficient roading units which forward bunched logs to a landing. One manufacturer, in order to increase the effectiveness of his machine, has installed an additional winch and fairlead which allows the unit to skid a greater number of logs per trip.

The large-size wheel skidder class (over 30,000 pounds in weight and developing over 160 horsepower) includes comparatively few makes and models when compared with the smaller classes. Included in this class are two rather distinct types. One type utilizes a diesel engine to drive the unit (Fig. 7–10), while in the other type each wheel is individually powered by its own electric motor. The energy to drive the "electric wheels" is provided by a diesel-generator power unit (Fig. 7–11). Generally the large-size wheel

Fig. 7–10. Large-size, diesel-powered wheel tractor. (Courtesy of the Kenworth Manufacturing Co.)

skidders are used on Pacific Coast, Alaska, and overseas operations which involve roading.

A recent innovation to tractive skidding involves an articulated hydraulically operated grapple which may be installed on rubber-tired skidders. This attachment provides a means whereby logs are clasped between the jaws of the grapple during the travel phase of the skidding operation. Logs may be skidded without the use of chokers and choker setting procedure. A turn of logs may be steered over or around obstructions such as stumps or residual trees. When reaching the landing the turn may be positioned on top of any logs which already are on the landing area and thereby more effectively utilizing landing space. The boom which supports the grapple is fitted with a roller fairlead which allows logs which are inaccessible to the grapple to be winched into position and skidded in the conventional manner. Figure 7–12 shows the grapple mounted on a rubber-tired wheel skidder.

Fig. 7–11. Large-size, diesel-electric-powered wheel tractor. (Courtesy of R. G. LeTourneau, Inc.)

Roading

The combination of two types of tractive skidding to move logs or tree length sticks from a stump location to a landing is termed roading. In principle a roading operation utilizes one type of tractive skidding to gather or bunch individual turns, and another type, usually larger and wheel mounted, combines two or more of the individual turns into one large turn for further transportation to the landing. The initial log movement from the stump areas usually is made by a relatively slow but maneuverable skidding unit over a comparatively short skidding distance. At the bunching landing, where the roading turn is prepared, a larger, faster, tractive skidding unit is used to haul the large turn to the landing over a bulldozed tractor road. Tractor roads for a roading operation should be laid out and constructed in such a manner which will allow the large roading tractors to operate at maximum efficiency when traveling both to and from the landing.

While no definite ratio of bunching distance to roading distance

Fig. 7–12. Skidding grapple mounted on a medium size wheeled skidder. (Courtesy of the Esco Corporation and the Beloit Corporation.)

is applicable to all conditions, the bunching distance generally is much shorter than the roading distance. Bunching with horses and roading with a small size wheel or crawler tractor may involve a distance ratio which is much different when large crawler tractors bunch turns for a large roading unit such as a large wheel-mounted tractor. The proper balance between the number of bunching units and the roading unit must be given due consideration so that maximum production will be achieved.

The capabilities and limitations of a roading unit should be carefully considered in connection with its performance under local operating conditions. Circumstances which might hinder a roading unit's performance are excessive adverse grades, unseasonal rainfall, and the degree to which the soil on the route provides adequate traction and flotation.

An efficient roading operation makes use of the same route both when skidding a **turn** and returning empty. Normally the roading unit will not have any difficulty when loaded; however, it may encounter difficulty returning via the same route when unloaded. A favorable grade when in the loaded condition is an adverse grade when in the empty or unloaded condition and may be beyond the climbing ability of the unit. The alternative involves providing a go-back route on terrain which, although of greater length, contains stretches of adverse grade which are within the capability of the unit.

LANDINGS

Normally, logs are subjected to more than one method of transportation before they reach their final destination. The initial log movement, from the stump location, usually is by means of some form of tractive skidding, and terminates at a landing.

Landings are prepared by clearing the area of all standing and down timber to facilitate unhooking the logs. Landings must be large enough to adequately accommodate the activities associated with receiving logs from the skidding operation and loading them onto trucks for further transportation. The actual size of a landing is dependent upon the size and number of skidding units, size of loader, and number and size of trucks being used on a particular operation. Naturally, a larger landing area will be required where tree-length or long logs are being logged in contrast to a short log operation. Furthermore, should log delivery or production schedules be such that log sorting at the landing is necessary, adequate landing area must be provided.

Consideration should be given to preparing landings which will provide an efficient and safe interchange of logs between the initial and subsequent log transportion activities. The side slope of the landing must be such that logs will not roll or slide when unhooked; consequently, side slopes should be limited to 10 percent. They should be large enough so that skidding output will not be restricted and provide adequate space for the necessary maneuvering of equipment. They should be well drained so mud does not accumulate and debris can be pushed away from the working area. Consequently, landings should be located on benches, not in draws.

SUGGESTED SUPPLEMENTARY READING

1. ALLIS-CHALMERS MANUFACTURING COMPANY. 1964. *Fundamentals of logging.* Allis-Chalmers Manufacturing Company, Construction Machinery Division, Milwaukee, Wisconsin. This handy reference booklet contains much useful information for both students and operators. The section on logging methods is particularly appropriate.
2. FORBES, R. D., and A. B. MEYER (eds.). 1955. *Forestry handbook.* The Ronald Press Company, New York. Section 16 contains data relative to timber harvesting production together with a description of both land and water transport.
3. Pacific Logging Congress. 1940 to 1966. *Loggers' handbook.* Portland, Oregon. These handbooks contain many articles which describe tractive skidding methods in the Intermountain, Redwood, and Pacific Northwest Regions.
4. SCHILLINGS, PAUL L. 1969. A technique for comparing the costs of skidding methods. U.S. Department of Agriculture, Forest Service, Intermountain Forest and Range Experiment Station, Ogden, Utah. (USDA Forest Serv. Res. Pap. INT-60). A series of tables shows costs of skidding as determined by various types of equipment, distances for skidding, percent of slope, logs per MBF, and MBF per acre.
5. SILVERSIDES, C. R. 1964. Developments in logging mechanization in Eastern Canada. H. R. MacMillan Lectureship address delivered at the University of British Columbia, Vancouver, Canada. No. 34 in the University's Lecture Series. The lectures by Mr. Silversides cover several mechanical logging systems. While the presentation is slanted toward pulpwood production the principles are applicable to harvesting small sawtimber.

8

Cable Yarding

Cable yarding is the movement of logs from stump site to a **landing** by a machine equipped with multiple drums or winches which operates from a stationary position at the landing. The machine is termed a **yarder, donkey** or hoist. The logs are yarded by reeling in a wire rope mainline on a **drum** barrel. The **choker** looped around the log is hooked to **butt rigging** attached to the end of the **mainline.** The butt rigging is pulled out to where the logs lie by another cable termed the **haulback** or outhaul line. The **lead,** or direction of movement of the lines, is controlled by blocks hung on stumps or trees. For the **high lead** system the mainline lead block is hung high above the ground on a **spar tree** or a steel **tower.** The **skyline** systems drag or carry the logs suspended from a carriage which rides on a cable stretched between a head spar at the landing and a tail spar at the back end of the yarding road. If the haulback line is too large to be pulled out manually for threading through the blocks, a light **strawline** cable, usually of $\frac{3}{8}$-inch diameter, is provided. The modern yarder is mounted on rubber-tired wheels or crawler tracks for mobility along the road between landings.

Cable yarding originated in 1881 with the invention of the Dolbeer steam donkey in the California redwood region. Line pull was applied to a manila rope by wrapping it around a vertical spool and pulling the rope off manually. In 1883 the Washington Iron Works of Seattle built the first logging donkey which stored the line on a drum as well as applying pulling force. Wire rope replaced manila rope. The cable was pulled out to the logs by a horse. Later a haulback line drum was added to the donkey and, as horsepower and line size increased, a strawline drum. Internal combustion engines began to replace steam in the nineteen-twenties.

The first cable yarding system was the **ground lead,** with the mainline lead block hung on a stump, and the log dragged along the ground. About 1905 the high lead was developed and since 1915 this has been the most commonly used yarding system in the West Coast regions of the United States and Canada.

Skyline or cableway yarding was invented in Michigan in 1883 and introduced to the West Coast in 1904. Since 1908, when the **slackline** system was invented, loggers have teamed with equipment manufacturers to develop many skyline systems. When truck roads replaced railroads in the nineteen-thirties, the popularity of skylines declined. But since 1949, when the first skyline crane was imported from Switzerland to North America, interest in skyline yarding has revived. The development of variable **interlocking-drum** yarders for **running skyline** systems, and of labor-saving **grapples** to eliminate choker-setting, has resulted in increasing use of skyline yarding in the nineteen-sixties.

Cable yarding predominates in the Pacific Coast regions of Alaska, British Columbia, Washington, and Oregon, There the topography is broken, the slopes steep, and, compared with other regions, the log size and the volume per acre is large, and clearcutting is practiced. In other North American regions, cable systems are used where the terrain is unsuitable for tractors, because of steep slopes, rough topography, or swamps.

Concern for soil compaction and consequent erosion due to tractor skidding is encouraging more use of cable yarding. American cable yarding equipment is used in a number of the developing countries, notably in the Philippines, Taiwan, and Sabah, North Borneo. Japanese cableway systems are used to log the mountain forests of that country. Skyline cranes developed for logging alpine forests in Switzerland, Austria, and Italy are used in the Himalayas and other mountainous countries.

Advantages of Cable Yarding

Cable yarding systems have characteristic advantages and disadvantages compared with tractive systems. Among the advantages are the following:

Cable systems can yard logs over ground on which a tractor cannot operate. Swamps, mud, rocks, steep slopes and broken topography can be logged by a cable system designed for the conditions.

Cable systems are available to operate in any direction—upslope, downslope, or along the contour. Tractors or animals can operate efficiently only when skidding logs downslope or on level ground.

Cable yarding can be done in any weather. On some soil types tractor skidding is limited to dry months.

Cable yarding is less damaging to the soil than tractor skidding. The crawler tractor disturbs a greater percentage of the surface area and, on the skid roads compacts the soil, which reduces water infiltration rate and accelerates runoff and erosion.

Since the yarder operates from a stationary position, a heavier machine, with more power than is feasible in a skidding tractor, can be used. Logs can be yarded at line speeds much faster than a tractor can travel.

Disadvantages of Cable Yarding

Disadvantages of cable yarding systems include the following:

Yarding distance is limited by the line capacities of the yarder drums. There is no physical limitation to tractor skidding distance although there is an economic limit.

The highlead, which is the most commonly used cable yarding system, is only used where **clear-cutting** is practiced. While the log turn tends to move in the direction of the lead, the logs may slide or roll sidewise as well. The tractor can be guided to avoid residual trees or patches of immature trees. The skyline systems give more positive control of the turn than the highlead, and some of them are effectively used for yarding thinnings and other partial cuts.

With the exception of grapple yarding, cable systems require larger crews than tractors, resulting in higher labor costs per unit of production. However, this may be offset, to some extent, by higher depreciation, maintenance, and repair costs of tractors.

The costs of moving, rigging-up and changing yarding roads are fixed per acre, regardless of the volume per acre logged. Consequently, the cost per unit of volume increases as volume per acre decreases. Tractors are more economical where the volume per acre cut is relatively small.

BASIC CABLE YARDING EQUIPMENT

The basic equipment required for cable logging includes the yarding machine, the wire rope lines and **rigging** attachments, and blocks through which the lines are threaded for control of direction of movement of the lines.

Yarders

The typical yarder is built on a rectangular frame of welded steel I-beams. It is powered by a diesel engine bolted at the rear of the frame, through a transmission at the engine output shaft. The transmission is usually either a 4-speed gear box or a hydraulic torque converter with a 2-speed gear to extend the gear range of the converter. A roller chain drive connects the transmission with the rear drum shaft. Other shafts forward are driven by gear trains meshing with the rear shaft gears. The drums float on bronze bushings on the shaft, and do not rotate until pushed into contact with a ring of friction blocks which are keyed to the shaft. On modern yarders a compressed air cylinder and piston controlled by the operator through a hand-operated air valve engages the drum. A representative 4-drum yarder is shown in Fig. 8–1. This machine is powered with a 200-horsepower diesel engine and a torque converter and has air drum friction controls and foot brake levers which actuate band brakes on the mainline and haulback drums. It is designed for operating a slackline skyline. The skyline drum brake is air-controlled. For highlead yarding only three drums, for mainline, haulback, and strawline, are needed. Yarders are manufactured in a wide range of sizes, powered with engines from 40 to 725 horsepower. The largest yarders have twin diesel engines.

The newest development in yarder design is the infinitely variable interlock. The interlocking mechanism permits mainline and haulback line drums to be interlocked to pay out one line at the

Fig. 8–1. A 200-horsepower 4-drum yarder, with air friction controls (center) and foot brake levers below, in the factory yard. (Courtesy of the Skagit Corporation.)

same line speed as the other line is reeled in, to maintain constant tension on both lines without the use of brakes or slipping devices. In highlead yarding the interlock is used to keep the butt rigging and chokers clear of the ground when outhauling from the landing and to tightline to raise the log turn to avoid hangups. The interlock is used to control the descent of the turn when yarding down steep slopes. The interlock is essential for the efficient operation of running skyline systems. Each yarder manufacturer has his own patented interlocking mechanism. All of them use a powershift transmission with a hydraulic torque converter and either a three- or four-speed gear. The standard models of interlocking yarders offered in 1971 were powered with diesel engines of from 335 to 525 horsepower. (An interlocking yarder is shown in Fig. 8–14.)

Yarder Mounts. In the past yarders were mounted on heavy timber sleds made of two logs hewn by hand on top with the ends sniped in a sled-runner curve. The runners were joined with hewn cross-timbers and steel tie rods. **Fairleads** with sheaves or rollers, alined with the drums so the lines will spool properly, are mounted on a cross timber on the front end of the sled. By stringing out the mainline through a multiple block purchase, the sled-mounted donkey can move itself cross-country over any terrain. Using mainline and haulback lines, it can move itself sideways for loading on truck or lowboy for transportation on the truck road. Wood sleds are still used, but donkey sled building is a vanishing skill. Yarder sleds are also fabricated from steel plates. The hollow sled runners serve as storage tanks for fuel and water.

In recent years the trend has been toward mounting the yarder on wheels for greater mobility on the truck road. The first wheel mounts for the larger yarders had 20 wide steel wheels. The machine was towed by a crawler tractor. Rubber-tired wheel mounting is the most common today. Three types of wheel mounts are available: trailers which are towed by truck-tractors, motor trucks, and

Fig. 8–2. Self-propelled yarder and steel tower mounted on rubber tires (Courtesy of the Skagit Corporation.)

undercarriages powered by the yarder engine (Fig. 8–2). Self-propelled crawler mounts of both tractor-type and military tank-type are also used. The tractor donkey or triple-drum yarder is a drum set mounted on the rear of a crawler tractor (Fig. 8–5). It can move over bulldozed tractor roads and is the favored yarder for cold-decking at a distance from the truck road.

Wire Rope

The wire rope used for yarder lines is usually 6x19 Seales construction, preformed, of improved plow steel. It is made up of 6 strands of 19 wires each, with either a hemp core or an independent wire rope core for greater strength and resistance to crushing. The Seales patent construction strand has 6 larger wires on the outside to resist abrasion, 12 smaller wires inside for flexibility, and a large wire in the center around which the other wires are laid. In making preformed wire rope the strand is given a permanent set in a preforming head situated just ahead of the die through which the strands are closed into rope. Preformed rope is easier to splice and does not have to be wrapped with wire to keep the end of the rope from raveling out as does non-preformed rope. The diameter of the mainline and haulback is matched to the line pull of the yarder. Following is the range of line diameters used with yarders of 200 to 500 horsepower: mainlines $1\frac{1}{8}$, $1\frac{1}{4}$, or $1\frac{3}{8}$ inches; haulback lines $\frac{3}{4}$ or $\frac{7}{8}$ inch; strawlines $\frac{3}{8}$ or $\frac{7}{16}$ inch. The wire rope choker is usually $\frac{1}{8}$ inch less than the diameter of the mainline, so that if the system is over-stressed the choker will break rather than the mainline. Wire rope handbooks which give breaking strength and other data on all the many constructions and grades manufactured are available from the wire rope companies.

Butt Rigging. The butt rigging which connects the mainline and haulback is a combination of clevises, chain links, and swivels which prevent the lines from twisting. The **butt hook** for the choker is hung from the swivel by links, clevis and another swivel. Figure 8–3 shows the type of butt rigging most commonly used in highlead yarding. In smaller timber or with the more powerful yarders longer butt rigging with three choker hooks are sometimes used.

Chokers. The most popular type of choker consists of a short length of flexible wire rope with a forged steel knob or **ferrule**

Fig. 8–3. Butt rigging for highlead yarding.

socketed or swaged to each end. A choker hook with a slot for the knob slides freely on the rope. One knob is passed under and around the log and inserted in the choker hook. The other knob is hooked into a slot in the butt hook, which is attached to the butt rigging. The "standard" size choker hook weighs 14 to 25 pounds. The length of choker used for highlead yarding in the Pacific Coast regions is 26 to 30 feet. The knobs, and other wire rope fittings, are variously attached to the end of the rope with poured babbit metal, with wedges or pins, or by swaging in a hydraulic press. Modern fittings are rifled to fit the rope strands.

Blocks

The typical logging block consists of a manganese steel sheave pressed on a steel shaft running in anti-friction bearings inserted in the two shells or sides. The shells are connected by a yoke. The haulback block yoke is held by a pin which can be released to thread the yoke through strap eye. Lead blocks to be hung on the spar tree have swivel and clevis attachment. Blocks are hung with straps made of a short piece of wire rope with an eye splice or a swaged "D" in each end. Logging blocks range in size from rigger's blocks with 6-inch-diameter sheaves weighing 21 pounds through haulback blocks with 9- to 16-inch-diameter sheaves to highlead blocks with 30- to 36-inch-diameter sheaves weighing 700 to 1540 pounds. A wide variety of blocks for moving, rigging, and loading are available.

Skyline Carriages

The **skyline carriage** is designed for the specific system with which it is to be used. Carriages for the North Bend and slackline systems have two sheaves 14, 16, or 21 inches in diameter which ride on the skyline. They are enclosed in steel plates triangular in shape with a clevis at the bottom for attachment of lines. Carriage weights range from 710 to 1600 pounds. Fall blocks used with the North Bend system have 12- to 20-inch-diameter sheaves and weigh from 220 to 770 pounds. The carriages used in other skyline systems are described and illustrated with the systems. The skyline is suspended on the spar tree by a large block or by a **tree shoe** or **jack**. The latter has a semi-circular hardwood shoe enclosed in steel plates with a rigging shackle at the top for the strap eyes.

THE HIGHLEAD SYSTEM

The highlead is the most commonly-used cable yarding system. Figure 8–4 illustrates the rigging and operation of the highlead. The

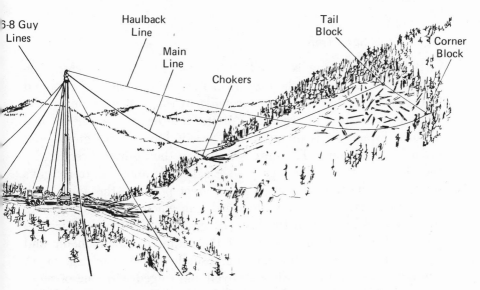

Fig. 8–4. Highlead yarding system. (Courtesy of the Washington Iron Works.)

principle of the highlead is the elevation of the mainline lead block on a spar tree or steel tower to exert a vertical component of force to lift the turn of logs over stumps, windfalls, or other obstacles encountered during yarding. The logs drag their full length along the ground as long as there is no obstacle to their forward progress. On level ground the effective lift decreases as the distance from the spar increases. When yarding downslope there is no lift when the turn is higher than the lead block. The highlead is most effective when yarding upslope as a vertical component of force is always present. The economical highlead yarding distance depends upon the height of the lead block above the ground, the power and speed of the yarder, the topography, and the direction of yarding with reference to the slope. The customary maximum external yarding distance, perpendicular to the slope, is 700 to 800 feet in the Pacific Coast regions. The **setting**, which is the area to be yarded to one landing, is usually rectangular in shape so that the greatest yarding distance from the long corner will not exceed 1,000 to 1,200 feet. The highlead is best suited to yarding clear-cut settings, since it affords no control of lateral movement of the turn when yarding parallel to the slope. Any understory trees left standing on the setting by the cutting crew are usually knocked down in the course of highlead yarding.

Rigging the Spar Tree

Before the setting is felled the tallest, straightest, sound tree growing on or near the landing is marked with a big "X" chopped in the bark. If the tree is not growing in the best position on the landing it is raised or "**jumped.**" If the tree has to be moved only a short distance it is jumped. The tree is topped and guyed and sawn off at the butt. The butt is pulled in the desired direction by the rig-up donkey, the rear guys slackened, and the front guys tightened. The tree is jumped in an erect position, the angle with the vertical varying from a backward lean when the butt is moved, to a forward lean when the forward guys are tightened. A tree to be raised is felled and bucked at the required top diameter and trucked or yarded to the landing, usually by crawler tractor. A "dummy" tree on the landing is guyed and rigged with one or more raising line lead blocks. The butt of the spar to be raised is anchored to stumps so it will not slide forward. Buckle guys and a raising line block are attached to the spar at about two-thirds of its length from the butt. Then the tree is raised by the rig-up donkey, with a multiple block purchase, usually a three-part line, which the logger terms a "one-and-a-half block" purchase. When the spar has been raised to the vertical the guylines are anchored to stumps, and the spar is rigged in the same manner as a standing tree. In the Douglas-fir region the practice of raising spar trees instead of rigging standing trees has increased as operations moved up to higher elevations where suitable trees are scarce. In the hemlock-true fir forest type a Douglas-fir tree will often be trucked from a lower elevation and used repeatedly as a spar tree.

In converting a standing tree to a spar, the first step is to limb and top the tree. The **high-climber** or head rigger climbs the tree with spurs similar to a telephone lineman's spurs, but with a longer point to bite into the thick bark. The tree is encircled by a manila rope with a wire rope safety core, which is fastened to a wide leather belt. As he climbs he cuts off the limbs with an axe or a chain power saw with a short blade. At the point selected for **topping**, usually 24 to 30 inches diameter on the larger trees used with the more powerful yarders, he fells the top with a power saw making an undercut and backcut in the same manner as a tree is felled at stump height. Topping is one of the most spectacular feats in logging. As the top separates and falls, the bole sways back and forth. Although high-climbing appears to be a dangerous occupation the accident frequency rate is low because the high-climber is conscious of the hazards and is always alert to avoid an accident. As he descends the tree after topping, he chops the bark off with an axe at the points where guylines and straps will be hung. This is a safety measure to prevent pieces of bark from being rubbed off during yarding and falling on the men working on the landing below.

The high-climber ascends the tree a second time with a **pass-line** or rig-up block and a section of strawline for the pass-line. He hangs the block by a wire rope strap just below the top of the spar. Thereafter, he ascends the spar in a **"bos'n's chair"** hoisted by the pass line which is powered by the strawline drum on the rig-up donkey. The rigging is hoisted by the pass-line. Figure 8–5 shows a spar tree rigged for highlead yarding. Six top guylines, usually of the same diameter as the yarder mainline, are placed 16 feet below the top of the spar. The guyline is wrapped with a chain at a dis-

Fig. 8–5. Spar tree rigged for highlead yarding. Yarder is a Carco triple drum and loader a Northwest crane equipped with heel boom and grapple. (Courtesy of the Pacific Car and Foundry Company.)

tance from the eyesplice a little longer than the circumference of the tree. The chain is fastened to the pass line and hoisted. The high-climber shoulders the free end of the guyline around the spar and shackles the eyesplice to the bight of the guyline. The end of the guyline on the ground is pulled around a notched stump and tightened by the rig-up donkey mainline, and spiked in place with railroad tie spikes. Additional wraps are taken around the stump, tightened and spiked. The guy stumps are spaced about 60 degrees apart in a circle about the spar, at a distance of not less than the height of the spar. The **bull block** or mainline lead block is hung on a heavy wire rope strap a few feet below the top guys. Recommended safety practice is to use 4 tree irons or plates between the guylines and the wood, with hooks at the bottom to hold the strap. A safety strap connects the block with a guyline so that if the main strap failed the block would slide down the guy away from the landing. The haulback lead block is hung on another strap about 6 feet below the bull block. Three or four buckle guys are hung at about one-third of the height of the spar from the top to resist bending stresses induced by hard pulls on the mainline. If the spar is to be used for a skyline system, a tree shoe or jack for the skyline is hung between the top guys and the bull block.

For yarding direct to tidewater from the coastal forest of Alaska an **A-frame** mounted on a log raft is used in place of a spar tree. The A-frame is made of two tree-length logs. The yarder is also mounted on the raft, which is towed between settings by a tugboat. The A-frame is used with slackline and interlocking skidder skyline systems as well as the highlead system.

Steel Towers

The use of portable steel towers instead of spar trees for highlead yarding has expanded rapidly in the Pacific Coast regions in recent years. Nine companies are currently manufacturing towers in lengths of 90, 100, 110, or 120 feet. The tower is made of steel tubing or of welded steel box sections. The fairlead sheaves for the lines are permanently installed in the tower. A drum is provided at the base of the tower for each of the six guylines. The guyline runs from the drum through a sheave at the top of the tower to a choker or sling placed around a stump. Each guyline is independently tightened by a drum, which is then dogged or locked. The tower is raised or lowered by either a multiple block purchase or a hydraulic arm. The tower and yarder may be mounted on a tractor, on crawlers, on rubber-tires, either self-propelled or on a trailer, or on surplus Army tank-retriever trucks. The tower may also be mounted on a steel sled which is towed by the tractor donkey which sets on the sled for yarding. The tractor-mounted tower (Fig. 8–6) lowers to an angle of about 30 degrees to the horizontal for moving. The self-

Fig. 8–6. Highlead yarding with a 90-foot Hyster Porta-Tower and Hyster triple drum yarder mounted on a crawler tractor. (Courtesy of the Hyster Company.)

propelled rubber-tired tower unit lowers the tower to a horizontal position for moving (Fig. 8–7). The tower is telescoped to shorten the length.

The advantage of the steel tower is the saving in labor cost of moving and rigging. An experienced crew can lower the tower, move to the next adjacent landing, raise the tower and set the guylines in as short a time as two hours. To raise and rig a spar tree requires from one-and-a-half to two days. The saving in labor cost more than compensates for the relatively high capital investment in the steel tower. The tower has the further advantage over the spar tree of

Fig. 8–7. A Skagit self-propelled 110-foot tower unit with the tower lowered and telescoped for moving on the truck road. (Courtesy of the Skagit Corporation.)

being quickly moved to positions on the landing which will give the most efficient lead for each quarter of the setting to be logged. The tower obviates the need for a high climber, and men with this skill are increasingly difficult to find in logging communities.

The Yarding Operation

The highlead "side" is composed of the men and machines who yard and load the logs from a given setting. The yarding crew consists of the hooktender, who is foreman of the side, a rigging slinger, 3 or 4 chokermen, the yarder engineer or "donkey puncher," a chaser and sometimes a signalman or "whistle punk." The loading crew varies with the loading method used. (See Chapter 9.) The hooktender plans the layout of the yarding roads or paths along which logs will be yarded. He may decide to start on the square lead, perpendicular to the truck road, or on the fairlead parallel to the road. A haulback tail block is hung by a strap on a stump or tree at the end of the first road and a corner block is hung several yarding roads away. A diagram of the layout of yarding roads is given in Fig. 8–8. If topography necessitates, side blocks may be hung to keep the haulback line from rubbing on the ground or logs. The strawline is pulled

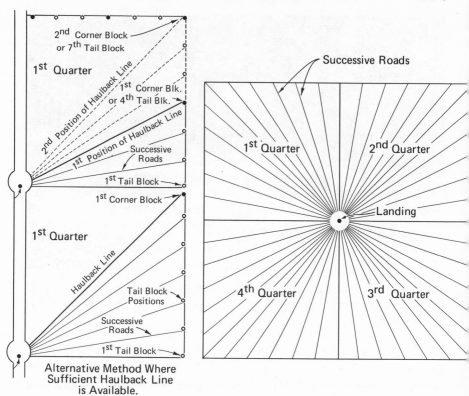

Fig. 8–8. Diagram of layout of yarding roads for highlead operation. (Courtesy of Isaacson Iron Works.)

out by hand from the spar tree to the corner block, then to the tail block, and back to the spar. The strawline and haulback line eye splices are connected by a "Hindu" hook or by a **molle,** a strand of a wire rope which is wrapped around itself to form an endless wire rope connection. Reeling in the strawline pulls the haulback line out through the corner block and then through the tail block, and back to the spar. The haulback is unhooked from the strawline and **clevised** to the butt rigging, which is attached to the end of the mainline. Reeling in the haulback pulls mainline, butt rigging, and chokers out along the yarding road.

When the butt rigging is over the logs selected to be choked to make up the turn, the rigging slinger signals the donkey engineer to stop. This may be done either by radio whistle signal, which blows an air horn on the yarder, or by shouting to the signalman, who transmits the signal through an electric wire which has been strung out from the yarder. The use of the radio signal from a lightweight transmitter carried by the rigging slinger eliminates the signalman from the crew. The signal horn is audible to all members of the crew so they can anticipate the movement of the lines. The rigging slinger tells the chokermen which logs to choke to make up the turn and assists them in setting the chokers. Usually two chokers to a turn are used. When all members of the rigging crew are in the clear, well away from the lines or the path of the turn, the rigging slinger signals to go ahead on the mainline with one blast of the horn. One blast also means "stop" when any line is moving. Two blasts means to reel the haulback and a series of short blasts means to slacken a line. When the turn arrives at the landing the chaser directs the engineer where to drop the turn by hand signals. He then unhooks the chokers, gets in the clear, and signals to reel in the haulback line. This operation is repeated until all the logs which can be reached from the first road are yarded. Then the haulback line is removed from the tail block and the block hung at the end of the second yarding road. Reeling in the haulback pulls the mainline out to the new road. When all roads up to that terminating at the corner block have been yarded, coils of strawline are sent out and threaded through the first corner block, which now becomes the tail block and a new corner block which has been hung. When the yarding of the first quarter-circle has been completed, the lead blocks are "swung" to the opposite side of the spar. (The mainline sheave of the steel tower swivels, so the operation of swinging the block when changing quarters is eliminated. This is another labor-saving advantage of the tower.) When the second quarter has been yarded the yarder is moved to the opposite side of the spar tree so as to face the remaining half-circle. When the third quarter has been yarded the lead blocks are again swung back to the side of the spar from which the first quarter was yarded.

Production on a highlead side varies with the yarding distance, the volume per acre cut, the topography and spar height, and yarder power and speed. The average daily production in old-growth timber in the Douglas-fir region is approximately 70,000 board feet.

MOBILE YARDER-LOADERS

The use of mobile machines which can yard as well as load is increasing in the western regions. The mobile yarder-loaders or mobile loggers, as they are termed, do not require as large a crew as the conventional highlead. Ability to vary the lead by swinging the boom to avoid residual trees during yarding gives them the capability of yarding partial cuttings. Also, the mobility of the machine permits it to yard perpendicular to the contour along parallel yarding roads and windrow the logs along the edge of the truck road. When a log truck arrives the machine can quickly be converted to a loader. The mobile logger is well adapted to logging small tracts of timber, to salvage logging of fire-killed or insect-infested timber, and to yarding road rights-of-way. The low lead to the mainline sheave in the end of the boom does not give the lift of the highlead and the efficient external yarding distance downslope or on level ground is reduced. However, the mobile logger effectively yards up steep slopes for distances comparable to that of the highlead.

The Jammer

The forerunner of the modern mobile loggers was the Idaho Jammer, which came into widespread use in the northern Rocky Mountain region following World War II. The **jammer** is mounted on a used motor truck. A vertical mast supports a pivoting pole boom which is set in a socket at the base of the mast. The top of the mast is guyed. A two-drum hoist, located between the mast and the truck cab, powers the haulback and the skidding line, which is also used as a loading line. The jammer operates along a network of parallel contour roads spaced 600 to 800 feet apart. The jammer is usually used where soil conditions or steep slopes are unsuitable for tractor skidding.

Mobile Loggers

The first mobile logger designed for West Coast logging was the Skagit SJ-8 which was introduced in 1951. It was followed by the smaller SJ-4 and a series of other models and sizes. The SJ-4R has a gooseneck boom supported by a vertical gantry which is braced during yarding by back guylines. The mainline-loading line fairlead is at the end of the boom and the haulback fairlead on top of the gantry. The machinery deck revolves full circle and the self-propelled rubber-tired undercarriage is stabilized by four hydraulic outrigger jacks during yarding and loading. For yarding the boom is raised to a fairlead height of 32 feet above the road surface. For loading the boom is lowered until the upper or heeling segment of the boom is horizontal.

To save the time and labor of setting guylines for mobile yarders, the guylineless tower was invented. The boom is supported during yarding by a stiff-leg pushing against the ground. The Skagit GT-4 "jillpoke sizzer" boom model folds the upper or heeling segment of the boom down for yarding (Fig. 8–9). The mainline fairlead is at

Fig. 8–9. Yarding with a Skagit GT-4 guylineless tower with "jillpoke sizzer" boom. (Courtesy of the Skagit Corporation.)

the top of the lower segment of the boom and the haulback fairlead is on top of the vertical gantry. For loading, the sizzer boom segment is raised, and the mainline runs over a sheave at the end of the boom. The GT-5 has a longer boom which is supported by a hinged stiff-leg. Both machines are mounted on tank-type tracks. Another recent development is the "snorkel" boom. The snorkel is a telescopic boom extension which is advanced to give the boom a reach of 60 or 75 feet, depending upon the model. When the log has been tonged or grappled, the snorkel is retracted and the log heeled for loading. The snorkel is especially well suited to logging narrow strips of timber such as a road right-of-way.

Mobile Cranes

Some makes and models of the mobile crane heel boom loaders described in Chapter 9 are manufactured with drum sets with line capacities which make them suitable for yarding as well as loading. One paper company logging on steep ground in the Cascade Mountains has replaced some highlead sides with Washington Trakloaders. The maximum yarding distance is 850 feet on the lower side of the truck road. A 4-man crew yards and loads an average of 9 truck loads a day with the Trakloader. This is a production rate of 10,000 board feet per man-day, as compared with the average rate of 8,000 board feet per man-day produced by the highlead sides. The Trakloader can also be used for slackline skyline yarding by installing a skyline sheave on a boom extension and a skyline drum in the drum set.

The mobile crane with a lattice boom is used for skidding as well as loading in the Western pine region. In one operation in Idaho a crane with a 57-foot boom and a grapple skidded up to 150 feet below the road, and 100 feet above, on slopes of 30 to 70 percent. The crane operator working alone decked 200 logs totaling 40,000 board feet in volume along one-third of a mile of road per day. The "Logger's Dream" and the "Big Stick Highlead," used for cable yarding and logging in the South, are described in Chapter 11 on pulpwood production. The "Logger's Dream" is also illustrated in Chapter 9 by Fig. 9–2.

Cone Yarding

An operation in Florida has for the past seventeen years successfully ground yarded hardwood logs from swamps with a metal cone. The cone is approximately six feet long and four feet in diameter at the base. The mainline is threaded through the nose of the cone. When the chokers are set on a turn of three or four logs and the mainline hauled in, the ends of the logs nestle inside the cone. The cone prevents hang-ups and keeps the ends of the logs clean.

The yarder consists of an A-frame and two-drum winch mounted on rubber tires. Drum capacities are 2,000 feet of $\frac{5}{8}$ inch mainline and 4,000 feet of $\frac{1}{2}$ inch rehaul line. The turn of logs for the cone is assembled by a "ground hog skidder" with a single drum winch and engine mounted on skids. The drum capacity is 500 feet of $\frac{1}{2}$ inch cable.

The operating procedure is to string out the lines from the A-frame yarder along a cleared yarding road for a distance of 2,000 feet in a direction perpendicular to the truck road. The tail block is hung on a tree about 10 feet above the ground. The groundhog is skidded out by the yarder rehaul line to the tail tree. The groundhog line is threaded through another block hung on the tree and pulled out by two men perpendicular to the main yarding road. Tongs are set and one log at a time pulled in by the groundhog. When enough logs for a turn have been assembled, they are attached with chokers to the end of the mainline, and yarded with the cone to the truck road. Signals between the chokermen and the yarder operator are by bell code with an old-fashioned hand-crank telephone. The telephone is also used for voice communication. The groundhog moves along the yarding road and pulls in logs from a strip 500 feet on each side. About 46 acres are yarded at each setting (Altman, 1965).

SKYLINE SYSTEMS

A skyline, in logging parlance, is a wire rope suspended between supports, termed tree shoes or jacks, elevated on spar trees or towers. Movement of the turn of logs is imparted through a carriage which rides on the skyline. The single span skyline systems require only two supports, a head tree at the landing and a tail tree at the back end of the skyline road. If the single span skyline is anchored at both ends it is classified as a tightline system. If one end is wound on a drum on the yarder, so the skyline can be lowered and raised, it is termed a slackline. As deflection or sag in the skyline is required for adequate working strength, the distance the single span system can yard is limited by the necessity of maintaining clearance between the skyline and the ground. The multi-span systems use intermediate supports at breaks in the ground profile and thus are limited in length only by the line capacity of the yarder or snubber, as the machine which operates the gravity systems is termed. The carriage is open on one side so it will travel past the intermediate support jacks or saddles. Systems with the capacity of yarding logs laterally to the carriage and then carrying them longitudinally clear of the ground are termed skyline cranes. Systems in which the carriage rides on the haulback line with no standing skyline used are termed running skylines. The diameter of wire rope used for standing skylines varies with the system, the deflection obtainable, and the length of span, and ranges from $1\frac{1}{2}$ to $2\frac{1}{4}$ inches. The wire rope used for skylines is made of the highest tensile strength grade of steel and has an independent wire rope core.

Standing Skyline Systems

The single span **standing skyline** systems, also termed tightline systems, use a cable stretched between a head spar and a tail spar tree, and anchored to stumps at both ends. The single span tightline systems include the north bend, the modified north bend, which is sometimes referred to as the south bend, and the interlocking skidder. The heavy skyline is usually stored and moved between settings on a large reel mounted on a used flat-bed truck. In some operations the skyline is dragged along the truck road by a tractor.

North Bend. The north bend, the simplest in rigging of the tight-line systems, is illustrated in Fig. 8–10. The mainline runs from the bull block (8) through a **fall block** (13) to an attachment to the carriage (6). The haulback line and the butt rigging is attached to the fall block. The haulback tail block (14) is positioned to pull the fall block down to choke the turn of logs. The logs drag along the ground until their forward progress is halted by a stump, windfall or other obstacle. Then the resultant of the forces in the two segments of the main line, one in the direction of the spar tree, the other upward to the carriage, raises the turn over the obstacle. If more lift is desired, by holding the brake on the haulback line and continuing to reel in the mainline, the turn can be lifted vertically. The attachment of the end of the mainline to the carriage tends to confine the turn to a narrow path beneath the skyline.

The north bend system is commonly used for swinging from highlead-yarded cold decks located at twice the yarding distance, or 1,400 to 1,600 feet from the landing on the truck road. It is the most popular swing system because no special donkey is required and any highlead yarder with sufficient drum capacity can be used. The north bend has been used for yarding where difficult topo-

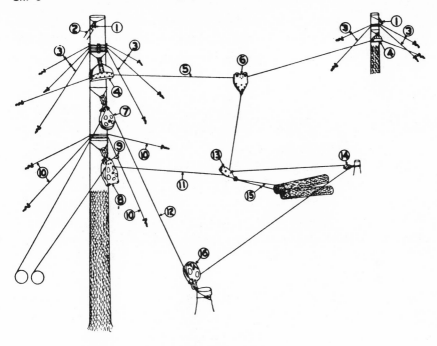

(T) Tail-Hold to Stump
(1) Pass Block
(2) **Pass Line**
(3) Top Guys
(4) Tree Shoes
(5) Tight Sky Line
(6) Carriage
(7) Head Tree Haul-Back **Block**
(8) High Lead Block
(9) Safety Strap
(10) Buckle Guys
(11) Main Line
(12) Haul-Back Line
(13) Fall Block
(14) Haul-Back Tail **Block**
(15) Chokers
(16) Haul-Back Corner **Block**

Fig. 8–10. North Bend Skyline System. (Courtesy of the Division of Safety, Department of Labor and Industries, State of Washington.)

graphic conditions call for a skyline system, but the special machine required by other skyline yarding systems was not available. The time required to rig up the tight skyline is a disadvantage in using the north bend for yarding. The north bend operates best upslope, or on moderate downslopes.

South Bend. For swinging down steep slopes the north bend system is modified to a south bend by threading the mainline through the fall block, thence through a sheave in the carriage, or a depending block hung from the bottom of the carriage, and back to the fall block where the end of the line is shackled. Thus, there are two parts of line between the fall block and the carriage to one part between the fall block and head spar. This gives more lift to the turn, but puts more load on the skyline. As the line wear running through two sheaves is more severe, a flexible tag line is usually used on the end of the mainline.

Interlocking Skidder. The interlocking skidder system (also termed the Lidgerwood system for the inventor) is a tightline system which requires a special yarding machine and carriage. The yarder has a heel tackle drum for tightening the skyline, a skidding line drum, a receding line drum and a slackpulling line drum, all three of which can be interlocked so they will operate together. The larger skidders also have a transfer line drum and strawline drum which are used to rig a second skyline while the first skyline is in use for yarding. Rapid change is effected by slacking the multiple-sheave heel tackle to lower the first skyline to the ground, releasing it, and threading the head end of the second skyline through the carriage and attaching it to the heel tackle. Tightening the heel tackle raises the skyline to the desired height.

The operating principle of the interlocking skidder may be visualized by referring to Fig. 8–11. The chokers hooked to the end of the skidding line are lowered to the ground by reeling in the slack pulling line. This line is threaded through the carriage so that the pivoted skidding line sheave is raised into contact with idler sheaves, feeding the skidding line through in an action similar to that of a laundry wringer. The brake is set on the receding line to hold the carriage in position during slack pulling and when reeling in the skidding line to raise the log turn to the desired height. Then the running lines are interlocked so that the receding line will pay out at the same rate the skidding and slack pulling lines are reeled in.

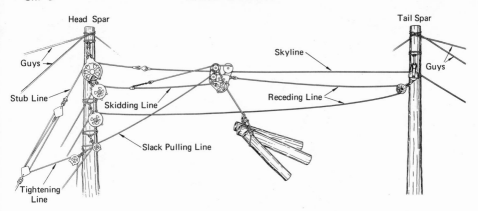

Fig. 8–11. Slack-pulling interlocking skidder system with Berger carriage. (Courtesy of the Smith-Berger Manufacturing Corporation.)

The interlocking skidder is suitable for yarding over rough, broken topography, upslope or downslope, wherever skyline deflection can be obtained. Popular in railroad logging days, its use declined when the motor truck replaced the railroad. In recent years the big skidders have been used mainly in Alaska for yarding the coastal fringe of timber from a raft. The skidder was revived in 1960, in a miniature size compared with the previous skidders, by the Longview branch of the Weyerhaeuser Company to do tree selection yarding in young growth stands. A yarder and a 56-foot steel tower, made from a ship's spar, are mounted on a 200-horsepower long-base six-wheeler truck. The yarder has skidding, receding, slackpulling, skyline, and guyline drums. The skidders are powered with 110 to 145 horsepower engines. The carriage is a Berger "miniature" model. Operating on spans averaging 660 feet in length, these skidders have proved to be efficient tree selection yarding machines with minimal damage to the residual stand. Revival of interest in the skidder system has lead the yarder manufacturers to develop machines with new mechanical principles in the interlocking system, and the use of these machines in old-growth as well as second-growth timber is increasing.

Slackline System

The slackline system, which requires a special yarder with a large skyline drum, equipped with powerful brakes is illustrated in Figure

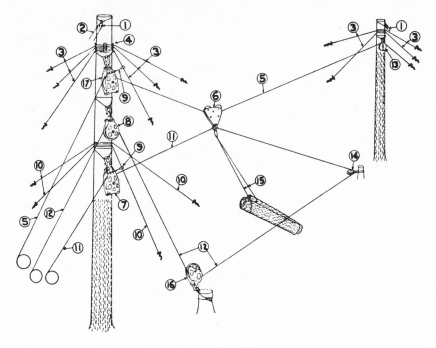

(T) Tail-Hold to Stump
(1) Pass Block
(2) Pass Line
(3) Top Guys
(4) Tree Plates
(5) Slack Sky Line
(6) Carriage
(7) Skidding Block
(8) Head Tree Haul-Back **Block**
(9) Safety Straps
(10) Buckle Guys
(11) Main Line
(12) Haul-Back
(13) Tree Jack
(14) Haul-Back Tail Block
(15) Chokers
(16) Haul-Back Corner Block
(17) Skyline Block

Fig. 8–12. The slackline system. (Courtesy of the Division of Safety, Department of Labor and Industries, State of Washington.)

8–12. The skyline runs from this drum through a large block on the head spar (17), thence through the carriage (6) and a tail tree jack (13) to a tail hold on a stump. The mainline, haulback, and butt hooks are all attached to a clevis at the bottom of the carriage. In operation the carriage is spotted above the turn of logs to be choked by the haulback. The skyline is slackened to bring the carriage and attached chokers down to the logs. When the chokers have been set the skyline is reeled in, raising the turn to the desired height. The haulback drum is released and the mainline reeled in. Arriving at the landing, the turn is lowered to the ground for unhooking by slackening the skyline. If desired the turn can be lifted clear of the ground. However, to reduce the stress in the skyline and permit a larger turn to be yarded, the usual procedure is to lift the front end of the log, leaving the rear end to drag on the ground. Road changing is accomplished by releasing the tail hold, reeling in the skyline until the end is at the landing, and then pulling the skyline out along the new road with the haulback line.

The slackline system is particularly well adapted to yarding down steep slopes or across canyons. Having the skyline spooled on a drum expedites rigging-up and road changing. Yarding distance with the larger slackline yarders is commonly 1,200 to 1,500 feet. Disadvantages are the cost and weight of the special yarder required. In addition to the yarding crew a rig-up crew is needed to rig tail trees ahead.

Running Skyline

The recent development of improved variable interlocking yarders by the yarder manufacturers has revived interest in the running skyline. For many years loggers have occasionally rigged a running skyline with their high lead equipment to yard down steep slopes or across narrow draws. The system was termed a Grabinski or a scab skyline. The butt rigging is hung from a traveling block or a light carriage which rides on the haulback line. The haulback runs from the head spar out along the yarding road, through a tail block hung on a tree well above the ground, and back to the rear of the carriage. The end of the mainline is attached to the front of the carriage. With a non-interlocking yarder, holding the brake on the haulback drum, and continuing to reel in, the mainline raises the carriage to free the turn of logs from hangups. When the brake is released the turn lowers and yarding continues as in high-leading. With an interlocking yarder, when the carriage is raised to the desired height, main and haulback drums are interlocked to maintain equal tension on both lines. The haulback line pays out at the same rate as the mainline reels in. This keeps the carriage elevated throughout the inhaul run. Figure 8–13 illustrates the running skyline system as used with a highlead yarder and steel spar.

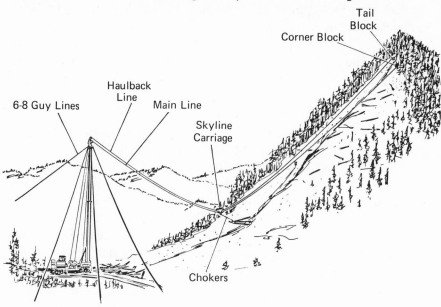

Fig. 8–13. Running skyline system operated by a highlead yarder and steel spar. (Courtesy of the Washington Iron Works.)

The development of mobile cranes with interlocking drum sets and of special running skyline carriages, beginning in 1967, has stimulated wide acceptance of the running skyline as a standard yarding method. The system offers labor-saving advantages over other cable yarding systems. It is well suited for the use of a yarding grapple instead of chokers. Only two men are required in grapple yarding— the machine operator and a spotter who is stationed where he can see the logs to be yarded and directs the operator by voice radio. The crane boom swings in an arc to help position the grapple over the log. At the landing the grapple is opened to quickly release the log. As some of the cranes require no guylines, and others only two rear guys, time is saved in rigging-up. Road-changing time and labor may be saved by attaching the tail block to an old crawler tractor. Changing to a new yarding road is effected by simply moving the tractor. Yarding is usually done perpendicular to the contour and the logs are windrowed along the road. Loading is done later by a mobile loader.

An example of the machines designed specifically for running skyline operation is the Washington 108 Skylok Yarder shown in Fig. 8–14. It has a boom 50 feet in length and is mounted on crawler

Fig. 8–14. Washington Model 108 Skylok Yarder operating a Skylok carriage and yarding grapple on a running skyline. (Courtesy of the Washington Iron Works.)

tracks. It is powered by a 320-horsepower diesel engine and the patented "Vari-lok" drive gives infinite ratio interlocking of the drum set. The four rectangles at the tip of the boom are floodlights for night operation. The Skylok carriage may be rigged for either grapple operation, as illustrated in Fig. 8–14, or for chokers. The two-line grapple is opened and closed by the two mainlines, which are reeled in simultaneously to yard the turn.

Yarding Grapples. Yarding **grapples** are of two general types. The single-line grapple with automatic cycling is illustrated in Figure 8–15. The grapple claws are held open by springs until the grapple

Fig. 8–15. Yarding grapple of automatic cycling type operated from a running skyline. (Courtesy of the Skagit Corporation.)

is dropped on the log to be yarded. When the grapple is lifted by raising the running skyline, the jaws close. The single-line grapple can be used with any machine suitable for running skyline operation. The two-line grapple is shown in Figure 8–16. The grapple is opened by the holding line pulling a short tag line which is fastened to one leg of the grapple. When line pull is applied to the closing line, the grapple jaws close on the log. Both holding and closing lines are reeled in simultaneously to yard the log. The two-line grapple requires a special yarder, such as the Skagit GT-5 Guylineless Tower illustrated (Fig. 8–16) or the Washington 108 Skylok Yarder previously described.

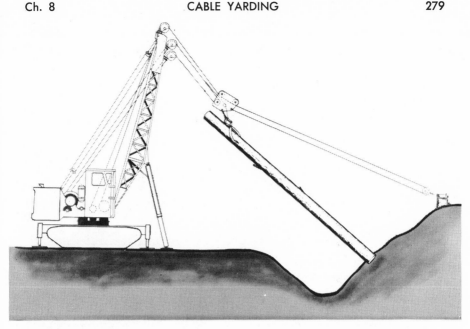

Fig. 8–16. Running skyline with a two-line yarding grapple. The three lines are the skyline, which also serves as the haulback, the holding line, and the closing line. (Courtesy of the Skagit Corporation.)

Skyline Cranes

Systems which transport logs through the air suspended from a carriage riding on a cable are widely used for timber extraction from alpine forests in Europe. They are termed, in German, "Seilkräne" or "Cable Cranes." The first European system to be introduced in North America in 1949 was marketed as the "Wyssen Skyline Crane." The words **"skyline crane"** have since become, in the United States, a generic term for systems using a standing skyline which combine lateral and longitudinal yarding capability. American-made skyline crane systems using radio controlled carriages were introduced in 1958. The skyline crane may be a single span system, or a multispan system, using an open-side carriage which can pass over intermediate skyline supports. Logs are yarded laterally to the skyline road for distances of from 75 to 250 feet, depending upon the system and the operating conditions, and then transported along the skyline road. Single span systems may only lift the front ends of the logs or may carry them suspended clear of the ground. The multispan systems carry the logs free of the ground. Some systems are designed to carry the logs downslope by gravity with the carriage controlled by a line from a snubber stationed at the upper end of the skyline road. Other systems haul the loaded carriage upslope with gravity return of the unloaded carriage. Others use a haulback line and may operate either upslope or downslope.

Skyline crane systems may be divided into two general types, determined by what line is used for lateral yarding. The European systems and some American systems use the mainline or snubbing line for both lateral yarding and longitudinal control of the carriage. The carriage is clamped to the skyline or held by the haulback line and means provided to pull slack in the mainline to enable the butt hook to be carried manually out to the logs to be hooked. The other type does the lateral yarding with a hoisting or tong line spooled on a winch drum in the carriage. One system has an engine in the carriage to power the winch; another powers the tong line drum by lines from the yarder. The mainline controls the carriage travel.

Multispan systems permit longer skyline roads than do the single span systems, due to the problem of obtaining adequate deflection in long skyline spans. Intermediate supports are placed at breaks in the ground profile to maintain deflection. A skyline road with a convex profile can be logged, whereas the single span requires a concave profile. To provide intermediate support, the shoe or jack on which the skyline rests is hung from a leaning spar or a vertical spar with the shoe pulled out from the spar by a guyline, so that the turn of logs will clear the spar. Where the cutting unit is fan-shaped, the shoe may be hung from a block riding on a line suspended crosswise of the skyline roads.

The Forest Service has encouraged skyline crane logging because of the reduction in soil disturbance from yarding and from truck road construction due to the wider road spacing. In an experimental operation in a Sitka spruce stand on slopes averaging 77 percent, it was found that mineral soil exposed on the skyline crane setting was 6.4 percent and on highlead settings 15.8 percent. It was estimated that soil disturbance due to landing preparation and road construction was 3.3 percent with skyline crane yarding and 9.8 percent with highlead yarding (Ruth, 1967). In a mixed conifer stand in eastern Washington, soil disturbance from skyline crane yarding was found to be 5.4 percent and from tractor skidding 22.2 percent (Wooldridge, 1960). Another advantage is that normally the logs are dragged to the skyline for short distances parallel to the contour, whereas the longer highlead roads cross the contour. The velocity of runoff along the skid trail, and consequently of erosion, is thus lessened with skyline crane yarding.

Disadvantages of the skyline crane are the large capital investment in equipment and the high cost of rigging up, especially if intermediate supports are required. A large volume of timber on the strip is needed to make the multispan system economical. Lost production time due to the maintenance and repair required by some of the more complicated skyline crane carriages has resulted in increased costs in some operations. The higher yarding costs with skyline cranes are offset to some extent by savings in road construction cost. The layout of multispan systems must be carefully engineered to achieve satisfactory performance. Logging with multispan skyline cranes is currently limited to situations where avoidance of soil disturbance is a prime consideration or where road construction and maintenance is unusually expensive and wide road spacing is desirable. Examples are municipal watersheds and tracts with unstable soils, such as slides or slumps. Timber areas interspersed with rock cliffs which could not be logged with more conventional systems, and other areas which have in the past been classed as inoperable, offer possible opportunities for some of the skyline crane systems. Most of the multispan skyline crane operations have been on public lands where the higher yarding costs are recognized in the stumpage appraisal.

Wyssen Skyline Crane. The Wyssen, which is made in Switzerland, is a multispan gravity system operated with one line from a snubber stationed at the upper end. The Wyssen snubber is a diesel-powered one-drum machine mounted on a steel sled. A large wing fan with vanes acts as an air brake to control the descent of the carriage. The snubbing line runs through the carriage and also acts as a hoisting line for lifting the turn of logs up to the carriage. Communication between the choker setters, the chaser at the landing, and the snubber operator is by telephone or radio. When the carriage is pulled up from the landing by the snubbing line, a hydraulic pump builds up pressure to overcome the spring tension on the cable clamp. When the carriage stops, a timing device set for a predetermined time interval, usually 20 seconds, bleeds off the hydraulic pressure and sets the clamp on the skyline clamps. This releases the hook at the end of the snubbing line to which the chokers are hooked. The snubbing line is slacked and the hook descends to the ground. When the chokers have been set the snubbing line is reeled in and the turn is hoisted. The hook locks into the carriage and releases the cable clamp. The snubbing line is payed out and the carriage rolls down the skyline. When the carriage is stopped on reaching the landing, the timing device again sets the cable clamp, releases the hook, and the turn is lowered to the ground.

The operating crew of 5 men includes the snubber operator, 2 chokermen, a signalman with a portable telephone stationed in view of the chokermen, and a chaser at the landing. The yarding strip width ranges from 150 to 250 feet each side of the skyline. In British Columbia, over distances of 4,500 to 6,000 feet, a Wyssen averaged 4 trips an hour. The Wyssen carriage is made in five model sizes with load capacities ranging from 3,300 pounds on a $1\frac{5}{16}$ inch diameter skyline to 22,000 pounds on a $1\frac{7}{8}$ inch skyline.

Baco Cable Crane. The Baco Cable Crane was the second Swiss system to be used in the United States. It is similar in principle to the Wyssen system except for the carriage mechanism. The carriage is hauled 3 meters beyond the point at which it is to be spotted. Then the snubbing line is slacked and as the carriage rolls back a system of levers actuates the cable clamp and disengages the hook at the end of the snubbing line. Hoisting the turn and re-engaging the hook releases the clamp. The Baco is manufactured in 2-, 3-, 5-, and 10-ton rated load capacities.

Radio-Controlled Carriages. Two types of radio-controlled carriages are made by the Skagit Corporation. The Skycar-type carriage contains an engine and winch drum on which a $\frac{7}{8}$-inch tong line is spooled. The tong line does the lateral yarding. The main or **snubbing** line is fastened to the carriage. The Skycar was designed to carry the log turn downslope on a multispan skyline. The Bullet-type carriage contains an engine which powers slackpulling sheaves to feed the main or load line through the carriage to lower the butt hook to the ground. The carriage is clamped to the skyline during lateral yarding. The Bullet is usually used to yard upslope on a single span skyline.

Controls for the mechanisms in the carriage are operated by solenoid valves actuated by radio signals. A radio receiver and a voice loud speaker is installed in the carriage and on the snubber or yarder. The rigging slinger and the chaser each carry a light portable transmitter on the belt or in the hip pocket, which is operated with one hand. The transmitter blows the signal horn on the yarder as well as controlling the carriage and is used for voice communication in emergencies. Another transmitter is installed on the yarder.

Skycar Type. The RCC-20 Skycar and the RCC-15 Torpedo carriages contain a 95-horsepower diesel engine, fuel tank, a winch for the tong line, and the radio equipment. The operating weight of the Skycar is 8,140 pounds and the rated load capacity 35,000 pounds. The Torpedo weighs 5,000 pounds and the load capacity is 30,000 pounds. The snubber usually is equipped with hydraulic retarders to provide ample braking power to control the descent of the loaded carriage.

The $1\frac{3}{4}$-inch skyline rests on head and tail tree shoes, and on intermediate support jacks hung from "gin poles" or spars leaning perpendicular to the slope and at an angle to the plane of the skyline so the turn will not brush against the spar. The location of the intermediate spars depends upon the ground profile, but generally average 1,000 feet apart. Skylines up to 5,000 feet in length have been used. When the carriage has been hauled up the skyline by the snubbing line to the desired spot, the rigging slinger sends a radio signal to the snubber operator to stop the snubbing line and set the brakes, and to the carriage controls to lower the tong line to the ground. When the turn of logs has been choked, he transmits the signal which reels in the tong line to raise the turn clear of the ground. Another signal sets the drum brake to hold the turn suspended. On signal the snubber operator releases the snubbing line brake and the carriage rolls along the skyline by gravity (Fig. 8–17). The carriage travels at speeds up to 70 miles per hour between support jacks, but it is slowed to pass over the jacks. Arriving at the landing, the chaser transmits signals to stop the carriage and to release the tong line brake, to lower the turn to the ground.

Usually a clear-cut strip 300 to 400 feet in width is yarded. Due to the precise control which the rigging slinger has of the speed at which the hoisting line is lowered and the elevation of the carriage above the ground, the butt hook on the end of the hoisting line may be carried laterally by the chokerman without exerting undue effort. A steep lateral slope restricts the distance the line can be pulled uphill, but increases the distance on the downhill side. The upward

Fig. 8–17. Logging with the Skagit RCC-20 Skycar in Mt. Baker National Forest, Washington. Upper lines are skyline and snubbing line. Lower lines are guylines from intermediate spar. (Courtesy of the Skagit Corporation.)

lead to the carriage facilitates hoisting the turn without hangups. Both rectangular cutting units with separate landings, and fan-shaped units radiating from a common landing, are yarded.

The Torpedo has also been used for yarding upslope as a single span. In one operation, where the rugged topography limited high-lead production to 50,000 board feet a day, the same equipment, with the addition of the Torpedo carriage, produced 70,000 board feet a day on spans of 1,200 to 1,400 feet. The mainline of the tractor donkey was used as the skyline, and the haulback line as the running line to the carriage.

Bullet Type. The Skagit RCC-10 Bullet and the RCC-13 Tracer carriages contain a 24-horsepower butane engine and fuel tank, radio controls, and an air compressor and air tank to operate the slackpulling sheave brake and the skyline clamp. The load line runs through the carriage around slackpulling sheaves, and out through a fairlead at the rear of the carriage. The rated load capacities of the RCC-10 and RCC-13 are 20,000 and 30,000 pounds, respectively.

Carriage weights are 3,000 and 4,000 pounds. The system is operated from a mobile logger with a boom or a mobile yarder with a steel tower (Fig. 8–18). Usually a live skyline reeled on the main drum of the yarder is used.

Fig. 8–18. Yarding a sanitation-salvage cutting unit with a Skagit radio-controlled Bullet carriage to a Skookum tower with a Carco triple-drum mounted on a Caterpillar tractor. Hydraulic knuckleboom grapple loader at right. (Courtesy of the Pacific Car and Foundry Company.)

The operating sequence in thinning or sanitation-salvage operations is as follows: The strawline is pulled out along the skyline road manually. The skyline is pulled out through a shoe on a tail tree by the strawline, and is fastened to a tail hold at the base of another tree. The skyline is tightened and the carriage runs by gravity down the skyline. When it reaches the desired position for yarding, the rigging slinger transmits a radio signal to the yarder engineer to stop the carriage, and to the carriage to set the skyline clamp. Another signal accelerates the butane engine and the load line feeds down to the ground. The rigging slinger and the chokerman carry the end of the load line laterally and hook the pre-set choker. On signal the load line is reeled in by the yarder to yard the log to the skyline road. When a turn of several logs is assembled, it is raised to the desired height and signals are transmitted to the carriage to set the brake on the slackpulling sheaves, and release the skyline clamp, and to the yarder to reel in the load line. Usually only the front end of the turn is raised and the other end drags along the ground. At the landing the turn can be lowered by slackening the skyline or by setting the skyline clamp and releasing the slackpulling sheave brake.

The Bullet has been used for thinning and sanitation-salvage in immature Douglas-fir stands since 1961. Skyline roads are commonly 800 feet long and lateral yarding is done from 100 to 150 feet on each side. Three or four roads are yarded to each landing. Due to the precision with which the carriage can be spotted to lead the laterally yarded log between standing trees, damage to the residual stand has been held within acceptable limits. Where the cut is 10,000 to 20,000 board feet per acre the production of the 4-man yarding crew is 30,000 to 35,000 board feet per day. The Bullet is also used for skidding partial cuttings in the ponderosa pine region. In British Columbia the Bullet has been used for swinging down-slope from a cold deck to a lake, a distance of 2,800 feet. A custom-made Powell snubber with a hydraulic retarder was used to control the descent of the carriage and the turn was carried in the air.

BALLOON YARDING

No new logging system has received more publicity in journals and attracted more interest among loggers in many countries than balloon yarding. The prospect of being able to fly logs through the air for long distances at high speeds, without having to rig up a skyline crane system, is intriguing to loggers operating in rough terrain or in tropical swamps. Foresters see in balloon systems the possibility of logging some of the millions of acres of West Coast forest land now classed as inoperable. Examples of such areas are watersheds with unstable soils, areas where the timber grows on benches interspersed with cliffs, or where steep, rocky terrain and relatively low volumes per acre make logging with conventional systems uneconomical because of high road costs. The advantages of balloon yarding in reducing soil disturbances and log breakage and the saving in road mileage due to wider road spacing also stimulated the interest of forest officers in the system. Experience with balloon yarding has demonstrated that other advantages are reduction in the amount of slash left on the ground and the safety of the yarding crew. The Forest Service has encouraged the development of balloon logging by making negotiated timber sales specifically for this system and conducting research in the system by the Seattle Forest Engineering Research unit of the Pacific Northwest Forest and Range Experiment Station.

Barrage-Type Balloon

The first balloon yarding tests were made in Sweden in 1956 and 1957, and in British Columbia in 1963. The balloon used was the single-hull type, well-known as the British barrage balloon of the Second World War. The configuration of this balloon is elliptical, tapering toward the tail, with a vertical tailfin beneath and a horizontal tailfin on each side (Fig. 8–19). In one Canadian operation two balloons were used, one being tethered vertically above the other. In 1965 a 75,000-cubic-foot barrage-type balloon was tested in Oregon. The balloon had a static lift capacity of 2,400 pounds, but log turns weighing 4,500 pounds were flown as additional dynamic lift due to inhaul speed and wind was obtained. The average cycle time per turn for outhaul, hooking, inhaul and unhooking was 6 minutes for an average yarding distance of 1,200 feet. The single-hull balloon is available in England in sizes up to 150,000 cubic feet.

A 100,000-cubic-foot barrage-type single-hull balloon was placed in logging service in British Columbia in 1967 by Balloon Transport, Limited. A specially designed hydraulic drive infinitely variable interlock yarder with drum capacities of 17,000 feet of line and average line speeds of 2,640 feet per minute was used. This system

Fig. 8–19. British barrage-type single-hull balloon flying a personnel gondola. Log turns are flown in the same manner. (Courtesy of the Forest Engineering Research, PNW Forest and Range Experiment Station, U.S. Forest Service.)

was used successfully to salvage windthrown patches and areas inaccessible to conventional logging methods in the Seymor River watershed which supplies water to the city of Vancouver, British Columbia. Yarding distances ranged up to 5,000 feet from the truck road. To solve the problem of getting the members of the choker-setting crew out to their work place, a personnel transport gondola was invented. The gondola is a steel drum weighing one ton, with accommodations for the crew in the upper section, and space for rigging in the lower section (Fig. 8–19). The balloon is left flying except when it is brought down every two weeks for inspection and replenishment of helium gas loss. The balloon is constructed to withstand winds of up to 100 miles per hour. A yarder which will operate the system at speeds of 40 miles per hour was being designed in 1968.

Spherical Balloon

Yarding with an American-made 250,000-cubic-foot capacity spherical balloon was initiated in Oregon by the Bohemia Lumber Company in 1967. This balloon is 82 feet in diameter and 87 feet in height, with a static lift capacity of 10,000 to 12,000 pounds. The static lift varies with air temperature and altitude. Depending upon the weight of the rigging and cables the balloon can lift and transport turns of logs weighing 5,000 to 9,000 pounds. The balloon consists of a spherical envelope, which is filled with helium gas and an external cone-shaped skirt-ballonet. The ballonet is pressurized by the wind blowing through a flap-valve system around the base of the skirt. The varying air volume in the ballonet takes up the volume change in the gas, which changes with altitude and temperature. The spherical balloon system is non-directional and will work regardless of wind direction. The spherical balloon flying a turn of two logs is shown in Figure 8–20.

The Bohemia Lumber Company began experimenting with balloon yarding in 1964 using a 75,000-cubic-foot twin-hulled Vee-balloon. The Vee-balloon has two football-shaped hulls joined at the nose and spread at the tail by a horizontal fin, which has a vertical fin on the upper side. The configuration provides dynamic lift in addition to static lift in a directional wind. In 1966 trials with a 175,000-cubic-foot Vee-balloon 162 feet long and 88 feet wide at the tail began, but were terminated in 1967.

The method of rigging for yarding with the spherical balloon is illustrated in Figure 8–21. The **tether** line is about 500 feet in length and the tong line 100 feet. The position of the forward tail block is changed as yarding progresses to bring the tong line and chokers down to within reach of the next logs to be yarded. Due to the mass of the balloon being so much greater than the mass of the turn of logs, when the haulback line is slackened, and the mainline reeled in, the turn is lifted quickly into the air. Then the main and haul-back drums are interlocked for constant tension on the lines to keep the turn flying at the desired height. Communication between the chokermen and the yarder is by two-way voice radio. When the turn reaches the landing the balloon is pulled down by the mainline until

Fig. 8–20. Raven spherical balloon flying a turn of two logs. (Courtesy of Raven Industries.)

there is enough slack in the tong line to unhook the turn. Balloon yarding requires a special yarder with large drum capacities and high line speeds. As no spar or tower is needed, the lines feed through fairleads on a bracket at the front of the yarder. For moving the balloon between settings, and to the bedding ground for weekly replenishment of helium loss, the tether line is transferred to a winch, which in the Bohemia operation is on an M-4 tank.

Balloon yarding does not yet appear to be competitive with conventional yarding methods in situations where the latter can be used. The promise of the balloon is for logging areas which would other-

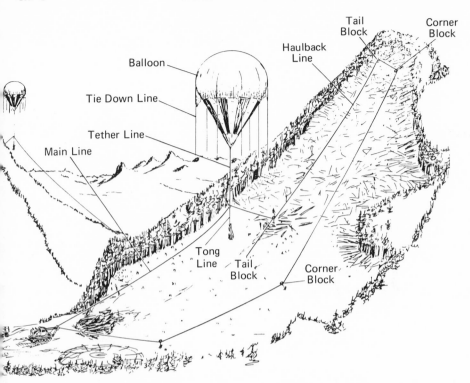

Fig. 8–21. Balloon yarding system with spherical balloon and track-mounted Washington Aero Yarder. (Courtesy of the Washington Iron Works.)

wise be inoperable. Much interest in balloon logging has been manifested in southeastern Alaska. The ability of the balloon system to yard long distances would reduce the mileage of costly truck roads to be constructed, and in some areas the balloon could reach from tidewater to timber line. In 1970 plans were being made for balloon logging in Alaska and in Idaho.

SUGGESTED SUPPLEMENTARY READING

1. O'LEARY, JOHN E. 1969. *Skyline logging*. Symposium Proceedings, School of Forestry, Oregon State University, Corvallis, Oregon. Every aspect of skyline yarding is covered in twenty-nine papers presented at a three-day symposium.
2. STILLINGS, FRED I., and WILLIAM M. GERSON, 1966. Report on an administrative study of skyline logging. 24 pp. U.S. Department of Agriculture Forest Service, Division of Timber Management, Region 5, San Francisco, California. Cost study of a skyline crane operation in the Klamath National Forest.

9

Log Loading
and Unloading

INTRODUCTION

Logs which have been skidded or yarded to a road are loaded on a wheeled vehicle, most commonly a motor truck and trailer, for further transportation. The output of a logging production unit is often determined by the capacity of the loading machine. For efficient operation the number of trucks must be balanced with the loader capacity to avoid delay waiting to load. Where skidding to a landing is done by tractors, the number of tractors used is determined by the loader capacity. Where yarding is by a cable system, the loading is limited by the yarding output.

A wide variety of loading machines and methods is available to the logging operator. As the high-capacity loaders are expensive, ranging in price to $100,000 or more, the machine used is sometimes determined by the capital investment the operator can afford. In the smaller operations the ability of the loader to be converted to a road construction machine may be a factor to consider in selecting the type of loader. For example, the crane or shovel loader and the forklift loader can be converted to an earth-mover or a road construction material loader by substituting a bucket for the log grapple or forks. In terrain requiring cable yarding, it is often desirable to have a loader which can double as a mobile yarder to log areas, adjacent to a road, which are too small to justify rigging up the conventional yarder.

The degree of mobility required of the loading machine is determined by the type of logging operation. Where the logs are brought to a landing for loading, stationary or crawler-mounted loaders may be used. Where the logs are left scattered along the edge of a truck road by the skidding tractors, rubber-tired mounting is desirable for high mobility during loading. The trend is toward rubber-tired mounting of the loader working at landings in order to save time in moving between landings, or to serve two landings alternately when yarding output at a single landing is less than loader capacity.

Logging timber stands where the yield per acre is small, such as salvage of fire-killed, insect-killed, or diseased trees, and in small scattered tracts, may call for self-loading log trucks. The hydraulic loader mounted behind the truck cab has become popular in such operations.

Where the major transportation system involves hauling logs on more than one type of wheeled vehicle, transferring or reloading is required. The term **transfer** is used when the entire log load is lifted from the truck and lowered onto a railroad car. The term **reload** is used when the logs are unloaded from the truck to the ground, and later loaded individually, or in bunches, on the car or another truck. At the log delivery point the operation of unloading or dumping is performed.

Log loading methods may be classified as either stationary or mobile. Characteristic of the stationary methods is that they have to be rigged at each landing and unrigged when loading at the landing is completed. The crosshaul, gin pole, A-frame, and single-tong guyline systems are today used only by logging operators lacking the capital to invest in more efficient loaders, and whose volume of production is small. The spar tree boom loading methods have been rendered virtually obsolete by the high labor cost of rigging them. The advent of the steel tower for high-lead yarding further contributed to the demise of the tree boom. These obsolete methods will be described briefly as they are still used in tropical countries where wage rates are low.

Today most logging operations use some type of mobile loader. They require no "rigging-up" other than to lower hydraulic outriggers to stabilize rubber-tired crane-type loaders. They are able to move about under their own power, and are ready to start loading as soon as they are within reach of the logs. They require fewer men on the loading crew than stationary methods. They are safer because they exercise better control over the movement of the log. They involve larger capital investment, but load logs at lower cost per unit of volume.

STATIONARY LOADING METHODS

Hand Rolling

The earliest method of loading was to roll logs onto a wagon or sled with a peavey. The **peavey** is a hand tool with a steel hook which swings freely about a bolt threaded through two lugs on a pointed steel shank. A stout wood handle five or six feet long is inserted in the shank. To roll a log, the hook is jabbed into the side of the log with the shank resting on top and the end of the handle pushed up to provide the necessary leverage. On sloping ground, a rollway was built of poles above the height of the vehicle to be loaded. The logs were skidded to the upper end of the rollway and rolled onto the vehicle. On flat ground, the logs were rolled up two pole skids extending from the ground to the log bunks on the vehicle. This laborious method has been obsolete for a long time.

Crosshaul

The **crosshaul** evolved from the hand rolling method. A rope was placed around the log in such a manner as to roll it when the free end of the rope was pulled. It was the earliest method used in loading railroad log cars with a steam donkey in the Douglas-fir region, where the method was termed a "parbuckle." The crosshaul is still used in some small operations lacking other loading equipment. The principle of the crosshaul is to roll the log onto the vehicle with a line from any available power source. As a portable method, two pole skids are laid at about a 45° angle from the ground to the vehicle bunks. Two lines or chains are anchored to the bunks, passed under and over the log to be loaded, and then hooked to a single line or chain leading across the truck to a power source on the opposite side of the vehicle. The power source may be animals, a tractor drawbar, or a winch. The one-drum winch used to power the cross-haul line might be on a tractor, a hoist on a sled, or on the truck. In the last case, the truck becomes a self-loading mobile unit. Pulling on the line rolls the log up the skids and onto the vehicle. The cross-haul is used for loading trucks, wheel or crawler wagons, and sleds. As the lines must be placed in turn around each log, or several small logs, it is a slow, low-production method. When used as a stationary loading method, logs to be loaded are skidded to a bank or platform at or above the elevation of the truck bunks.

Gin Pole and A-Frame Loading

The simplest stationary loading method which lifts the log, rather than rolling it, is the gin-pole or the A-frame using a single loading line. A **gin pole** is a pole or small tree erected in a leaning position so the load line lead block is centered over the vehicle to be loaded. Two back guys hold the gin-pole in position and, for safety, a snap guy is set opposite the back guys. The snap guy prevents the pole kicking back when the stress on the loading line is suddenly released. In small timber a standing tree is sometimes rigged as a gin-pole by topping and hanging the lead block and back guys, leaving the latter with the proper amount of slack. Then a shallow felling undercut is made and followed by a back cut, leaving as much holding wood as possible, until the tree leans. It is restrained from falling by the guys. The **A-frame** is made of two poles separated far enough at the base to give lateral stability with only one back guyline used and fastened together at the top with several wraps of wire rope or chain. The A-frame is often mounted on a timber sled carrying a small hoisting engine. For short moves the sled is pulled by a tractor. For longer moves the loader pulls itself onto a truck. The A-frame may also be mounted on the rear of a used flat-bed truck, with a winch powered from the truck drive line to provide a relatively inexpensive mobile loader.

The single loading line may be powered by a one-drum towing winch on a tractor, or the line may be hooked to the drawbar of a tractor and the log raised and lowered by shuttling the tractor forward and back. Usually **tongs** are used to grab the log. The tongs are set as close to the balance point of the log as possible. The single line and tongs system necessitates man-handling of the log to position it over the truck load. A safer method is to use a **crotchline** and

end hooks, but this requires an additional man on the loading crew. The two end hooks are each **shackled** to a line 1.2 to 1.4 times the half length of the logs to be loaded. The other ends of the two lines are shackled to the end of the single loading line to form the crotchline. Wire rope straps may be used instead of hooks. The strap need not be looped around the log, but merely placed against the end of the log. The horizontal component of force developed when the crotchline is tightened results in the strap biting into the end of the log enough to lift it. The end hooks usually have a short length of manila rope attached by which the men on the ground can guide the log into position on the load and pull out the hooks after the log is loaded. Gin-pole loading is adapted to small timber which is tractor-yarded to a landing. It is used by small operators who do not have the capital for investment in a separate loading machine. It is a low production method.

A vertical standing tree may be used instead of a leaning gin-pole by hanging the loading line lead block on a long strap. Pulling the loading line in a direction perpendicular to the road when the log is hoisted pulls the lead block away from the tree until it is centered over the truck.

Guyline Loading

Where a spar tree is used for yarding, the loading line lead block may be hung from a **jack** on a guyline centered over the road. The guy is anchored to the guy stump and to the spar tree by cables. The loading line runs up through the block, then to another block on the spar tree and down to a loading drum. In tropical countries where low yarder production can be handled by this loading method, provision for a loading drum on the yarder obviates the need for a separate loading machine. Either tongs or a crotchline is attached to the end of the loading line. For better control of the log two loading guys may be rigged. Two tongs are used and a two-drum loader is required. The loading procedure with the single-tong method is similar to that described under Gin Pole and A-Frame Loading.

Spar Tree Boom Loading

Two types of loading booms are used with spar trees when the modern crane loader is not available. The rigging of the **McLean**

boom, referred to by loggers as the "hayrack," is shown in Figure 9–1. The boom poles are a little longer than the logs to be loaded.

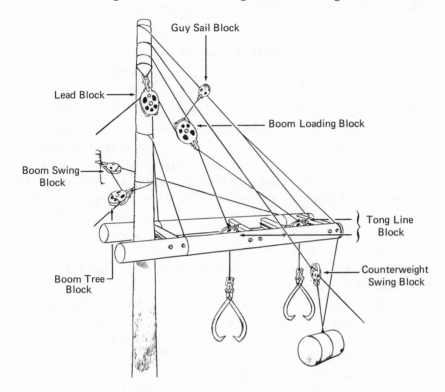

Fig. 9–1. McLean boom. (Courtesy of the Young Corporation.)

The block purchase afforded by the boom loading block doubles the lifting force and enables two tongs to be used with one loading line. A second drum powers the boom swing line. When the swing line is slackened, the boom is swung in the opposite direction by the counterweight, which is rigged at a safe distance to a block on a guyline. The boom gives more positive control of the log than do the guyline loading systems, and eliminates the hard work of carrying tongs out to the logs. The boom drops the tongs over the log and the tong man has only to spread the points.

The **heel boom** is rigged in a similar manner, except that only one tong is used, and the bottom of the boom is lined with steel rails. The log is tonged off-center and the end of the shorter segment of the log is heeled against the rails. The heel boom saves one man on the loading crew, as the McLean boom requires two tong men for efficient operation. The heel boom was widely used in the West Coast regions before the advent of the crane loader. The crane retains the principle of heeling the log against a boom.

MOBILE LOADERS

Truck-Mounted Loaders

The earliest mobile loaders used for loading log trucks were home-made. A hoist and a pole boom were mounted on an old truck which was no longer suitable for log hauling. The hoist was operated through a power take-off from the truck drive line, or by a separate engine. The boom was either an A-frame mounted on the rear end of the truck, or a swinging pole pivoting from or near the base of a vertical mast. The latter type is that used on the Idaho **jammer** which, in both home-made and factory-made versions, is popular in the northern Rocky Mountain region for cable skidding as well as loading. For loading with the A-frame the loader truck must be positioned perpendicular to the road, opposite the accumulation of logs to be loaded. A log is tonged or hooked and hoisted to clear the truck. The truck then backs under the log, which is lowered to its desired position. With single tongs, man-handling of the log to swing it into alinement with the truck is done by a man standing on the load. With end hooks, the swing of the log is controlled by two men on the ground by ropes attached to the hooks. Since the boom of the jammer pivots in a horizontal plane, the loader sits on the road, behind the truck, and picks up logs from either side of the road. The boom and mast can be mounted so that it is self-centering over the log truck when the log is hoisted.

Probably the first factory-made truck-mounted mobile loader to gain wide acceptance was the "Logger's Dream" manufactured by Taylor Machine Works, Louisville, Mississippi (Fig. 9–2). Since the larger models are designed for cable skidding as well as loading, the Logger's Dream was also a pioneer among "mobile loggers." It is used in both hardwood and pine regions, especially in the South, and in tropical countries. The unit consists of a hoist and a boom mounted on any truck of the customer's choice. The boom is made of two parallel lengths of tubular steel, strengthened with a truss on the upper side. The boom pivots in a vertical plane from the rear end of the truck and is guyed to the front end. The boom lengths range

Fig. 9–2. "Logger's Dream" loading pine logs with crotchline and end hooks in Alabama. Log is guided with ropes by the men on the ground. (Courtesy of the Taylor Machine Works.)

from 22 to 26 feet, depending upon the model. A two-drum hoist powered by an industrial engine, or a one-drum hoist powered from the truck engine, are mounted on the truck frame. The hoist operator's seat and control levers are between the hoist and the truck cab. The lines run through sheaves suspended from the top of the boom.

Tractor-Mounted Loaders

When crawler tractors became available, the proverbial ingenuity of the logger was exercised by converting an old skidding tractor to a mobile loader by mounting some type of boom on the tractor and using the winch drum to operate the loading line. Manufacturers of tractor accessories developed several types of loading attachments. The hydraulic knuckle boom and the forklift are today the most widely used tractor-mounted loaders. The hydraulic boom is described under self-loading trucks in this chapter.

Forklift Loaders

The newest log-loading device is the **forklift** mounted on the front end of a tractor. Some manufacturers term this attachment a grapple, but the general name forklift is used here to avoid confusion with the log grapple used with a heel boom crane. The principle of the well-known forklift truck widely used in industrial plants, warehouses, and on shipping docks has been adapted to log handling, both at the landing and in the log yard at the mill. Probably the first use of the forklift principle in the woods was in the Drott Skid-Loader mounted on an International crawler tractor which could embrace a bundle of pulpwood in its grapple arms, skid it to the roadside using the grapple as a sled, and lift it to load the bundle onto a truck. The development by the major tractor manufacturers of the earth-moving bucket loader gave an impetus to the forklift log loader. The same hydraulic mechanism is used to operate both, and bucket and fork are interchangeable. Figure 9–3 shows log

Fig. 9–3. The "Cat 980" Log Loader, manufactured by the Caterpillar Tractor Co. (Courtesy of the NC Machinery Co.)

Fig. 9–4. Loading Southern pine logs on a Taylor trailer with a Taylor Tree Loader. (Courtesy of the Taylor Machine Works.)

forks on a wheel tractor designed originally for earth-moving. Figure 9–4 shows a forklift loader designed especially for logging.

A typical fork lift consists of two straight tines projecting horizontally from vertical uprights attached to lift arms which are powered by hydraulic lift cylinders. The fork can be tilted forward or backward by tilt cylinders. An optional curved top clamp, with either one or two members, is raised and lowered by a third cylinder. In operation the tractor moves perpendicular to the axis of the log to be loaded, with the fork tines lowered to the ground, and slides the tines underneath the log near its center of gravity. Tilting the fork back cradles the log between the tines and the uprights. If a top clamp is used it is lowered to grip the log. The lift arms are raised to the desired height for loading and the tractor moves to the side of the truck. The top clamp, if used, is raised and the fork tilted forward until the log slides off the tines onto the truck load. The tractor backs away from the truck, turns to the next log to be loaded, and repeats the cycle. Another type, the Libu pivoting timber fork, pivots 90° to carry the logs longitudinally, thus reducing side sway in moving over uneven ground. To even the ends of a group of short logs from loading they can be moved forward to a vertical position and tapped against the ground.

Log fork attachments are available for all sizes of both crawler and wheel tractors. The crawler tractor is preferred on muddy landings. The development of the center pivot steering or articulated wheel tractor gives promise of increased use of fork lift log loading. The fork can be positioned by pivoting the front wheels through an arc from 60° to 90°, depending upon the make of tractor, without moving the tractor. Fork lift loading requires larger landings than crane loading for maneuvering space. For efficient operation the landing should be smooth and flat. The fork lift loader provides a versatile unit for the operator with limited equipment, as he can replace the log forks with an earthmoving bucket and use it for road construction.

Crane Loaders

The standard mobile **crane** or power shovel used in the construction industry is adapted to log loading by the attachment of a special loading boom. The products of all the well-known manufacturers of earthmoving and hoisting cranes are in use as log loaders. The popularity of the crane loader led to the development, by yarder manufacturers, of crane-type mobile loggers which can be used for cable yarding as well as loading.

Basically, the crane consists of a machinery deck mounted on an undercarriage and a boom. The machinery deck carries, from back to front, a power unit, a multiple-drum hoist unit, a travel and swing unit, and an operator's cab and controls. The power unit is usually a diesel engine driving a gear transmission through a hydraulic torque converter. The hoist unit has a main or loading line drum, a boom hoist or topping drum, and optional drums such as closing or haulback drum, and tagline or utility drum. The machinery deck sits on trucks with rollers which ride on a turntable with a large bull gear. A vertical swing pinion engages the bull gear and rotates the machinery deck full circle around a king pin. The self-propelled crane has a travel gear which drives crawler tracks or rubber-tired wheels. The crane turntable may also be mounted on a motor truck. The loading boom is attached to the front end of the machinery deck, and a removable counterweight at the rear end. Rubber-tired cranes are usually equipped with hydraulic outriggers which are extended to the ground during loading to take the weight off of the tires and increase lateral stability.

Cranes are rated for size by the maximum lift capacity in tons, or by the size of the earthmoving bucket it is designed to handle. Sizes used for log loading range from $\frac{1}{2}$ or $\frac{3}{4}$ yard for small timber to 2 or $2\frac{1}{2}$ yard for large timber. One crane manufacturer offers a range of sizes of 15 to 45 tons capacity in self-propelled rubber-tired models, and up to 60 tons truck-mounted, for log loading.

Crane Loading Booms. The most widely used type of log loading boom used on cranes or converted shovels is the heel boom of either the gooseneck (Fig. 9–5) or the hinge type (Fig. 9–6). A typical

Fig. 9–5. Tong loading with Albin gooseneck heel boom mounted on a 40-ton-capacity Lima crane. (Courtesy of the Albin Manufacturing Company.)

gooseneck boom used with a 1- or $1\frac{1}{2}$-yard power shovel is built in two segments. The lower segment, 9 feet in length, bends at an angle of 52° with the 16-foot-long upper heeling segment. The lower segment is hinged to the front of the crane. The upper side of the gooseneck carries a gantry with sheaves for the boom hoist line which raises or lowers the boom. The outer end of the upper segment carries a swiveling fairlead with one or two sheaves, de-

Fig. 9–6. Washington TL-6 Track Loader with hinge boom and grapple, mounted on self-propelled rubber-tired carrier, in Alaska. (Courtesy of the Washington Iron Works.)

pending upon whether tongs or grapple legs are used. In the example cited the length from the fairlead to the gooseneck is 19 or 20 feet and the heeling height 13 to 20 feet. In operation the log is grasped by tongs or grapple between the near end and the center of gravity, and the end heeled against the under side of the upper segment of the boom. The log will slide back no farther than the gooseneck bend.

The hinge type heel boom is straight, and is hinged to a vertical support mounted on the front end of the crane. The boom hoist gantry is an extension of the vertical support. A cab guard projects on each side of the support to stop the log should it slide along the heeling portion of the boom. On a typical hinge-type boom for a 1-yard machine, the hinge is 13 feet above the ground and the distance from fairlead to hinge is 27 feet when the boom is horizontal and 19½ feet when the boom is raised to a 45° angle. For a given size of crane, the hinge type boom has greater lifting capacity than the goose neck as the log may be heeled closer to the machinery deck.

Where air or hydraulic tongs are used, the conventional boom and dipper stick of the power shovel, with the addition of a toothed bar on the underside of the dipper stick for heeling the log, is retained (Fig. 9–7). The tongs replace the shovel bucket at the end of the stick. For loading with crotchline and end hooks in the western pine regions, the crane is equipped with a long lattice-type boom, similar to the construction hoist. The log does not touch the boom.

Loading Tongs. Conventional log loading tongs are made of two forged steel curved legs, hinged near the upper end, with clevises for attachment of the loading line, and sharp points on the inner side of the lower ends of the legs. They engage the log by the bite of the points penetrating the bark and sapwood. To set the tongs on a log, the legs must be separated manually by a man on the ground and positioned on the log. The log is lifted, swung, and lowered onto the truck. The tongs are released by a man on the truck by pulling out one point and swinging the leg parallel to the axis of the log to release the other point. The top loader customarily stands on a platform on top of the cab of the log truck and signals the truck driver to back or go ahead. Handling the larger size tongs required in big timber is hard physical labor. If a secure bite is not obtained, or the sapwood is not sound, the tongs may pull out or tear a slab off the log with consequent hazard to the loading crew. They also leave holes in the log which causes waste in peeling veneer. Tong holes are not a problem in a sawmill as they do not extend deeper than the slab. Logs too large in diameter for the tongs are lifted by placing a wire rope strap around the log and hooking the two eye splices in the ends of the strap over the tong points. Tong loading is illustrated in Fig. 9–5.

Power Tongs. The Berger-Rees **air tongs** were introduced in 1949. They are attached to the front end of the **dipper stick** on a power shovel. The tongs are opened and closed by an air cylinder mounted between the tong legs. They grasp the log by closing the tong legs by air pressure. The air tongs are curved slightly to engage a portion of the circumference of the log and have no sharp points to penetrate the wood. The tong air cylinder is controlled by a valve in the operator's cab. Thus, manual handling of the tongs on the ground or on the truck is eliminated. The shovel operator performs all the operating functions involved in handling the log while loading. Since the dipper stick can be moved forward or backward to position the log on the load, shuttling the truck back and forth during loading is obviated. For this reason the dipper-stick power shovel is well adapted to preloading trailers. In 1957 Berger hydraulic tongs were introduced. They operate in the same manner as the air tongs, but use an oil hydraulic cylinder to open and close the tongs. Two types of Berger hydraulic tongs are now manufactured. Type "A" tongs have the cylinder mounted horizontally above the tong legs. Type "C" tongs have a vertical cylinder enclosed in a triangular-shaped housing (Fig. 9–7). For decking logs in the mill yard, hydraulic tongs are used in pairs at the ends of a spreader bar, which is suspended from the lattice boom of a crane.

Fig. 9–7. Berger hydraulic tongs, on the end of a power shovel dipper stick. (Courtesy of the Smith-Berger Manufacturing Corporation.)

The advantages of loading with a power shovel with a dipper stick may also be obtained by attaching conventional sharp-pointed tongs to the stick. Pushing the tongs against the log to be loaded opens them. A line attached to one of the tong legs pulls the point out and turns the tongs to release the other point.

Grapples. The labor-saving and safety features of the air tongs led other manufacturers of logging equipment to develop competitive substitutes for the conventional tongs for use with the heel boom crane loader. In 1953 the ESCO **grapple** was introduced. The sheave-type grapple opens the legs to engage the log by a line running through a sheave at the upper end of one leg, above the pivot point, thence through a sheave in a framework above the grapple and back to a dead end on the upper end of the other leg. Reeling in on the line pulls the two upper ends of the legs together and the lower ends apart. The grapple cradles the log, lifting it from below the center line. The grapple and log are lifted by a hoist line clevised to the top of the frame, imparting a scissors-like action to grip the log. Often a third or tag line is attached to the side of the grapple to pull it back. Releasing the tag line suddenly casts the grapple beyond the reach of the boom. The height of the grapple frame necessitates a jib or elevated extension on the end of the heel boom in order to raise the grapple to the elevation of the heeling portion of the boom. Other makes of sheave-type grapples widely used in the western regions are the Albin-Ritz (Fig. 9–8), the Young, and the Mar "Grapple-tong." The use of the grapple has been stimulated by the demands of some plywood plants that their log suppliers use grapples for loading, to obviate holes in the logs made by con-

Fig. 9–8. Loading with a Ritz grapple and Albin gooseneck boom. Crane is a 45-ton Northwest mounted on a Pierce rubber-tired carrier. (Courtesy of the Albin Manufacturing Company.)

ventional tong points. Sheave-type grapples are also classified by the number of legs, as 2-point, 3-point, and 4-point. The 4-point is mainly used for decking logs, where it is desired to pick up several logs at a time.

Another type of grapple, the Washington-Priest "Tongrapple" was introduced in 1961. Sheaves are eliminated by the use of levers. The Tongrapple is opened by a line to a bar, which is connected by lever arms to the tong legs below the pivot. The hoist line is clevised to levers connecting the upper ends of the legs above the pivot. One drum operation of both lines is permitted by the use of a companion hydraulic "Line-shortener" which keeps tension on the opening line and avoids slack when the log is hoisted. A divider on the drum separates the two lines.

In the western pine regions the grapple is also used for loading with a conventional construction crane with a lattice boom up to 60 feet in length. The log is not heeled.

Crotchline and End Hooks. Loading with a crotchline with end hooks is a commonly used method in the western pine regions. The loading machine is a mobile crane, usually mounted on rubber tires, with a lattice boom generally 40 or 45 feet in length. Two crotch-lines, each about 23 feet in length for loading the standard 32-foot log, are clevised to the end of the loading line. At the end of each crotch line is an L-shaped end hook, with a spike point and cup pointing up from the lower arm of the L. A hand line of manila rope is attached to the hook. While the loading line is being slackened, the two loaders pull the end hooks out to the log to be loaded. The point of the hook is held against the end of the log. Tightening the loading line pulls the spikes into the wood. As the log is hoisted it is guided into position on the truck by manipulating the hand lines. The end hook is released from the log by a jerk on the hand line. This is a fast system for timber of the size logged in the pine regions, but requires an additional man on the loading crew.

Self-Loading Trucks

A number of manufacturers offer hydraulic loaders for permanent mounting on the log-hauling truck. A turnpost is mounted at the rear of the truck cab, and a loader operator's seat and controls on top of the cab. The loader is powered by the truck engine through a power take-off. A boom on top of the turnpost swings through an arc of 180° to 320°, depending upon the make of loader. Two types of booms are available. The straight boom uses a loading line running from a winch drum at the base of the turnpost through sheaves in the turnpost and boom to a fairlead sheave at the end of the boom. The boom is raised, usually through an angle of 75° from the horizontal by a hydraulic cylinder. A hydraulic outrigger on each side of the truck in the plane of the turnpost gives stability during loading and relieves truck springs and tires. A hydraulic grapple at the end of the loading line permits one-man operation.

An example of the newer and more widely used type, the hydraulic **knuckle-boom**, is illustrated in Fig. 9–9. The 4-axle self-

Fig. 9–9. A self-loading truck with a Ramey hydraulic loader in a salvage logging operation in Oregon. (Courtesy of Ramey Self Loaders, Incorporated.)

loading truck is loading logs in a salvage operation in Oregon. The way in which the knuckle-boom imitates the action of the human arm is more clearly shown in Fig. 9–10. Although the boom in this illustration is mounted on a tractor, similar booms are used on self-loading trucks. The boom arm has two segments of approximately equal length. The part of the boom bolted to the tractor is the

Fig. 9–10. Prentice hydraulic knuckle-boom loader mounted on International crawler tractor lifting two hardwood logs with hydraulic grapple. Lake States region. (Courtesy of Prentice Hydraulics, Incorporated.)

turnpost; and the bank of levers, which is partially obscured by the operator, is the hydraulic controls. The attachment of the first segment to the turnpost corresponds to the shoulder. It swings in a horizontal plane and raises in a vertical plane by a hydraulic cylinder mounted underneath the boom. The junction of the second segment with the first corresponds to the elbow. A hydraulic cylinder on top of the first segment pivots the second segment downward in a vertical plane. A hydraulic grapple with a self-contained rotor attached to the end of the second segment corresponds to the wrist and fingers. The boom can be lowered to pick up logs below the elevation of the road and raised as high as necessary to load the peaker log. This hydraulic loading unit may be mounted on a tractor or on a tractor-drawn trailer for transporting logs to the truck road. There the loader reloads the logs from the trailer onto a truck.

A log heeling attachment is available for some makes of hydraulic knuckle boom loaders. The log heeler is a toothed bar extending below the upper segment of the boom. The log is grasped by the grapple short of the center of balance and raised against the heeler. The heeler obviates finding the center of balance of the log, and is said to give more positive control of the log than is possible without the heeler. The Hy-Hoe Logger offers a hydraulic heeler which is attached to the end of the boom and retracts against the boom when handling short logs or "tops," or extends through a 90° arc by a hydraulic cylinder. The Hy-Hoe Logger is mounted on a rubber-tired carrier.

An older type of self-loading truck is equipped with two side arms which are raised by lines from a winch on the truck to toss the log up onto the truck bunks. One make of this type is aptly named the Timber-Tosser. The logs are rolled onto the arms from a pile by hand. It is a slow method of loading compared with the hydraulic loader, but the capital investment required is much smaller.

The self-loading truck is popular with the operator whose production does not justify investment in a separate loader. The truck driver can load without help; and, if truck unloading facilities are lacking, he can unload the truck at mill or log yard, and stack the logs. He can also reload into a gondola railroad car. The self-loading truck is especially well adapted to logging scattered patches of timber too small to justify moving in a separate loading machine. The self-loading truck is best suited to loading short logs, usually 16 or 18 feet in length, although small-diameter logs or poles of longer length can be handled. Self-loaders are used on 2- or 3-axle trucks without trailers, or on 2-axle trucks with folding reach "hop-on" semi-trailers. A full trailer may be loaded alongside the truck and coupled to the truck for hauling. The disadvantage of the self-loader is that it reduces the payload of the truck. For example, a hydraulic knuckle-boom self-loader with a rated maximum lift of 10,000 pounds weighs 3,000 to 3,500 pounds, and one with a lift of 16,000 pounds weighs 7,000 pounds.

THE LOADING OPERATION

The customary sequence of loading a truck tractor and semi-trailer with a heel boom crane in the West Coast regions is as follows: The tractor, with the semi-trailer loaded on the tractor bunk, backs to the landing. The member of the loading crew on the ground hooks a wire rope strap on the trailer to the tongs or grapple. The trailer is hoisted and the truck moves ahead of the trailer reach. The trailer is lowered to the ground and the reach hooked to the tractor. The driver connects the trailer air and water hoses, while the loader adjusts bunk chock blocks or stakes. The driver returns to his cab where he remains to shuttle the truck back and forth so the logs will be properly positioned lengthwise. He is usually directed by the loader standing on a platform on top of the cab, by voice signals, stomping on the platform, or a "peanut" whistle, a small air whistle. The bunk tier of logs long enough to span the distance between bunks is first loaded. Then successive tiers are added, and finally the peaker log. Then the truck driver puts on the binder chains. In some operations the truck moves clear of the landing before putting on the binders.

The crew organization varies with the loading system. For loading with tongs, a "second loader" on the ground sets the tongs on the logs selected by the "head loader," who stands on the truck cab platform. When the log is loaded, the head loader walks along the log and releases the tongs. Communication with the crane operator is by hand signals. Sometimes a "third loader" is employed to mark the ends of the loaded logs with a branding hammer or paint spots, or both. The brand indicates the ownership of the logs and the property from which they were logged. The third loader also chops off the occasional limb which survived yarding. For grapple loading the crane operator assumes the duties of a head loader in selecting the logs to be loaded. A second loader is employed to perform the other tasks incident to loading.

The legal load limitations on highway trucks govern not only the total load but the load distribution between tractor and trailer, and between wheels on both sides. It requires experienced judgement on the part of the man who selects the logs to be loaded in turn to obtain a safe, well-balanced payload without underloading and without exceeding legal limits. The tiers of logs above the stakes and the peaker must be well "saddled" to avoid rolling or shifting. The safety of the trucking operation depends upon the skill of the loading crew, as well as that of the truck driver.

UNLOADING LOGS

Where the truck load of logs is to be dumped into water, at mill pond or booming and rafting ground, the most commonly used unloading system is an adaptation of the crosshaul or parbuckle loading system. The ends of two wire rope straps are fastened to a brow log, the top of which is level with the truck bunks. The straps are passed under the logs, above the reach pole of the trailer, and hooked to the unloading line. The unloading line runs through a block at the apex of an A-frame, in a vertical plane with the brow log, to a winch drum which is usually powered with an electric motor. The straps are tightened against the load. Then the binders are removed. State safety codes forbid removing the binders before the unloading straps are in position and taut. If the load is to be strapped in a bundle with steel bands, this operation is performed after the unloading straps are taut and before the binders are removed. Further winching in on the unloading line rolls the load off the truck and over the brow log into the water.

Where a large volume of logs is to be handled, the **bridge crane,** similar to the bridge crane used in factories and sawmill dry storage sheds, is the most efficient means of unloading logs into water. The bridge extends from a support on the landward side of the truck runway to a piling structure in the water, leaving water space for a load of logs. The crane runs on steel rails on the bridge. To unload, the crane is centered above the truck, and the entire load is lifted by two wire rope straps on a spreader bar or by hydraulic grapples or tongs. The crane then moves out over the water and lowers the load until the tension on the straps is relieved by the buoyancy of the logs in the water and the automatic hooks joining the straps release. The same principle of raising and lowering the entire truck load of logs is sometimes used with a heavy A-frame which projects over the water. Lowering of logs has the advantage over dumping of avoiding breakage and reducing the accumulation of bark and debris which must be dredged periodically from a shallow dump. Strapping of bundles of logs prevents loss of sinkers, and enables a greater volume of logs to be stored in a given water space.

Railroad cars equipped with log bunks are generally unloaded into water by a mobile unloading machine running on a parallel track. The logs are pushed off the car by a hydraulically powered arm which pivots from an A-frame extending from the unloader. The "jillpoke" unloaders of the steam donkey and railroad logging days which pushed the load from the car as the train moved, are now obsolete. Gondola cars, which have side and end walls but no roof, must be unloaded by lifting out the logs. This is usually done by a crane with a grapple.

The entire operation of lifting truck or rail car loads and lowering them into the water can be performed by one man with the hydraulic tong. The ESCO hydraulic tong is illustrated as an example (Fig. 9–11). It is offered in capacities of 30 tons for unloading highway trucks, and 50 and 60 tons for off-highway trucks and rail cars.

Fig. 9–11. Unloading a truckload of logs with an Esco hydraulic tong operated from an Ederer bridge crane. (Courtesy of the Esco Corporation.)

The four L-shaped arms completely enclose the load so that un-bundled as well as strapped bundles of logs may be handled. The full operating cycle, from open to closed to open, is approximately 20 seconds. An 85-ton-capacity ESCO hydraulic log grapple is oper-ated from a crane mounted on a barge to provide self-loading and unloading of large bundles of logs in Alaska.

An aerial photograph of a log dump and booming ground at tidewater is shown in Fig. 9–12. The logs are brought to the dump

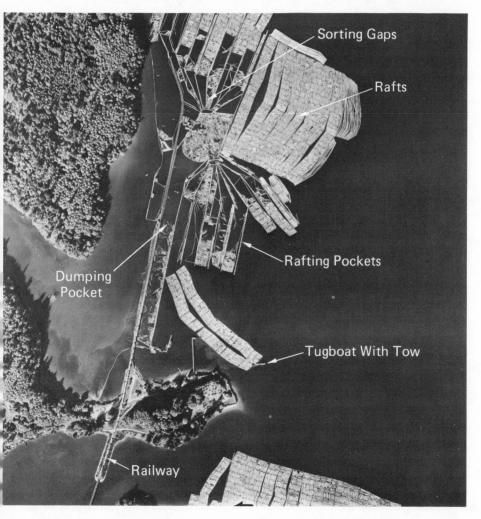

Fig. 9–12.   Aerial view of a log dump and booming ground. (Courtesy of Carl M. Berry, Photogrammetric Engineers.)

by a railway which runs from the bottom of the picture to near the top. The rail cars are dumped by an unloader traveling on a parallel track into the long rectangular pocket adjoining the railway. The logs are pushed by a boom boat to the sorting gaps, which appear as triangular-shaped areas in the photograph. The logs are sorted by species and grade and poled into the rectangular pockets where they are made up into flat rafts. Completed rafts, ready for towing to the mills, are seen to the right of the pockets, and at the lower edge of the picture. A tugboat has two rafts in tow between the peninsula at lower left and the photo center. This dump was designed to handle about one million board feet of logs a day.

Stackers

The development of the log **stacker** which can lift an entire truck load of logs has revolutionized unloading and log storage at the mill or plant. The stacker is similar in principle to the forklift tractor with the modifications necessary to lift much heavier loads. The stacker has two straight forks to carry the load and two curved arms or tusks to clamp the logs from the top. Some stackers have push arms or kickers to push the logs off the forks. Some users prefer kickers; others find tilting the forks downward sufficient to discharge the load. The pioneer stacker was the R. G. LeTourneau electric wheel log stacker, which was first placed in service in 1953. The machine has individual electric motors in each drive wheel, and each lift component is powered by an electric motor. It is offered in 2-wheel drive or 4-wheel drive. A diesel-powered generator furnishes the electric current. The LeTourneau stacker is manufactured in seven models with lift capacities ranging from 20 to 50 tons. The Wagner Lumber Jack, with conventional automotive gear drive and hydraulically operated lift components, entered the stacker market in 1959. The FWD-Wagner Lumber Jack is available in models with lift capacities of 20 to 65 tons. The Kenworth stacker is offered in three models with lifts of $17\frac{1}{2}$, 40 and 60 tons. The Kenworth stacker is mounted on a tricycle wheel configuration for maneuverability and three-point stability. All of the well-known makes of forklift wheel tractors are used for sorting and stacking, and for unloading in the smaller operations where one machine must perform all the operating functions in the log yard. If the capacity is insufficient to lift the entire truck load, the load is "split" to remove a few logs at a time. The stacker unloads by lifting the entire log load from truck or rail car. The 30-ton-lift stacker is designed to unload legal highway truck loads, and the larger stackers to unload off-highway trucks or rail cars. At one large pine sawmill the stacker reduced unloading time from the 20 minutes required for dumping into the log pond to less than 2 minutes. Unloading with a stacker saves the log breakage experienced in dumping. Scaling and grading logs on the ground can be done more quickly than on water. Most of the exterior of the log can be seen for grading, without rolling the log over as must be done in water.

Since stackers became available, many sawmills, pulpmills, and plywood plants have abandoned log ponds and converted to dry land handling and storage. In addition to unloading, the stacker spreads the load on the ground for scaling and grading, sorts the logs by species and grades, carries them to storage piles, and transports logs from storage to the mill in-feed deck in accordance with the mill's requirements. Although the capital investment in stackers and log yard is large, the labor saving from reduction in the number of men required in water handling has reduced the cost per unit of volume. Concern to avoid water pollution from log dumps has also had an influence.

The stacker is also used by logging operators in sorting yards. As an example, one logging company which sells logs to a sawmill, a pulpmill, and a plywood plant, and for export, first hauls with off-highway trucks to a sorting yard at the terminus of their private road. The logs are sorted by market destination, stacked in bins holding a highway truck load, and reloaded by the stacker for the highway haul to market. Another example is found in a paper company tree farm. Tree lengths from second growth thinnings are trucked to a sorting yard adjacent to salt water. The trucks are unloaded by a stacker and spread for marking. The tree lengths are bucked into logs for maximum value, and scaled. They are sorted by a forklift tractor and deposited in bins, where the bundles are strapped with steel straps. The bundles are carried by the stacker to the water. There the bundles are pushed by a boom boat into rafts. The rafts are towed by tugboat to market destination.

Log Yard Operation. Following is typical log yard operating procedure: when a truck load of logs arrives at the yard, the stacker moves to the side of the load, inserts the straight fork tines under the load and lowers the curved top clamp or tusks. With the load securely held by the stacker, the truck driver can safely remove the binders. The stacker lifts the load to clear the bunk stakes, and backs away from the truck (Fig. 9–13). It then proceeds to a scaling

Fig. 9–13. Unloading a truckload of ponderosa pine logs with 30-ton-lift FWD-Wagner Lumber Jack. (Courtesy of the Nelson Equipment Company.)

bay and spreads the logs out in one tier. A metal tag bearing a serial number is tacked to the end of each log. This number is recorded in the scale book together with gross and net scale, and these data are later recorded on a computer punch card. Where sorts by grade are to be made, the side of the log is grade-marked in paint with a spray gun. While tagging and scaling are being done, the stacker works at sorting or stacking elsewhere. The stacker returns to the scaling bay, sorts the logs by pulling all the logs of one grade and species onto the fork tines with end of the tusks (Fig. 9–14). When a stacker load is accumulated, it is transported to the storage area and stacked or decked. The place where a row of logs is decked is variously termed an aisle, a bay, or a raceway. While the larger stackers can deck logs up to a height of 20 to 24 feet, such height is feasible only for large-diameter logs. Small logs are usually decked up to a height of about 12 or 13 feet. In yards for logs of 32 to 40 feet length, pole skids or stringers are used under the decks and in scaling bays. Where the logs handled include some

Fig. 9–14. Picking up logs from the sorting bay with a Kenworth Stacker. (Courtesy of the Totem Equipment Company.)

short logs 8 to 12 feet in length, they are sorted and stacked on the ground without the use of skids. When the stacker driver is informed of the species and grade which the mill or plant wants, he goes to the deck, scoops up a load, and carries it to the mill in-feed. The metal tag is removed and the number recorded, and the mill log intake data sent to the computer. Thus, a continuous inventory of the logs in the decks is maintained.

In most log yards a smaller stacker or a forklift wheel tractor is used to supplement the large stacker which is required for unloading an entire truck load of logs at one lift. The smaller machine sorts and decks, while the large stacker unloads and feeds the mill. In some log yards the stackers are equipped with two-way radio. The log requirements of the mill are communicated to the stacker driver by radio from the mill log deck. Colored signal lights are used by one mill as another means of communication.

The size of the log yard varies with annual log input, the volume to be stored in decks, the height of the decks, and the number of sorts. For efficient operation of the yard, ample space must be provided for maneuvering the stacker. For each 10 million board feet of annual input, 0.4 acre working area and 10 acres storage area for decks 12 feet in height, and 6 acres for 20-foot decks are generally required. Careful planning of the layout of the log yard is essential to efficient operation. Log yard runways are usually surfaced with a wearing course of gravel, crushed rock, or asphalt. The heavy wheel loads of stackers necessitate a good base course of gravel, the required thickness depending upon the bearing strength of the underlying soil. As the yard is nearly level, a good drainage system is important. At some tidewater mills a thick layer of bark is used satisfactorily as a log yard surfacing material on soft ground.

During the dry season the log decks are sprayed with water to inhibit insects and checking and to abate fire hazards. One system uses agricultural sprinkler heads spaced 40 feet apart connected by aluminum pipe laid along the top of the deck. A mist system developed by the St. Regis Paper Company, Tacoma, Washington, uses plastic heads between the decks at a spacing of 36 feet, which spray a fine mist. The mist system is reported to produce higher relative humidity and lower air temperature than the sprinkler system, and to wet the ends of the logs better. Less water drips down to the aisle surface along which the stacker operates. The stacker operator experiences no difficulty in driving through the mist.

A representative log yard which supplies a plywood plant and a sawmill in the Douglas-fir region is shown in Fig. 9–15. When the aerial photograph was taken, the 41-acre yard contained a log inventory of 18 million board feet. The average daily log input was 752,000 board feet in 65 truck loads. The gravel surface of the yard appears as a gray tone in the photograph and the heavy black lines are oiled roads. The loaded log truck enters at upper right and proceeds along the right-hand oiled road to one of the 17 sorting bays bordering the road. The logs are lifted off the truck by a 65-ton lift LeTourneau and spread out in one layer in a bay. The type of log is paint-marked and the logs scaled for volume. Sorting is done by two Caterpillar 966B wheel tractors. The logs of a given type are picked up with the forks and deposited in one of the 15 bins which lie in a curved row paralleling the row of bays. Each bin contains a different type of log. When a fork load has accumulated in a bin, it is picked up by one of the two 40-ton lift Kenworth stackers, carried to one of the 58 aisles, as the straight rows of decked logs are termed, and stacked. Any type which coincides with the type being run through the debarker at the time is transported from the bin direct to the debarker, which is located at bottom center in the photograph, over the log pond. The debarked logs

Fig. 9–15. Aerial photograph of a 41-acre log yard with 18 million board feet of logs decked. (Courtesy of the Weyerhaeuser Company.)

drop into the water, and are moved to the jackladder of the plant or mill. The Kenworth stackers keep the debarker supplied with logs from the decks.

In some log yards the "layout" system of sorting is used instead of the "bay-bin" system. The truck loads of logs are spread out on the ground in a continuous row. The sorts by log type are piled on the ground in a herringbone pattern on each side of the row. A time study of both systems in one log yard showed a 30 percent saving in sorting time with the layout system. The distance the markers and scalers walked in a day was reduced 60 percent. The layout system was advantageous to the truck drivers as they drove to the unloading machine instead of looking for an empty bay.

RELOADS AND TRANSFERS

Logs to be transported by common carrier railroad are trucked to a place at a railroad siding termed a reload or a transfer. At the reload the logs are unloaded from the truck to the ground and later reloaded, singly or in sorted groups, on the rail car. At the transfer the entire load is hoisted from the truck and lowered intact on the car.

Reloading is customary where the log trucks haul on public roads or highways, as the truck bunk width is limited to 8 feet and the rail car bunks are wider. The load capacity of the rail car is greater than that of the highway truck. For example, the staked log cars furnished by the Northern Pacific Railway are 8 feet $5\frac{1}{4}$ inches width between the stakes in the western pine region, and 9 feet $6\frac{3}{4}$ inches in the Douglas-fir region, and the load capacity of each is 80,000 lbs. When gondola cars are 8 feet inside width, the load capacity is 100,000 lbs. The reload is often combined with a log sorting yard and log stackers used for unloading, sorting, and loading bundles of one species and grade on the cars. Where sorting is not done at the reload, the cars are usually loaded by conventional cranes.

Where railhead can be reached by a private road, and off-highway trucks of the same bunk width and load capacity as the rail car can be used, the stacker lifts the load from the truck, which stops beside an empty car, and, when the truck has been driven out of the way, moves forward to lower the load onto the car. The capa-

Fig. 9–16. Unloading a railroad car with an FWD-Wagner Lumber Jack. Whiting car switching machine at the right. (Courtesy of the Nelson Equipment Company.)

bility of the stacker to load or unload a staked rail car is illustrated in Fig. 9–16.

Cable transfers powered by yarders were used before stackers were available. The tightline system, with a spreader bar suspended between two spar trees, was the method of rigging commonly used. Two steel shafts suspended from the spreader bar at balance points were guided under the truck load to function as bunks while the truck load was being hoisted, moved laterally, and lowered onto the rail car. As each empty car had to be spotted under the spreader bar, movement of the train during loading was controlled by a spotting line from the yarder. The loading spur was located on a slight grade so the train would roll by gravity when the brake on the spotting line drum was released. The bridge crane is well adapted to use as a transfer, as well as an unloader (Fig. 9–11).

SUGGESTED SUPPLEMENTARY READING

1. Volumes of the *Loggers Handbook* published annually by the Pacific Logging Congress, Portland, Oregon. New developments in logging technology appear each year. Also articles on logging management and the proceedings of the congress session of the previous year.
2. Volumes of the official proceedings of the Oregon Logging Conference. Published annually by the Oregon Logging Conference, Eugene, Oregon. New developments in logging technology in both Douglas-fir and ponderosa-pine regions are discussed each year.

10

Log Transportation

This chapter concerns itself with the movement of logs from landings, yards, or rollways to a location, which generally is adjacent to a mill, where logs are stored prior to being subjected to some form of conversion. Normally this subsequent log movement is made over a much greater distance and in most cases at greater speed than the initial log movement, which involves some form of tractive or cable skidding.

Historically, the earliest method of log transportation made use of water as a means of transporting logs.

Forest railroads date back to 1852, when they were used to transport white pine logs in upper New York State. Since that time and up to well into the first quarter of the present century, forest railroads were the backbone of the logging industry. Initially, pole roads were used to engage the flanged wheels of the log cars which were drawn by animals and later by steam locomotives. The train or stringer railroads next were developed, where the rails, first wooden and then steel, were secured to cross ties comparable to the method used on the common carrier railroads. The final development, occurring in 1885, was the geared locomotive capable of negotiating 40-degree curves and 6 percent adverse grades. Two distinct types of geared locomotives were available—the center-shaft and side-shaft. The Climax and Heisler were center-shaft types and the Shay the side-shaft type. The center-shaft type transmitted power to the driving wheels by means of an articulated shaft and gearing arrangement positioned under the boiler. The side-shaft type transmitted power through a side-mounted sectional shaft which was geared to the right hand driving wheels. The individual sections of the shaft were connected by universal joints. This feature together with two slip joints allowed the shaft to lengthen or shorten when negotiating curves.

The motor truck entered the logging industry in the Douglas-fir region in 1913, but did not supplant the logging railroad until the nineteen-thirties. Since its introduction to the logging industry, the truck has all but eliminated other means of log transportation.

While natural waterways still are used to transport logs, it is the motor truck which hauls logs from landings to log storage areas adjacent to water or log dumps for subsequent water transport. In turn, the motor truck has made it possible for the common carrier railroads to become an integral part of many log transportation systems by transporting logs to loading points or sidings.

MOTOR TRUCKS AND TRAILERS

In general, there are two general categories of trucks used for hauling logs. On one hand, there is the small truck, which never was intended to serve as a log truck. Included in this category are the standard or production line vehicles, which usually are fitted with a flat bed and the necessary accessories such as bunks and stakes. Normally, these vehicles may be lacking in power, braking capacity, proper framing or inadequate spring assembly to name a few of the possible areas of inadequate design for log trucks. The so-called small truck might haul a reasonable load of logs but the fact remains that it was not designed for that specific purpose. On the other hand, there is the vehicle which has the combination of components to supply the strength and power necessary to reduce the possibility of failure to a minimum. Trucks which are designed or engineered for a specific task tend to show a better unit of volume-per-mile cost than when little or no consideration is given to the selection of vehicle components.

It is acknowledged that many small logging contractors or gyppos do not have operations of a size which would allow them to acquire a hauling unit which would be ideal for their particular operating conditions. Of necessity, they must acquire a truck which is within their financial means even though it is not the most practical or economical hauling unit for their use.

Production Line

The production line gasoline-powered truck is widely used throughout most of the logging regions as a means of transporting logs. The basic unit consists of a two-axle truck, with the conventional two-axle arrangement, one steering axle and one driving axle with dual wheels and is generally used for small and short length logs. Figure 10–1 shows a truck of this type.

Fig. 10–1. Production line truck with flat bed, loaded with hardwood sawlogs, Pacific Northwest. (Courtesy of the College of Forest Resources, University of Washington.)

Tandem rear axles, 4- and 6-wheel drives, braking improvements, and sturdier over-all construction have provided the logger with a relatively light-weight truck, capable of hauling loads of 2,000 to 3,000 board feet, depending upon the species being hauled.

Tandem rear axles increase the load-carrying capacity of trucks. One type is merely mounted on helper springs behind the driving axle and come into use when the truck settles under the load. Tandem axles which transmit power to all four sets of rear wheels (4-wheel drive) have definite advantages in that tractive power is more evenly distributed as well as weight. When the latter type is incorporated in a hauling unit usually the truck frame is strengthened and stronger springs installed. These units are capable of hauling nearly double the rated capacity of the truck in a safe manner (Simmons, 1951). Figure 10–2 shows a 4-wheel drive truck.

Fig. 10–2. Four-wheel drive, three-axle log truck. Log bunks fitted with cheese blocks. (Courtesy of the Swanson Bros. Lumber Co.)

A three-axle truck with axles all under power (6-wheel drive) has definite advantages under those conditions where traction may be limited; such as on a landing or yard during rainy periods, on stretches of haul road having excessive adverse grade, on haul roads with little if any provision for drainage, and in those localities where surface rock is unavailable. The all-wheel drive may be the difference between being able to haul logs or shutting down the operation.

Custom-Built

In contrast to the production line type of log truck, the custom-built vehicle consists of a particular combination of truck components capable of performing under a given set of operating conditions. The custom truck is designed and built to provide a vehicle which will perform accordingly.

The custom-built truck may be classed into on-highway and off-highway trucks. The on-highway truck would be used in the case where logs must be hauled over a network of roads which could include at least four classes: spurs, secondary, mainline and public highways. The off-highway truck, as the name implies, would be used in those cases where it is not necessary to use the public highways. The entire haul from the woods to a mill or dump would be over private roads where public travel is either restricted or eliminated completely. While there are no weight, width, length and height restrictions on an off-highway haul, the on-highway haul must be carried on in keeping with the motor vehicle restrictions as set forth in the motor vehicle laws of the state in which logs are being hauled.

The off-highway truck must have the ability to haul large loads to justify the cost of the vehicle and the additional road construction cost of a pavement structure necessary to support loads much in excess of conventional on-highway loading. Certain large off-highway log hauling combinations would be in excess of legal axle loading when in an unloaded condition. Generally, these large combinations are used in the Douglas-fir and redwood regions.

The largest log hauler in existence has the ability to transport payloads up to 60 tons over rough unimproved roads. This unit has definite advantages in those areas which have heavy rainfall and no ballast or rock is available for road surfacing—a combination of conditions which exists throughout much of the tropical forest regions. Each 6-foot wheel has its own individual electric motor and reduction gear built within its rim. A diesel generator provides each wheel with the energy necessary to move heavy loads over difficult terrain. The unit is definitely an off-highway hauler. It is oversize in all respects and weighs 61,000 pounds when in an unloaded condition. Payloads of from 10,000 to 12,000 board feet have been reported by an owner logging in British Columbia, Canada.

Savings in fuel tax and license fees and a reduction in hauling cost per unit of volume are items which point to the advantage of an off-highway log-hauling operation.

ACCESSORIES

The basic truck must be equipped with certain items before it can be used to haul logs. The vehicle may be used as a single self-propelled motor truck carrying its load on its own wheels. Should the vehicle have the capability it may also be used in combination with a pole trailer having single or tandem axles or two-axle full trailer.

Log Bunks and Stakes

Trucks with an axle arrangement as shown in Fig. 10–1 which are used solely for hauling logs must be fitted with log bunks. Usually two wooden or steel log bunks are mounted across and securely attached to the truck frame. Should the truck also be used for hauling products other than logs a flat bed of wooden plank construction is mounted upon the truck frame with bunks placed on the flat bed and both attached to the frame. Stakes are installed in the ends of the bunks to help in holding the load. Bunks with integral stakes along with a safety tripping mechanism are very effective.

Chock blocks or "cheese blocks" are triangular shaped steel fixtures attached to logs bunks and provide stability to the first tier or bunkload of logs. They tend to prevent large logs from rocking or shifting position while being transported by means of trucks or railway cars. Blocks have been replaced by bunk stakes in those regions where large logs are becoming increasingly uncommon.

Trailers

Pole trailer and full trailers are drawn by truck tractors and trucks respectively. The truck tractor is so designed that it carries part of the weight and load of a pole trailer. A pole trailer may be either a single or tandem axle unit and is attached to the truck tractor by means of a metal, wooden, or fiberglass pole or reach. One of the more common types of log hauling units which allows the logger to haul maximum legal loads of long logs consists of a three-axle truck tractor and a two-axle pole trailer (Fig. 10–3). This combination requires that both the truck tractor and pole trailer be

Fig. 10–3. A three-axle truck tractor and two-axle semi- or pole-trailer. Log bunks fitted with integral stakes. (Courtesy of the Kenworth Motor Truck Company.)

fitted with log bunks and stakes. The same requirement for log bunks and stakes is necessary for full trailers (Fig. 10–4), except that each trailer requires two bunks and log stakes. Truck tractors and trucks are so designed that it is possible for the pole or full trailers to be loaded "piggyback" fashion (Fig. 10–4), upon the driving unit for the return trip to the woods. Transporting the empty trailer in this fashion saves wear and tear on the trailer which otherwise would have a tendency to bounce around on the road during the return trip. The added weight on the drive wheels allows the empty vehicle to climb grades more effectively. Also, the truck is more maneuverable and safer to drive.

Fifth Wheel

Truck tractors may be fitted with a weight-bearing swivel connection positioned over the tandem driving axles. This arrangement, commonly identified as a "fifth wheel," allows the vehicle to be combined with a lowboy semi-trailer or similar unit (Fig. 10–5).

Fig. 10–4. Truck tractor and full trailer loaded "piggyback." (Courtesy of the Swanson Bros. Lumber Co.)

Fig. 10–5. Truck-tractor fitted with a "fifth wheel," which is positioned on the frame between the rear axles. (Courtesy of the Kenworth Motor Truck Co.)

One company (Dwyer, 1956) has developed a truck tractor which may be used as a conventional truck tractor together with a pole trailer for log hauling, and also as a truck tractor in combination with a lowboy semi-trailer to transport equipment. The truck tractor is fitted with a conventional fifth-wheel which is permanently attached to the frame. Log bunks positioned over the fifth-wheel are detachable and may be removed in a matter of five minutes. Thus the truck tractor may be readied for either type of haul in a matter of a few minutes. The fifth-wheel truck tractor when combined with a conventional flatbed semi-trailer may also be used to haul lumber.

Pre-Loading Trailers

In those logging regions where semi- or full trailers are used in the log hauling operation, pre-loading has proven to be very effective. The pre-load trailers provide the means whereby an operation becomes more productive and more versatile.

Basically a pre-load system involves a hydraulic truck mounted component which allows the driver to detach or engage empty or loaded trailers at will, without assistance. Through the use of additional trailers a high degree of flexibility is achieved which, in turn, permits more efficient use of logging trucks by reducing to a minimum the length of waiting periods at both ends of a haul (Fig. 10–6).

Generally a pre-load operation is carried on in this manner: a truck upon returning to a landing with an empty trailer simply exchanges it for a loaded one and returns to the point of delivery. The exchange, leaving an empty trailer and linking up with a loaded one, may be made at what is termed an intermediate holding site, which can be a turnout, wide spot, shoulder, intersection, or spur road. The loaded trailers are hauled and positioned in these holding sites by "snap" trucks. These trucks perform a shuttle function—constantly striving to move either loaded or empty trailers from and to locations where they can either be moved out loaded or positioned at the landing for loading. The "snap" trucks generally are older vehicles which remain in the operational area and would therefore not be subject to licensing and insurance as in the case of trucks which use public roads.

Fig. 10–6. Pre-loading semi-trailer with heel boom loader. Steel spar in background. (Courtesy of the Page and Page Company.)

Air Brakes

Trailers, of all types, should have brakes of the proper size and capacity to assist the truck tractor in exercising control over the hauling combination when it is in the loaded condition. Air brakes are the conventional type associated with either full or pole trailers. Air to activate the brakes is provided by an air compressor driven by the engine. The compressed air is accumulated in a reservoir or storage tank and applied to the brakes by a system of brake lines and brake chambers. Air hose is used to connect the braking system on the trailer to the compressed air source on the truck or truck tractor. The hose is fitted with a quick lock-unlock coupling which facilitates engaging or disengaging the air supply, depending upon whether the trailer is in the hauling or return (piggyback) condition.

Water for Cooling Brakes

The usual method of dissipating the heat generated, when a more or less continuous braking action is necessary, is accomplished by applying water to the service brakes of a log truck. While this does not increase the total braking capacity it does prolong the effectiveness of good brakes and tends to prevent what is known as "brake fade." A supply of water is carried on the truck and applied to the brakes by a hose system. Vehicles in combination where the trailer is carried piggyback during the return trip have the same type of coupling system for the water supply as for the air supply.

Retarders

These units are used to retard or slow down a vehicle when traveling down a steep grade. There are three basic types of retarders in general use that depend on the rear axle or gear train to help hold the vehicle on down grades. One type uses water as the retarding agent while another type uses oil to develop the braking effect. The other is an electric retarder which utilizes field coils in much the same manner as a large generator operates.

In addition to the retarders there are two types of engine brakes which are used in conjunction with diesel engines and provide a retarding effect. One type operates by means of a butterfly valve in the exhaust pipe. This arrangement tends to make a low-stage air compressor out of the engine. The other type mounts on the rocker arm housing of certain makes of diesel engines and is considered to be the more effective of the two. The latter type converts a power producing diesel engine into a power absorbing air compressor.

Binders

The motor vehicle laws of the individual states require that loads of logs be cinched tightly with at least two binder chains or wire rope wrappers. The binders are cinched or tightened around a load of logs by means of a steel load binder, which consists of a leverage mechanism, whereby the truck driver is able to tighten the binders and thereby prepare his loaded vehicle for travel.

LOG HAULING REGULATIONS

Hauling logs by motor truck over public highways places the trucker under motor vehicle laws and regulations of the state in which logs are being hauled.

Each state has specific laws relating to size and weight limits. Violation of these limits may result in a citation and fine. Included in the size restrictions are width, height and length dimensions of the loaded vehicle. There is a further maximum length restriction for combinations of vehicles; such as, a truck tractor and a pole trailer hauling unit. When tandem axles are used there is a minimum spacing between axles specified in most states. Gross weight is stated in pounds per inch of tire width, pounds per axle, and pounds per set of tandem axles.

The gross weights of log trucks used separately or in various combinations with pole or full trailers are identified according to an engineering oriented system. Should the entire payload be carried on a truck chassis the gross weight is identified by the letters GVW and means gross vehicle weight; however, if the payload is carried partly on a truck tractor and partly on a pole trailer, the gross weight is identified by the letters GCW and means gross combination weight. In the case where one unit of payload is carried on a truck chassis and another payload unit is carried on a following full trailer the gross weight is identified by the letters GTW and means gross train weight. A truck tractor-pole trailer hauling unit followed by a full trailer also would be considered a train and the gross weight identified as GTW.

Certain states have special regulations which apply specifically to hauling logs or pulpwood. For example, Maine and Washington have a 10-percent weight tolerance regulation which permits a log trucker to exceed by 10 percent the gross weight limit in effect for trucking commodities other than logs. California issues a special 30-day permit for one-log loads regardless of weight. Montana issues a special permit which allows for a 9-foot bunk width whereas 8 feet is the maximum width of loads when not hauling under permit.

Oregon, presently the leading state in log production in the country, has the following size and weight restrictions under which truckers haul logs (Table 10–1). A 3-axle truck tractor in combi-

TABLE 10–1. State of Oregon Motor Vehicle Restrictions with Special Application to the Truck Hauling of Logs

Maximum height	$12\frac{1}{2}$ ft
Maximum width	8 ft
Maximum length (truck tractor and pole trailer)	55 ft
Maximum axle load	19,000 lb
Minimum tandem axle spacing	40 in.
Maximum tandem axle spacing	6 ft
Maximum tandem axle load	34,000 lb
Maximum gross weight (truck tractor and pole trailer)	76,000 lb
Maximum tire load (lbs. per inch of tire width)	550 lb

nation with a 2-axle pole trailer is most commonly used for log hauling.

Load Weight Control

The log trucker operating on public highways is constantly confronted with the problem of hauling an economical payload on his vehicle without exceeding the legal load limit. The problem is intensified by the rather wide range in weights of logs per board foot. Unit weights vary with log size, position of the log in the tree, time since felling, and tree species. For these reasons, no two loads of the same log scale volume will weigh the same (Pearce, 1954).

Overloads are costly to the log trucker in that fines are imposed when legal load limits are exceeded. Underloads are also costly to the log trucker. Failure to load up to the legal limit for which the vehicle is licensed results in loss of revenue and higher hauling costs per unit of volume.

This problem of overloads and underloads has led to the development of various types of weighing devices which either indicate the weight of the payload or gross vehicle weight. Still another type determines the weight of individual logs.

Bunk scales provide the trucker with a means whereby he can determine the weight of his load. These scales are permanently attached to the vehicle and are available in two types; hydraulic and compressed air. The hydraulic bunk scale consists of a "pressure pad" of two steel plates, with a thin film of oil between them, connected to a dial gauge by tubing. Two "pressure pads" are mounted between the log bunks and frame of a truck or trailer, one on each side of the bunk center. The air type also is positioned between log bunk and frame. It weighs the load by lifting the log bunk by means of two steel lifting chambers actuated by compressed air. The chambers are connected to dial gauges which are calibrated in both pounds of load and pounds per square inch pressure (Pearce, 1954).

Portable truck scales or "jump scales" are used to determine gross vehicle weights of log trucks. This type allows more than one truck to be weighed while bunk scales being attached permanently to a vehicle, can only supply weight data regarding the vehicle on which they are installed. Portable scales are positioned between the road surface and the tire contact area and are an effective means of controlling the weights of loaded log trucks (American Pulpwood Association, 1961).

INTERNAL COMBUSTION ENGINES

Trucks and truck tractors used for log hauling are powered either by gasoline or diesel engines. Basically, the main difference between the two types of engines, other than the fuel requirement, is the method by which the fuel-air mixture is ignited after it has been compressed by piston action within the cylinders. The gas engine utilizes a spark produced by its own ignition system to ignite the fuel, whereas the diesel engine depends entirely upon the heat of compression to ignite the fuel. Hence, the diesel does not need an electrical ignition system as required by the gas engine.

All gasoline and all but one make of diesel engines used in trucks and truck tractors for log hauling, operate on the four-stroke-cycle principle; intake, compression, power and exhaust piston strokes, which requires two revolutions of the crankshaft during each cycle, or to put it another way, there is one power stroke to two revolutions of the crankshaft. A two-stroke-cycle diesel engine also is available to the industry and has gained wide acceptance. In contrast to the four-stroke-cycle diesel engine, the two-stroke-cycle engine completes a power stroke every revolution of the crankshaft. The two-cycle principle involves a compression stroke and a power stroke; in between or during these two fuel and air are introduced into the cylinder and exhaust gases are discharged.

Trucks and truck tractors used for the off-highway type of log haul are practically all powered by diesel engines. Diesel-powered trucks perform very effectively when hauling heavy loads of logs over roads which have stretches of sustained adverse grades. Equipped with one of the various retarding mechanisms to provide additional braking effect, the vehicle functions satisfactorily and safely on the type of road generally associated with off-highway log hauling.

Many diesel-powered log-hauling units also are used in connection with on-highway hauls. Improvements in metallurgy have contributed much to provide truck manufacturers with metals and alloys which are both light and strong. Reducing vehicle weight allows a greater payload to be hauled, which in turn reduces ton-mile costs.

Diesel engines are reputed to be the most efficient power source because they convert a higher percentage of heat energy from fuel into work. Diesel oil is lower in price than gasoline, which makes it cheaper to operate a diesel engine than it does to operate a gas engine. Consequently, gasoline-powered trucks are considered to be better suited for short hauls where fuel economy is not a major consideration. Maintenance cost usually is lower and gas engines are more easily serviced, which is of particular importance to the small operator who usually does his own servicing and much of the necessary repair work. The initial cost of a gas-powered hauling unit is usually less than that of a diesel of comparable size and power.

Results of a comprehensive study of log hauling costs, as related to road design, state that there is small difference between the cost of hauling logs with gasoline-powered trucks when compared with diesel powered trucks. The advantage of the lower cost of diesel fuel is offset by higher repair, lubrication, and fixed costs (Byrne, Nelson, and Googins, 1960).

TRUCK PERFORMANCE

Consideration of truck performance is essential so that a log hauling operation will provide a steady flow of logs from the woods to mill, log dump, or railhead at a cost which will justify the investment.

Several items should be considered and examined in an effort to select the most effective truck for a particular set of conditions. The distance logs are to be hauled and the time required to effect the log movement are closely associated with the type and condition of the roads which must be used. Steep grades and sharp curves tend to increase travel time which results in increased hauling cost. These considerations should be examined in keeping with local and state hauling restrictions so that an efficient and economical log hauling operation will be achieved.

Hauling logs is a somewhat unusual transportation operation in that a log truck is running empty one-half of the time. A truck used to haul logs should have the ability to haul heavy bulky loads over a rather wide range of road conditions: from the worst, adjacent to the loading point; to the best, the paved public highway. A log truck should be able to negotiate both adverse and favorable grades in a safe, efficient manner, and also be able to pull through a muddy stretch of road as well as achieve the posted speed limit during the empty or return trip.

Truck performance may be estimated according to procedures presented in Supplementary Readings 7 and 8. These procedures provide the means for determining the best combination of driveline components which should be incorporated in a log truck and allow it to operate effectively under certain operating conditions. The combined effect of a particular component mix may be compared to operating conditions which will be encountered during the log haul.

TIRES

While new developments in tire manufacture have resulted in better performance, the introduction of the wide-base single tire to replace dual tires is of major significance to the industry.

The advantage of large, wide tires for off-highway travel has been long recognized, particularly on earthmoving equipment; however, certain problems developed when this type of tire was subjected to high-speed highway travel. One of the major problems was the tendency of side sway on curves. Manufacturers have corrected this situation by incorporating stronger side walls, a lower tire profile, and developed a special truck tire and rim assembly.

One single wide-base tire of the 18.00–22.5 size replaces two 10.00–22 tires which is the size generally associated with log trucks hauling loads within legal limits. Several manufacturers are producing a wide base single tire and claim certain advantages over duals, which agree quite well with user experience. Besides providing better flotation and traction, the single tire has a less damaging effect upon both truck and road by providing a cushioned or softer ride. The tendency of rocks to lodge between duals is eliminated, as is the problem of the inside dual having to accept a disproportionate share of the load on those portions of a road having excessive crown (American Pulpwood Association, 1962).

It appears that a log truck rolling on single wide-base tires would have less damaging effect upon both truck and road than when dual tires are used. An additional gross load allowance for log trucks equipped with single wide-base tires may be a practical consideration.

WATER TRANSPORTATION

The earliest method of log transportation made use of natural waterways. Nearly every large stream in the forest regions of the United States has at some time or other served as a means of moving logs from forest to mill. The colorful log drives are now history even though a considerable amount of pulpwood still is moved in this fashion in the Northeast, the Lake States, and Eastern Canada. Considerable log volume is presently carried by water either in the form of log rafts or loaded on barges.

Rafts

While transporting logs in rafts used to be of major importance, this method of log transportation has been reduced to minor significance or eliminated completely in most parts of this United States. The main reasons for the transition are the introduction of truck hauling and the increased use of hardwood species which do not have the floatability generally associated with softwood species.

The Pacific Northwest, British Columbia and Alaska are the only areas where rafting still is an important medium of log transportation. In general, two types of rafts are used: the flat raft and the Davis raft.

Flat Rafts. Flat rafts are used where tows are made in harbors, bays, and on large rivers, where water action does not develop to the extent that logs are dislodged from the area within the boom sticks. While towing distance generally is not an important factor in flat raft operations, wind and wave action with resulting rough water conditions may restrict or curtail operations. Flat rafts are towed the length of Puget Sound in the State of Washington, a distance in excess of 130 miles, and also along the Columbia River for many miles.

Flat rafts are made up in booming grounds which are situated adjacent to a log dump where log trucks or common carrier railroads unload their loads of logs. Log sorting by species or grade may be accomplished at this point. The logs then are guided to various rafting pockets by boom men who are experts in the use of pike poles which they use for moving and positioning the logs. The rafting pockets are narrow lanes approximately 75 feet wide and from 800 to 1,000 feet long enclosed by boom sticks held in place by piling or piers spaced at about 70-foot intervals. The boom men first position boom sticks, which are 75 to 80 feet in length, across the far end and along both sides of the pocket, after which logs are poled into and stowed parallel to one another in the far end or first section of boom sticks. When the first section has been filled with logs, a boom stick is placed across the end of the section and attached to the outer boom sticks of the section. Thus a section of a flat raft is approximately 75 feet square. The procedure is continued until the rafting pocket is filled with 10 to 12 sections, making up a flat raft ready for towing. Log volume per section varies with log size and species and ranges between 15,000 to 20,000 board feet for small Douglas-fir logs to 100,000 board feet for old-growth logs.

As an added precaution against logs being dislodged from within a section of a flat raft during blustery weather, a cross boom or "swifter" may be placed on top of the logs and fastened to the side boom sticks.

Davis Rafts. Classed as ocean-going, Davis Rafts have the ability to operate in fairly heavy seas without an appreciable loss of logs.

The raft is constructed by means of wire rope passing over and under the bottom tiers of logs at each end and fastened to boom sticks on either side. In other words, the logs are laced firmly together, thereby making a floor of logs upon which additional logs are placed which in turn are secured with additional wire rope or chain. A Davis raft 70 by 150 feet in size will carry approximately 750,000 board feet of logs. This type raft is used in British Columbia and certain coastal areas of Washington and Oregon.

Benson Rafts. An earlier ocean-going raft called the Benson Raft made use of a 700-foot wooden cradle or frame to build up the raft. One side of the cradle was detachable and, when released, the raft would slide sidewise into the water. Logs were stowed in the cradle with the butts toward the center of the raft after which chains were passed around the logs and tightened. The raft ready for towing was cigar-shaped and very seaworthy. Rafts of this type were used to transport logs, piling and also deck loads of sawn material between Columbia River ports and San Diego, California.

Log Brands. While log marking or branding is no longer as necessary as it was in the days of the colorful river drives, it nevertheless is still required when logs are transported on navigable waters. Log marks or brands denote ownership in the same manner as cattle brands do in grazing country. In those sections of the country where logs are rafted and towed, notably the Pacific Northwest, log brands are registered and controlled at the state level. For example, the Washington Department of Natural Resources currently registers about 4,500 brands.

Barges

Transporting logs by loading them on barges is restricted to those locations where waterways and navigable rivers connect log delivery points and mills. The conventional log barge used in the industry is loaded and unloaded by means of a crane. In contrast, a new type developed in British Columbia is classed as a self-loading and self-unloading type.

Conventional Log Barges. This type is used to transport hardwood logs on portions of the Mississippi River and its tributaries, and also in some parts of the southern hardwood bottomlands. The largest barges in general use are 220 feet long, 36 feet wide and have a draft of 7½ feet. A barge of these dimensions, loaded with cotton-wood logs 22 inches and larger in diameter, would be able to hold approximately 200,000 board feet. However, log barges of somewhat smaller size, approximating 110 feet long and 20 feet wide, and having a 6-foot draft, are much more common. Barges are used to transport pulpwood in the Southeast, Eastern Canada and on some of the Great Lakes. Barges are constructed either of wood or steel and may be flat-decked or open, depending upon the system of loading or unloading used.

Self-Load, Self-Unloading Barge. A steel barge, 340 feet long, 64 feet wide, and 19½ feet deep, is performing very satisfactorily for a company in British Columbia, Canada. This barge, with unique self-loading and self-unloading characteristics, has a capacity of 6,700 tons or 1.5 million board feet of logs. The loading is accomplished by two 80-ton cranes which are mounted on pedestals and located so that the 90-foot booms may position logs throughout the effective length and breadth of the barge. The cranes may also be used to accomplish selective unloading; however, the log cargo normally is dumped by flooding two large side-tipping tanks which tilt the barge to approximately a 45-degree list.

Advantages of the self-loading barge lie in faster loading and turnaround time and in greater flexibility in picking up logs in isolated areas. This barge is able to accomplish in a matter of a few days what would take several weeks if the logs were transported by Davis or flat-raft.

Another feature of this operation is the inclusion of a 20-foot steel, diesel-driven boom tug which is carried on the barge at all times. The tug is lifted on and off the barge by one of the cranes and is used to push logs to the barge and position them for the cranes to insure an uninterrupted supply of logs in the loading area.

SUGGESTED SUPPLEMENTARY READING

1. ARD, HENRY N. 1960. Brakes and logs. *Loggers Handbook.* Vol. XXI, 1961. Pacific Logging Congress, Portland, Oregon. Compression brakes and retarders are discussed in this article.
2. CUMMINS, C. LYLE, JR., and G. S. HAVILAND. 1961. The Jacobs engine brake—a new concept in vehicle retarders; 387A. Society of Automotive Engineers, Inc., New York. The SAE publication covers the development of the Jacobs brake together with its effect upon the engine, driver, and vehicle. Vehicle performance also receives attention.

3. EKSE, MARTIN. 1965. Wide single vs. dual tires—a logging road test. Department of Civil Engineering, University of Washington, Seattle, Washington. This report, prepared for the Pacific Northwest Forest and Range Experiment Station, is the result of a study which compared performance characteristics of large single tires and dual tires.

4. FITCH, J. W. 1956. *Motor Vehicle Engineering Guide.* 3rd Edition. Arrow Lithographing Co., Chicago, Illinois. This book is an engineering guide for the selection, operation and design of motor vehicles. Methods for calculating vehicle performance including gradeability, speed, weight balance, turn radii and other vital factors are outlined in a clear, concise manner. The book is written in semitechnical language and is clarified by the generous use of charts and illustrations.

5. Peterbilt Motors Co. (no date). Truck engineering—a guide to truck specifications. Peterbilt Motors Co., Newark, California. Provides a means for an engineering-oriented approach to driveline component selection as related to truck performance.

6. Society of Automotive Engineers, Inc. 1952. Truck ability prediction procedure; HS 82. Society of Automotive Engineers, Inc., New York. This report describes an uncomplicated procedure for predicting truck performance using accepted considerations. It is designed to assist anyone concerned with the problem of truck selection.

7. Society of Automotive Engineers. 1951. Truck Ability Prediction Procedure. Society of Automotive Engineers, Inc., New York. This report simplifies the prediction of truck performance using accepted considerations. It is designed to assist anyone concerned with the problem of truck selection. It is not necessary to understand or even read the complete report to predict truck performance.

8. Thomson, J. G. 1963. Off-road transportation. Faculty of Forestry, University of New Brunswick, Fredericton, New Brunswick. This publication consists of a series of lectures which cover many aspects of transportation and related equipment.

11

Pulpwood Production: Organization and Cutting

INTRODUCTION

Pulpwood is round wood bucked to length from the stem of a tree for utilization as a source of fiber in a **pulpmill** producing **woodpulp** for subsequent manufacture into newsprint paper, paperboard, and other products. It is classified as **shortwood** if less than 120 inches in length and **longwood** if lengths are 120 inches or more. Bolts or sticks of shortwood range in length from 48 inches to 100 inches. Customary lengths cut in the pulpwood producing regions of the United States are 48 inches in the Northeast, 60 inches and 72 inches in the South, 96 inches in the West, and 100 inches in the Lake States. In eastern Canada standard lengths are 49 and 98 inches, or 50 and 100 inches, commonly referred to as 4-foot or 8-foot wood. The length required by the pulpmill is determined by the debarking, chipping, and handling equipment. In the groundwood mill the width of the grindstone face limits the pulpwood length.

Pulpwood Measurement

Short pulpwood is measured by the **cord** of 128 cubic feet of stacked roundwood. The cord originated with the custom of stacking

4-foot fuel wood in ricks 4 feet high and 8 feet long. The number of cords in a stack of any size is obtained by measuring the length of the stack and the average height, multiplying the three dimensions of the stack in feet, and dividing by 128. Wood longer than 4 feet, ricked 4 feet high and 8 feet long, is termed a unit. For example, a unit of 63-inch wood is 168 cubic feet or 1.3125 cords. Longwood is scaled in board feet in the United States, and in cubic feet in Canada, the same as sawlogs and veneer logs. Statistics of pulpwood consumption, including chips and logs as well as short pulpwood, are expressed in cords. The cubic content of solid wood in a cord varies with the diameter and length of the stick, with the straightness and smoothness, and with the bark thickness of unpeeled wood.

In eastern Canada pulpwood is also measured in **cunits** of 100 cubic feet of solid wood without bark. The average cubic content of wood per cord of 4-foot softwood is 85 cubic feet with bark, and 95 cubic feet peeled. Thus a stack of one cunit of peeled 4-foot wood 4 feet high would be 8.4 feet long, and occupy 134.7 cubic feet of space, or 5.2 percent more than a cord. A cunit of unpeeled 4-foot wood would be 9.4 feet long and occupy 150.6 cubic feet. Production data on pulpwood harvesting systems in Canada are customarily reported in cunits.

Weight scaling or purchased pulpwood is being practiced by an increasing number of companies. The loaded truck is weighed on platform scales at the delivery point. The net weight of the wood is obtained by deducting the tare weight of the empty truck. The wood may be paid for at a rate per unit of weight, at a ton-mile rate by which the price varies with the haul distance, or at a rate per cord, which requires applying conversion factors obtained from records of volume-weight relationships. In the latter case sample loads are measured as a check. Advantages of weight scaling include saving in truck turn-around time, saving in scaler labor, and safety as compared with measuring the dimensions of the load. A "Print-O-Matic" attachment automatically prints out the weights on a delivery ticket.

An electronic technique for measuring the solid volume of stacked pulpwood has been developed by the Ottawa Forest Products Laboratory, Ottawa, Canada. A photo of a pulpwood load is placed before a black-and-white television camera. The ends of the pulpwood bolts are seen as white and the air spaces between the bolts as black. The white areas are scanned by a video area evaluator. Radio-frequency pulses are fed to an electronic counter. The digital readout is in the desired units of volume. The measurement is made in less than 5 seconds with an accuracy within plus or minus 2 percent.

Pulpwood Consumption

The consumption of pulpwood by regions of the United States in the year 1969 is given in Table 11–1. These data are taken from the

TABLE 11–1. Pulpwood Consumption, 1969

	CONSUMPTION	
REGION	**1,000 Cords**	**Percent**
South Atlantic	21,286	34
South Central	18,337	29
West	12,810	20
Northeast	5,668	9
North Central	5,264	8
Total United States	63,365	100

Monthly Pulpwood Summary for October, 1970, issued by the American Pulpwood Association. Pulpwood consumption in 1969 was 5.0 percent more than in 1968, and for 10 months of 1970 the increase in consumption was 2.6 percent more than for the same period in 1969. In October, 1970, the pulpmills received 71.6 percent of their consumption in the form of roundwood, and 28.4 percent in the form of residues from sawmills and veneer plants. Softwood accounted for 74.5 percent and hardwood for 25.5 percent of the consumption.

The pulpwood industry offers to the owners of small tracts of timber land the major market for their wood. For example, in 1965 the American Pulpwood Association reported that 69.3 percent of the roundwood harvested came from farm woodlots and other privately owned lands. Paper company lands supplied 18.7 percent, federal forests 9.3 percent, and state forests 2.7 percent. Comparable data for later years were not available from the Association.

Production Unit Size

Pulpwood production is unique among major industries in being dependent upon a large number of very small producing units. Based on American Pulpwood Association data, part-time pulpwood producers, mostly farmers delivering less than 1,000 cords per year, produced 52 percent of the roundwood consumed in 1964. Full-time pulpwood producers accounted for 36 percent, and paper company logging crews the remaining 12 percent. The Association estimated that the number of men engaged in the pulpwood industry in 1964, including labor, producers, dealers, and foresters totaled 222,000.

A significant factor influencing the size of the American pulpwood producing unit is the eight-man exemption provided by the following section of the Fair Labor Standards Act of 1938 as amended on September 23, 1966:

Sec. 13(a) The provision of sections 6 and 7 shall not apply with respect to any employee employed in planting or tending trees, cruising, surveying or felling timber, or in preparing or transporting logs or other forestry products to the mill, processing plant, railroad, or other transportation terminal, if the number of employees employed by his employer in such forestry or lumbering operations does not exceed eight.

Section 6 of the Act as amended established minimum wages of $1.40 an hour during the first year from the effective date of the amendment, February 1, 1967, and $1.60 an hour thereafter. Section 7 regulates maximum hours and overtime pay. Prior to the enactment of the eight-man exemption, a twelve-man exemption was in effect under Section 13(a)(15), Amendment of 1949.

In a report to the Congress of the United States in January 1964 the Department of Labor stated that of an estimated 137,000 workers engaged in logging in 1963, approximately 87,000 workers came under the twelve-man exemption. In a "small-logging" survey of small logging camps and logging contractors primarily engaged in cutting timber and producing logs, bolts and pulpwood, the Department of Labor found that the average number of workers per establishment was 5 in 1,256 establishments in four Southern states and 4.9 in 354 establishments in the three Lake States.

Producer-Dealer Organization

The small size of the typical producing unit has led to the development in the southern and north central regions of the producer-dealer system of marketing pulpwood. The **producer** is an independent operator who acquires stumpage to cut, invests capital in equipment and employs labor. His operation performs all the functions of production necessary to deliver pulpwood to a concentration yard where he sells it at the market price. The small producer is a working member of the crew and usually drives the truck, which is the equipment item in which he has the largest investment.

The **dealer** is a middleman who buys from the producer and sells to the pulpmill company, or who acts as a commission broker for the company. He usually operates a **concentration yard** which is centrally located in a pulpwood producing area on a railroad or a barge waterway. He provides facilities for unloading the producer's truck, stacking in the woodyard, and reloading on rail car or barge for shipment to the pulpmill. He may also locate timber to cut, buy and sell stumpage, and advance working capital to the producer.

Concentration yards are also operated by pulp and paper companies in localities that are beyond economical trucking distance to the mill. The typical installation consists of a scale house, a railroad spur storage space, and a pulpwood loader which unloads trucks and loads rail cars (Fig. 11–1).

TYPES OF OPERATION

Short pulpwood operations are of three general types. The older and more common type is the "stumpwood" or "yarded wood" operation. The tree is **limbed** and **bucked** into pulpwood lengths where it is felled. The "**sticks**" as they are termed in the South, or the "**bolts**" in the terminology of the northern regions, are bunched at stump site or piled along trails by manual labor. The movement of the wood to the truck road, termed "**forwarding**" in the northeastern region and in eastern Canada and "**prehauling**" in the southern and Lake States, is done with a great variety of equipment. Prehauling is eliminated where favorable terrain permits the truck to be driven to stump site for loading the piled wood. The stumpwood type of operation is favored where tree selection cutting or thinning is done, or where low volumes per acre, small tree diameters, or small tracts make it the most economical type.

Fig. 11–1. A concentration yard in the South. The "Pulpwood Dream" unloads trucks and loads rail cars. (Courtesy of the Taylor Machine Works.)

The second type is the tree length or multiple-stick length operation. The felled tree is limbed and topped, and skidded to a landing or roadside where it is bucked into bolt lengths. In some Lake States operations, the tree is not limbed before **skidding** because in cold weather most of the branches will break off during skidding. If the tree length is too large for the skidding equipment used, it is bucked into log lengths before skidding. In clear-cut operations or where skidding of tree lengths can be done without damage to the residual stand, the trend appears to be away from stumpwood toward tree-length skidding. Studies made of the two types of operations under comparable conditions indicate that the tree-length type gives greater production per man-day.

The third and newest type is the "full tree" operation. This type developed with the invention of machines which perform all the tasks of **felling,** limbing, bucking, and loading mechanically. The Buschcombine processes the tree at the stump site and transports the wood to the road, where it is transferred to truck or pallet. Other processors or harvesters operate at the landing to which full trees are skidded.

Longwood operations which deliver logs to the pulpmill use the same logging methods as are used to produce sawlogs in the locality. They are usually mechanized with wheel or crawler tractor skidding and mechanical loading of the truck and trailer. Since larger machines and more investment in equipment is required for longwood operations, they are usually conducted by the larger companies. Pulpwood is also produced in multiple product logging operations, also termed integrated operations, together with sawlogs, veneer logs, poles and piling, etc.

Manual Handling

Short pulpwood production in the past has been characterized by the great amount of hand labor involved, and the relatively low volume production per man-day as compared with logging sawlogs.

Since a pulpwood bolt is usually of a size and weight that can be lifted by a man, hand piling and loading has been widely practiced. Manual handling at some stage of the operation is still common today. In the less mechanized operations, bolts bucked where the tree is felled are either bunched by hand for easier gathering for forwarding to the truck road, or piled in a pallet or on the bed of one of the many devices used for forwarding. Where cutting is done by piece work, the bolts are piled for cord measurement. On steep slopes the bolts may be pitched, rolled, or tumbled by hand to collection points. Trucks are sometimes loaded by hand, especially where farm equipment, which does not include mechanical loading devices, is used. While the chain power saw has greatly reduced the physical labor of felling and bucking, cutting pulpwood is still a manual operation (Fig. 11-2).

Fig. 11–2. Traditional pulpwood production system in the South: skidding by mule, bucking at the landing with a bow saw, and manual loading of the pulpwood truck. (Courtesy of the Hiwassee Land Company, Calhoun, Tennessee.)

Mechanization

The mechanization of pulpwood production with labor-saving machinery is expanding rapidly. This development is stimulated by rising wage rates, a declining supply of good woods labor, and the unwillingness of young men entering the industry to do hand labor. To attract boys reared in an age of machines to enter an industry they must be offered machines to operate. The mechanization of farming, with consequent reduction in the farm labor force, affects the pulpwood industry. Fewer men are available for seasonal woodswork and the farm tractors and trucks are used for part-time pulpwood and production. The trend in migration of young men to the cities from the farms and country towns, which are the traditional source of woods labor, depletes the labor supply. In eastern Canada where most of the pulpwood is produced in large company operations, union organization of woods labor has given an impetus to mechanization.

Workmen's Compensation rates and medical aid costs of industrial injuries are rising with consequent increase in industrial insurance costs. Claims for back injuries resulting from lifting heavy bolts of pulpwood in hand operations are of special concern to the industry. The mechanized operations with experienced crews generally have a lower accident frequency rate than the manual operations.

Two other important elements contribute to increasing the cost of pulpwood delivered to the mill. The long-term trend in stumpage prices for timber is upward. Taxation of trucks, tires, and fuel to support the public highway programs, as well as rising truck operating costs, affect the large volume of pulpwood which is hauled by truck. The use of larger capacity trucks, and of trailers, to reduce trucking labor costs is increasing.

To maintain the woodpulp industry in a competitive position in the market the cost of pulpwood must be held within economic limits. Increasing the productivity of labor by mechanization of pulpwood production appears to be the solution to this problem.

Research and Development

Research and development in labor-saving, cost-reducing systems for pulpwood production are being conducted by many agencies. Machinery manufacturers are developing new or improved machines. The design of a machine for the mass pulpwood producer market presents a challenge to the manufacturer. It must be manufactured to sell at a price the small producer can afford to pay and constructed so it can be maintained and repaired without elaborate shop facilities or skilled mechanics. Logging engineers or forest engineers for pulp and paper companies are doing operations research, experimenting with new machines, and working with the producers to improve the efficiency of their operations. The forest engineering laboratories of the United States Forest Service are engaged in research in new harvesting systems. Some of the forest schools in pulpwood-producing states engage in research in this field. In Canada the woodlands section of the Pulp and Paper Research Institute of Canada and the

Ottawa Laboratory of the Forest Products Branch, Department of Forestry, are actively pursuing research in pulpwood harvesting.

The American Pulpwood Association employs forest engineers in five regional offices who maintain field contact with pulpwood producers and machinery manufacturers. They report new developments in production equipment and methods in technical releases and trade journal articles. In 1967 the Association initiated a five-year pulpwood harvesting research project which is sponsored by six pulp and paper companies. By 1971 the number of sponsor companies had increased to fifteen, and the term of the project extended.

Labor Productivity

Pulpwood labor productivity in the South doubled during the period 1944–1964. In 1944 the rate was only two cords per man-day (Traczewitz, 1965). The power saw, the mechanized concentration yard, and the use of machines for prehauling or skidding and truck loading contributed to this increase in productivity.

For the year 1966 eleven paper companies in the South reported production rates in fourteen company pulpwood operations of from 2.16 to 6.99 cords per man-day. The average rate was 3.72 cords per man-day (Rolston, 1967). These operations were presumably well mechanized.

The effect of stick size and volume cut per acre on productivity is shown graphically in Fig. 11–3. The graphs result from· a production study made in South Carolina (Jarck, 1967). They also give comparison between types of operation and loading methods. They show that when the average stick size is under 8 inches diameter, prehauling stumpwood gives greater productivity than skidding tree lengths and bucking at the landing. For cuts of less than 5 cords per acre, loading by hand and the relatively inexpensive Big Stick truck-mounted self-loader was preferable to the hydraulic grapple.

The results of a study of four methods of producing 4-foot pulpwood in the northeastern region presents an interesting comparison of cord-feet moved per man-hour (Table 11–2). In all cases the activities included felling, limbing, bucking, and piling, and the production rate was 4 cords per man-day for all methods. Cord-feet is a measure of production related to distance the wood is moved. When bucking and piling at stump site was not done, the cord-feet produc-

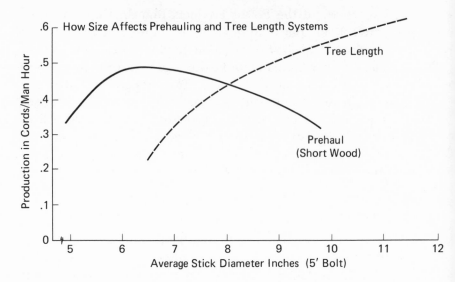

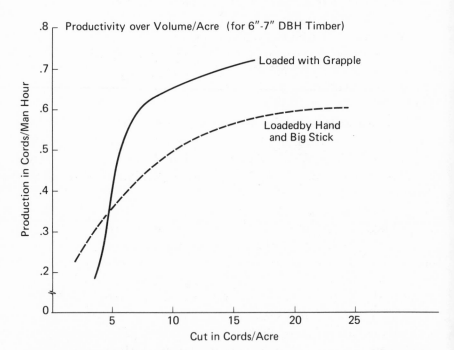

Fig. 11–3. The effect of stick size and volume cut per acre on productivity. (Courtesy of Pulpwood Production and Saw Mill Logging Magazine and Mr. Walter Jarck.)

TABLE 11–2. Cord-feet Production

Number of Men	Method	Distance, feet	Cord-feet per Man-hour
1	Hand	25	12.5
2	Horse	200	100
3	Tractor	400	200
5	Wheel skidder	800	400

tion per man hour of the tractor crew was 300, and of the wheel skidder crew 632 (Hazelton, 1966).

PULPWOOD HARVESTING SYSTEMS

Influences

A pulpwood harvesting system is the aggregation of men, equipment, and material to perform the total operation of converting the standing tree to pulpwood and delivering the wood to the pulpmill or concentration yard. It has been estimated that there are probably 200 or more individual pulpwood harvesting systems in the United States and Canada. With the invention and development of new machines for handling and processing pulpwood, the number of systems is increasing.

The wide diversity of pulpwood harvesting systems is due to the following principal influences:

1. Terrain, which varies from swamp and muskeg or dry plains to steep mountain slopes.
2. Climate ranging from the summer heat of the South to the sub-zero winter cold of the northern regions. The amount and duration of snowfall may be an important factor in determining the harvesting system.
3. Timber type, species and size, ranging from plantations to old growth, and from pure to mixed stands. Short pulpwood is generally cut from trees less than 20 inches in diameter at stump height.
4. The volume per acre cut, which may be as little as 3 or 4 cords per acre in partial cuts or thinnings, and the total volume to be

cut on a tract of timber to which equipment must be moved. The silvicultural method prescribed—intermediate cutting (thinning), partial cutting (selection or shelterwood), or clear-cutting—affects not only the volume per acre cut, but also the harvesting system. where damage to residual trees must be minimized.

5. The pulpmill specifications, which dictate acceptable species and size limits, and the wood procurement policies of the pulp company which determine operating periods and volumes delivered.

6. The supply of stumpage available to the producer and the size of the timber tracts in which he operates. Uncertain supplies of stumpage and small scattered tracts impose limitations on investment in labor-saving machines. Assured timber supplies and large tracts encourage mechanization.

7. The size of the producing unit and the capital available to the producer for investment in equipment.

8. The quantity and quality of the labor force and the wage scale. The use of labor-saving machines tends to increase as labor supply decreases and the wages rise. However, labor skills in operating and maintaining a given machine may determine whether the machine is used.

9. The major transportation facilities available. Land transportation facilities include roads, highways and common-carrier railroads. Water transportation facilities may be rivers, lakes or canals on which barges or rafts can be towed, or cargo ships operated. River driving, the oldest method of pulpwood transportation, is now practiced only in eastern Canada, and, to a limited extent, in the northeastern region.

Subsystems

Producing longwood is essentially a logging operation. Since logging is covered in previous chapters of this book, this chapter will mainly be concerned with the production of shortwood. The short pulpwood harvesting system is generally composed of the following first-order subsystems:

1. Preparation of the bolt or stick.
2. Movement of the wood from stump site to landing.
3. Loading on the vehicle used for major transportation.
4. Transporting to the delivery point.

The preparation of the bolt or stick involves three component or second-order subsystems:

1. Felling the tree.
2. Delimbing
3. Cross-cutting or bucking to bolt length.

Bucking may be done at stump site where the tree is felled, or at a landing adjacent to the truck road.

The bucking site chosen determines whether: (a) bolts or sticks are forwarded or prehauled to the truck road, or (b) tree lengths or multiple-stick lengths are skidded to the landing. Forwarding usually involves four components:

1. Piling, stacking or bunching by hand at stump site.
2. Loading the wood on the equipment used for forwarding.
3. Transporting the wood to the truck road.
4. Unloading alongside the road or reloading on waiting trucks.

The tree length system divides the preparation of the stick between felling and limbing at stump site, and bucking at the landing. Where **processors** equipped for delimbing are used at the landing, full trees, including limbs, are skidded. Skidding components include hooking, skidding and unhooking. Bunching of scattered tree lengths to expedite hooking a payload may be an additional component. After manual bucking the wood is stacked on the ground or in pallets ready for loading. If a bucking machine, termed a **slasher,** is used, the wood is dropped into a pallet or moved by a chain conveyor directly to a waiting truck or trailer.

Some of the subsystems or their components may be combined by the use of the multiple-function machines such as the Beloit Harvester, which limbs, tops, fells and bunches multiple-stick lengths; the Sicard Feller-Skidder which fells, bunches and skids tree lengths; and the Buschcombine which fells, limbs, and bucks, and prehauls the shortwood and transfers it to truck or pallet.

PREPARATION OF SHORTWOOD

Felling

Felling the tree is most commonly done with the light-weight, gasoline-powered one-man chain saw with a straight blade. The chain saw buyer in the United States is offered a choice of some sixteen makes and one to six models of each make. The most popular size used in felling pulp timber is the 5 to 7 horsepower saw with blade lengths of 18 to 24 inches, weighing 17 to 24 pounds. The felling procedure is that described in Chapter 6, except no wedging is needed. For felling small pine trees in the South some saw operators prefer the bow saw. The saw is held in the right hand, with the handle resting on the knee, and plunged into the tree butt to make a sloping cut. The left hand rests against the tree and pushes it over when nearly severed. The stump is about 4 inches high on the near side and 2 inches high on the far side (Fig. 11–4).

Fig. 11–4. A bow chain saw is used in the conventional method of felling trees for harvesting pulpwood in the South. (Courtesy of the International Paper Company, Southern Kraft Division.)

Limbing

The limbs of the felled tree are either chopped off with an axe or trimmed with a power saw. In hardwood or heavy-limbed species the power saw is preferred for limbing. If the tree is to be bucked where it is felled, the bolt lengths are measured concurrently with limbing, using a pole or rod. Limbing is the most tedious and hazardous part of pulpwood cutting.

Bucking

The bole of the tree is cross-cut to bolt or stick length with either a straight blade or a bow-type power saw. The plunge cut bow saw is preferred for bucking tree lengths at the landing as it tends to pinch less than the straight blade, and the chain can be kept clear of dirt and grit (Fig. 11–2). The blade and bow are interchangeable on the power unit. The bow saw is heavier than the blade, and the diameter which can be cut is limited by the size of the bow. By "boring" with the end of the straight blade, diameters larger than the blade length can be sawn.

In the southern pine region the Lowther wheelbarrow or bicycle-type circular saw antedated the chain saw. The saw diameter is 20 to 30 inches and the gasoline engine has 6 to 8 horsepower. The saw frame rides on two wheels with pneumatic tires. The sawblade is driven by a belt so it can be swiveled 90° from the horizontal plane for felling to the vertical plane for bucking. Limited in use to flat, brush-free ground, the wheelbarrow saw is now used mainly for bucking tree lengths at the landing.

Production Rates. A study of stick preparation with chain power saws in the Tennessee Valley found that an average time of 3 man-hours was required to cut one cord of 63-inch wood. The trees averaged 7 inches D.B.H., and the yield was 5 sticks per tree and 21 trees per cord (Schell, 1961). Another study of 24 full-time producer operations in Tennessee found that preparation of the stick required 39 percent of the total production time in pine and 31 percent in hardwood. The average man-day production was 1.75 cords in pine and 1.66 cords in hardwood for the entire operation of stick preparation, skidding or prehauling, and trucking (Walbridge, 1960).

In eastern Canada manual preparation of the bolt and piling at stump site typically requires 4.54 man-hours per cunit in 7-inch diameter spruce. In tree length operations, bucking and piling at the landing requires 2.48 man-hours per cunit (Silversides, 1964).

New Developments. Advances in metallurgy and design have enabled the manufacturers to reduce the weight and bulk of the chain power saw. An example is a saw with a 20-inch bar which weighs only 14½ pounds complete with bar and chain. New developments in hand-held power tools which are expected to be used in pulpwood cutting in the future are the rim saw and the wood-milling cutter. (Burkett 1967) The rim saw has the cutting teeth mounted on a rotating ring driven by a gear transmission. The rim saw cuts faster than the conventional chain saw because the teeth and depth gauge are stabilized in the cut. The rim saw is a much safer tool as the rotating rim is farther away from the operator and is shrouded on the operator's side. The rim saw shown in Fig. 11–5 has a rim

Fig. 11–5. Bucking with a pilot model of a rim saw. (Courtesy of the McCulloch Corporation.)

diameter of 16 inches and cutting capacity of 12 inches. The wood-milling cutter is similar to a twist drill and side cuts in the same manner as a metal-cutting end mill. A finger-like guide projecting at an angle to the rear of the cutter wedges the cutter into the wood. Only a force applied along the axis of the drive shaft is required to effect cutting. This promises to be the safest pulpwood cutting tool.

Slashers

In the larger mechanized pulpwood operations, a **slasher** is often used for bucking tree lengths at the landing. The components of the older type of slasher are an infeed mechanism of live rolls, a circular saw, and a chain conveyor which can either load a waiting truck, or dump the bolts into a rack or pallet. For mobility the unit is either mounted on a trailer for towing, or is self-propelled. This type of slasher requires an operating crew of 4 to 6 men and production is said to range from 12 to 32 cords per man-day, varying with the make of slasher. Production also varies with the efficiency of the skidding equipment in supplying the slasher, and the volume of timber tributary to a landing which determines the frequency of moves to other landings.

The demand for labor-saving machines led to the development of slashers which require fewer men on the operating crew. Representative of the newer type of slasher is the Pettibone Log Slasher, which cuts logs automatically to pulpwood lengths. A knuckle-boom loader with hydraulic grapple places the tree stem in the log conveyor and off-loads the pulpwood from the hopper in which it is collected to the haul truck or stack. One model is self-propelled at highway speed and is operated by one man. Bucking is done by a 48-inch circular saw. Another model is self-propelled at the job site by a hydraulic motor, but is towed between jobs. An option of chain saw or circular saw is offered. This model is operated by a crew of two men.

The SPP-1 Pulpwood Processor combines the bucking shear assembly used on the Buschcombine with a hydraulic knuckle-boom loader. Tree lengths are skidded alongside the processor, and the tree stem is placed in the processing assembly with the loader. The stem is propelled through the shear, automatically metering any length from 48 to 104 inches. The severed sticks drop into a wood receiver (Fig. 11–6). The shortwood is lifted from the receiver and loaded

Fig. 11–6. The SPP-1 Pulpwood Processor shearing a pulpwood length from a tree stem, which was placed by the hydraulic loader. (Courtesy of the Continental-Emsco Company.)

on trucks or into pallets with the loader. The processor is operated by one man. The machine is mounted on steel skids which fit the standard pallet truck for moving. It may also be dragged by a wheel skidder for short moves.

A new concept which enables one man to perform all the jobs on the landing with one machine is the Allis-Chalmers Buckmaster. The wood-handling and bucking devices are mounted on a 4-wheel industrial tractor. A group of logs is lifted with the tines of a forklift loader. A hydraulic plate pushing against the log ends aligns the ends and measures the stick length automatically. The logs are clamped firmly by a cable and bucked by a 48-inch circular saw. The bundle of sticks is raised with a tilting mast and loaded on the truck or into a pallet. Cycle time is said to be 60 to 90 seconds, and an operator in Wisconsin reports a production rate of 40 to 50 cords a day with his Buckmaster. A rear-mounted dozer blade is provided to straighten the log piles dropped by the skidder and to clear debris from the landing.

Portable Debarkers

In the Lake States and eastern Canada, where peeled pulpwood is produced, hand peeling is limited to the months of May, June and July. The use of portable or mobile debarkers permits both hardwood and softwood to be peeled at any time of the year, even when frozen. Three different types of debarking mechanisms are available. The rosser type rotates the bolt in contact with a rosser head or toothed drum or wheel. An example of this type is the Morbark debarker, which is powered by a 37 horsepower engine. Bark is blown away from the machine by a blower attachment. The unit is mounted on a trailer which can be towed by a light truck or tractor. The Morbark will debark all lengths of shortwood from 3 to 15 inches in diameter, and production is reported to be 15 to 20 cords a day with a crew of two men.

The mobile drum type debarks by tumbling the bolts against each other and against steel lugs on the inner surface of the drum. The unit is mounted on a truck and debarking can proceed while prehauling to the loading point, in the manner of the ready-mix concrete truck. Earlier models rotated the entire drum. In a later model the drum is stationary and a rotating axle with fixed arms carries the wood around. The production rate in 63-inch length southern pine is 12 to 15 cords a day with a crew of three men when they charge the drum from scattered piles of unpeeled wood. Loading the wood into the drum requires about two-thirds of the time. When operating at a central location the production rate is 4 to 5 cords per hour.

The cambium-shear type debarker has fingers or tools on the inside of a rotating ring which shear the bark off at the cambium layer. The pulpwood bolt is conveyed longitudinally through the ring. Several makes of ring debarkers are available for mounting on tractors or trailers. The Soderhamn Cambio Debarker feeds the bolt with three rotating spiked rollers on each side of the ring. The five debarking tools are rubber spring loaded against the bark surface. The ring or rotor is fitted with fan blades which eject the removed bark.

The cambium shear type does not remove any wood as the rosser type may do, but it is less efficient in the autumn and winter when the resistance of the cambium layer to shearing is higher than during the tree growing seasons of spring and summer.

In eastern Canada, where peeled pulpwood is customarily produced, several makes of machines which will process full trees have been invented. The processor incorporates a delimber, a debarker and a bucking saw or a shear. It is reported that some of the processors were abandoned after field trials of pilot models. Others are still undergoing field tests. It appears likely that rising wage rates and a decline in the woods labor force will encourage the development of processors which will find acceptance in the pulpwood industry. Some of the processing systems include the design of special skidders to supply full trees with limbs to the processor.

Tree Shearing

The invention of the hydraulic **shear** has made the most significant contribution to the mechanization of felling and bucking. The shear is mounted on a tractor for felling, is used in tree harvesters for felling and topping or for felling and bucking, and in slashers and processors. The cut is made by the scissors-like action of two steel blades, or by a single blade opposed by an anvil. The cutting blade is activated by a hydraulic piston. The angle of the cutting edge on the felling shear blade acts as a wedge for directional control of felling. The shear saves 1 percent of the wood volume in chain saw kerf, and 1 to 2 percent in stumps, as it permits the felling cut to be made flush with the ground. In eastern Canada the shear is termed a "guillotine."

Tree Harvesters. The first successful application of the hydraulic shear to pulpwood cutting was on the Buschcombine, a one-man machine which performs all the operations of harvesting pulpwood except truck hauling. The Buschcombine is a 70 horsepower, a 4-wheel drive, pneumatic-tired felling, bucking, and prehauling machine. It was invented by Mr. T. N. Busch, a forest engineer for the International Paper Company, and developed commercially by the Timberline Equipment Company. Following is the sequence of events in harvesting pulpwood with the Buschcombine, shown in Fig. 11–7.

Fig. 11–7. The Buschcombine, operated by one man, fells, delimbs and bucks trees into pulpwood-length sticks and deposits them in a cradle. Illustration shows a southern pine tree being felled with the felling shear. The bucking shear is at upper center. (Courtesy of the International Paper Company, Southern Kraft Division.)

The machine drives alongside the tree to be felled. The felling shear assembly attached to the end of a knuckle-boom type of arm is lowered to the ground. The curved anvil blade fits against one side of the bole, and the pivoting cutting blade is forced through the wood from the other side. The hydraulic piston speed is such that the blades will close to fell a tree with a maximum diameter of 19 inches in 7 seconds. The butt end of the tree is raised by the lift arm into processing position on the machine, and the debranching assembly closed around the stem. Travel is metered to cut bolt lengths of either 63 or 72 inches. Bucking is done by a single shear blade opposed by a fixed anvil. The bolts drop into a loader sling suspended from a Z-frame boom over the rear wheels. When a package of 1.1 cords of 63-inch wood or 1.26 cords of 72-inch wood has accumulated, the Buschcombine is driven to the truck road. The pivoting boom extends either to load the package on truck or trailer, or into a pallet.

The Buschcombine has a rated production of 2 cords of southern pine pulpwood per hour. One paper company which operated 18 Buschcombines in five southern states in 1966 obtained an average production of 2.21 cords per machine hour. The production record is held by a skilled Buschcombine operator in Florida who averaged 3.13 cords per hour and 19.8 cords per day for 234 days. This was clear-cutting in longleaf pine stands of 6 to 12 cords per acre, averaging 10 trees per cord (Busch, 1967). The accident frequency rate for the year, for a production of 58,812 cords, was only 26 percent of the national average for the logging industry compiled by the National Safety Council.

The Sicard Feller-Skidder was developed in Canada to enable one man to fell and bunch full trees. An articulated four-wheeled tractor has a hydraulic loading head and shear mounted in front (Fig. 11–8) and a grapple-bunk and apron over the rear axle. The tree to be felled is gripped by the loading head and sheared at $4\frac{1}{2}$ inches above the ground. The tree is swung over the cab by the loading head and the butt end deposited on the bunk. The grapple holds the tree on the bunk while the machine moves to the next tree to be felled. When a bunk load has accumulated the trees may be released to slide off in a bunch, ready for skidding with a separate wheel skidder, or skidded directly to the landing. The machine will fell trees up to 14 inches diameter on slopes up to 30 percent. Rated production is 5.25 cunits per hour clear-cutting timber averaging 15 trees per cunit. The grapple-bunk will hold 75 to 110 cords.

The Beloit Tree Harvester, illustrated in Fig. 11–9, uses hydraulic shears for both felling and topping. A mast, which carries the shears and a delimbing head, is positioned by a knuckle boom from a $\frac{3}{4}$-yard crawler crane base. The mast is extended against the standing tree, with the felling tool around the base of the tree and the hydraulic delimbing tool encircling the stem above it. The delimber is pulled rapidly up the tree by a line over a sheave on the top of the mast. At the desired top diameter the topping tool shears off the top while the delimber maintains a grip on the stem. The delimber is dropped two-thirds of the height to hold the stem while the felling shear severs it from the stump. The felling shear acts as a platform while the stem is swung to bunching position by the boom. The mast is tilted forward to guide the fall of the tree

Fig. 11–8. Closeup of the hydraulic loading head which grasps the tree (upper) and the felling shear (lower) of the Sicard Feller-Skidder. (Courtesy of the Pacific Car and Foundry Company.)

and the delimbed tree released. The rated hourly production of the Harvester is 5 cords. In stands of black spruce and jack pine averaging 8 inches in stump diameter, a productive rate of 55.4 trees per hour has been achieved. The tree lengths are laid down in bunches for later skidding to the truck road. The complete Beloit harvesting system includes a Beloit wheel skidder and a Beloit heel boom loader.

Tractor-Mounted Shears. The most recent development in mechanized felling is the mounting of hydraulic tree shears on crawler tractors or on wheel skidders. The shear unit consists of a fixed arm

Fig. 11–9. A Beloit Harvester delimbing a tree. The delimbing head is seen about half way up the tree. (Courtesy of the Beloit Corporation, Beloit Woodlands Division.)

or bar and a pivoting cutter blade actuated by a horizontal hydraulic cylinder. The blade is beveled to act as a wedge to fell the tree in a direction perpendicular to the axis of the tractor. Vertical movement of the front-mounted shear is by a hydraulic cylinder in front of the radiator. The vertical movement of the shear mounted on the rear of a wheel skidder is controlled by the winch line running through the fairlead. Pin connections for the shear frame and quick-connect hydraulic couplings permit the shear to be removed quickly to use the tractor for skidding. Rear mounting leaves the front-mounted blade on the tractor for clearing windfalls and heavy brush from its path.

In felling with the shear, the blade is retracted, the unit driven against the tree, lowered to ground level, and the butt sheared in a scissor action. Advantages of shear felling include labor-saving, salvage of wood volume left in the stump by hand felling, and safety, since the operator is protected by the tractor canopy and the wedge action of the blade ensures the direction of fall. Skidding is facilitated by the absence of high stumps, and by brush being trampled down by the felling tractor. Felling trees at ground level is also advantageous if planting machines are used for reforestation.

The Roanoke Tree Shear, illustrated in Fig. 11–10, is manufactured in three models for maximum stump diameters of 19, 22, and 26 inches, with jaw openings at the tips of 31, and 37 inches, and weights of 2300, 3000, and 3650 pounds. One of the safety features is a roll bar on the arm to deflect crooked or leaning trees away from the tractor.

Fig. 11–10. Roanoke Tree Shear mounted on International crawler tractor. (Courtesy of Industrial and Logging Equipment Sales, Raleigh, North Carolina.)

The Fleco Hydraulic Tree Shear is manufactured in five models for mounting on tractors of the Caterpillar D4, D5, and D6 sizes. The unit may be frame mounted or blade mounted. The maximum stump diameter cut is 26 inches. The weight of the unit ranges from 3,235 to 4,000 pounds. The Fleco has a vertical bar on the fixed arm which acts as a kicker to throw the log butt up so that the tree top hits the ground first. This is said to keep the butt up for faster choker setting when skidding tree lengths.

The Esco tree shear with a guillotine blade was introduced in 1970. Unlike the scissor shears, the thin steel blade closes parallel to the anvil, and is operated by twin hydraulic cylinders. An integral wedge provides controlled fall of the tree 90° to the right. A metal rim around the blade is designed to hook the butt of a sheared tree when it is hung up in order to pull it down. The entire unit can be raised to a vertical position to push logs for decking. The first model produced will shear a maximum diameter of 24 inches.

A test of the capability of the hydraulic tree shear in 30 inches of snow and subzero temperatures of −10° to −40°F. was made in the Prince George area of British Columbia in March 1967. The trees were pine and spruce with diameters of 7 to 28 inches, averaging 11 inches in pine and 16 inches in spruce. The stand volume was 45 cunits per acre and clear cutting was practiced. The trees were topped at 4-inch diameter and the tree lengths skidded to landings by rubber-tired tractors. The terrain was level to rolling with a maximum slope of 15 percent. A Roanoke Tree Shear mounted on a D4 Caterpillar tractor was used in the test. The operator worked a face toward the landing, cutting a swath 10 to 15 feet wide at each pass. A chain saw operator followed 150 to 200 feet behind the machine to buck the tops. In $61\frac{1}{2}$ hours spent in cutting, the Roanoke felled 4,183 trees or an average of 68 trees per hour. This time did not include service or down-time or travel between landings. The wood volume cut was 12.3 cunits per hour. Estimated costs showed a savings of 14 percent of chain saw felling cost. Skid trail cost was reduced 50 percent. The stumps were 8 to 16 inches lower than the hand-felled stumps.

SUGGESTED SUPPLEMENTARY READING

Suggested supplementary reading on pulpwood production is given at the end of Chapter 12.

12

Pulpwood Production: Transportation

SKIDDING PULPWOOD

Tractors

The tractive skidding methods described in Chapter 7 are used for skidding tree-length or log-length pulpwood. The smaller models of all the well-known makes of crawler tractors are used in the pulpwood industry. A single-drum winch and a bulldozer blade are usually mounted on the crawler tractor. A variety of accessories are used to elevate the front end of the turn to reduce dragging resistance. Skidding accessories include integral arches, rubber-tired arches or sulkies, steel pans, and home-made wood sleds or drays. Ground skidding from the drawbar of the tractor, or from the winch drum, is often done in the smaller operations.

The rubber-tired farm tractor with large rear wheels and small front wheels is widely used by farmer-producers. For skidding the spacing of the front wheels is widened to that of the rear wheels, instead of the narrow gauge used for crop cultivation or for towing pulpwood carts. For use on swampy ground or in snow the farm tractor may be converted to a semi-belt tractor by mounting idler wheels in front of the rear wheels and encircling drive wheel and idler with a rubber belt with steel grousers. The "ARPS Half-Track" illustrated in Fig. 12–1 is representative of such attachments. The

Fig. 12–1. Towing a pulpwood cart with an International farm tractor equipped with ARPS Half-Track in southern bottom land. (Courtesy of the ARPS Company.)

ARPS is available to fit 128 models of 8 makes of farm tractors. Dynamometer field tests have shown that the Half-Track increases the drawbar pull of the farm tractor as much as 488 percent. The increase varies with soil type and moisture content.

Another type of tractor used by some pulpwood producers operating in snow or swamps is the belt tractor. The Bombardier tractor has a bogie with 3 or 4 rubber-tired wheels, the number varying with the model, and a drive sprocket, all of which are encircled with a rubber and nylon belt reinforced with steel cable. Steel grousers are provided for added traction. The Bombardier may be equipped with a tilting platform and winch for forwarding shortwood, or with an integral arch for skidding longwood. A belt-type trailer is available for hauling pulpwood. The International Harvester "Track Skidder" has a steel track encircling 3 rubber-tired bogies. The grooved rear tire is the drive wheel. The belt tractor offers the advantages of low unit ground pressure, faster speed than the crawler tractor, and greater gradeability than the wheel tractor.

Wheel Skidders

The development of the four-wheel-drive rubber-tired tractor with articulated or center-pin steering by hydraulic cylinders, has revolutionized pulpwood skidding and accelerated the trend to tree-length systems. These machines are generally termed **wheel skidders.** With a speed of about three times that of the crawler tractor and ability to work in terrain formerly considered unsuitable for wheel tractor skidding, the wheel skidder has doubled economic skidding distance and road spacing. The center-pin steering gives the wheel skidder greater maneuverability between trees than front wheel steering. The front axle oscillates, enabling one wheel to rise as much as 35 inches, in some models, above the other. This tends to keep all four wheels in contact with the ground despite uneven surfaces. For operating in the alluvial sand and silt soil in the southern bottomlands, the wheel skidder is equipped with extra large tires. For operating in snow or on ice tire chains may be used.

Most wheel skidders are equipped with an integral arch, mounted over the rear axle, a single-drum winch behind the driver's seat, and a light bulldozer blade mounted in front. The blade is used for clearing skid trails and decking at the landing. Several methods of connecting the chokers with the winch line are used. One popular method is illustrated in Fig. 12–2. Four sliding butt hooks are threaded on the winch line. The choker cable has a swaged ferrule or knob on each end. One end engages the sliding choker hook and the other end is inserted in the butt hook. In pulpwood operations the wheel skidder usually skids longwood or tree lengths. Figure 12–3 illustrates skidding of a turn of nine or more pieces of pulpwood. Figure 12–4 shows how full trees are skidded. The wheel skidder has also been adapted to forwarding or prehauling short-wood.

The Garrett "Tree Farmer" is credited with being the pioneer wheel skidder. It was first used in logging in 1957. In 1969 the number of makes of wheel skidders manufactured in North America

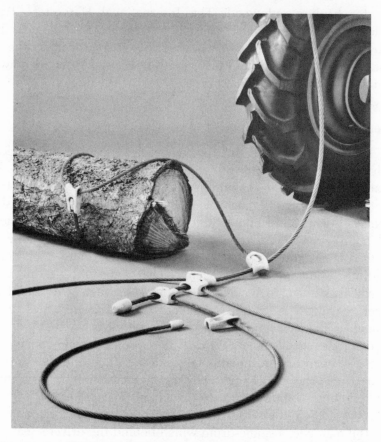

Fig. 12–2. ESCO rigging for wheel skidders showing sliding butt hooks threaded on the mainline and chokers with a swage ferrule on each end and a Bardon-type hook. (Courtesy of the Esco Corporation.)

had grown to 26. Five models of the "Tree Farmer," ranging in horsepower from 45 to 170, were offered. It was manufactured in Canada as the "Can-Car Tree Farmer," in Sweden, and in the United States.

While the use of the crawler tractor for skidding is declining, it is still an essential machine in the pulpwood operation for road construction, clearing skid trails and landings, and other bulldozer tasks. For some soil, terrain, and snow conditions producers still prefer the crawler tractor for skidding.

Wheel skidders have attracted more research interest than any other pulpwood production subsystem. One of the most compre-

Fig. 12–3. Skidding pulpwood with a "Tree Farmer" wheel skidder in Ontario. (Courtesy of the Canadian Car Fort William Division, Hawker-Siddeley Canada Ltd.)

hensive research projects, "The measurement of environmental factors and their effect on the productivity of tree-length logging with rubber-tired skidders," is being conducted by the Pulp and Paper Research Institute of Canada. A preliminary report indicates that volume per turn is the most important factor, that tree size and volume per acre are significant, but that variations in skidding distance over a normal range are relatively unimportant. Human factors, skidding crew, supervision, and planning, may account for more production variation than all other factors combined. (Bennett 1965)

Remote Winch Control. In many operations the tractor driver also does the choker-setting. If a number of scattered logs have to be bunched behind the tractor to make up a turn, the driver has to get down from the tractor to set the choker and then climb up on the

Fig. 12–4. Skidding full trees with a "Tree Farmer" wheel skidder in Ontario. Courtesy of the Canadian Car Fort William Division, Hawker-Siddeley Canada Ltd.)

tractor to operate the winch for each log. This is time-consuming and tiring. Radio controls to operate tractor and yarder winches from a distance have been used in Sweden for some time. In 1967 at least two American manufacturers introduced remote controls for hydraulically controlled tractor winches. The finger-tip control unit is compact and light in weight and can be carried in the pocket or strapped on the belt. It operates the tractor winch from distances up to 250 feet.

Hydraulic Skidding Grapples. The hydraulic grapple, which is widely used for loading logs, has been adapted to wheel skidding to dispense with chokers and choker-setting. The Esco skidding grapple, believed to be the pioneer in this field, is mounted on a knuckle-type boom which swings through an arc of 70 degrees to either side (Fig. 12–5). The tractor is backed to the log, the boom is swung to position the grapple over the log, and lowered to grab it. A supplementary winch is provided to pull in logs beyond the

Fig. 12–5. ESCO skidding grapple on Beloit wheel skidder. (Courtesy of the Esco Corporation.)

reach of the grapple. When a turn of several logs has been bunched behind the tractor, the turn is embraced by the grapple and the boom raised to the skidding position. To deck logs at the landing or alongside the truck road, the boom is swung to one side and the grapple opened to release the turn without stopping the tractor. Two models are offered, one with a boom reach of 9 feet and a grapple load diameter of 30 inches and the other with a reach of 12 feet which will handle logs 5 to 40 inches in diameter or three 14-inch logs. The ESCO skidding grapple is available for mounting on all makes of wheel skidders.

Several of the manufacturers of wheel skidders now offer their machines equipped with skidding grapples of their own design. A representative example is the Beloit Grapple Skidder, a 100 horse-power wheel skidder with a hydraulically controlled knuckle-boom which swings 90 degrees to either side with a horizontal reach of 12 feet and a hydraulic grapple which will hold a turn of up to 68 inches in diameter. Studies made in North Carolina found that production with the Grapple Skidder was 26 percent more than with the choker skidder with a chokerman and 62 percent more than when the skidder driver set the chokers. The swinging boom of the Beloit grapple skidder permits it to also be used as a loader to load tree lengths on a preloaded trailer. This permits operating with a production unit of three men—a faller, a skidder operator, and a truck driver.

Cable Yarding

Where topographical and ground conditions make tractor skidding difficult, cable yarding of pulp timber is practiced. Most of the cable yarding machines combine the dual functions of yarding and loading. In the South the "Loggers Dream," described in Chapter 9, is used for yarding swamps and steep slopes. In the mountains of the Tennessee Valley area, the "Big Stick, Highlead," an adaptation of the "Big Stick Loader," is popular for yarding bundles of pulpwood. The "Big Stick Highlead" is described under pulpwood loading. A variety of jammers, usually home made and mounted on used trucks, are used in localities where the terrain is steep or broken. The jammer has a mast and pivoting boom and a two-drum winch for in-haul and out-haul lines. The winch may be powered from the truck engine through a power take-off, or by a separate engine. In eastern Canada cable yarding of both longwood and bundles of bolts was formerly done. The bolts were hand-piled in stacks of 1 to $1\frac{1}{4}$ cords in parallel rows 60 to 66 feet apart, and up to 700 feet long. The lead was from a 44-foot A-frame on a sled.

Animal Skidding

Animal skidding with mules in the South and with horses in the northern regions is not yet obsolete. As an authority in pulpwood production has stated, "No machine has yet been invented that can operate on all topography, in small timber, partial cutting and scattered stands as efficiently as the mule" (Walbridge, 1960). In the Tennessee Valley a mule will skid 6 to 8 cords a day of multiple-stick or tree lengths. Skidding is done by a single animal and a driver, who does the hooking and unhooking with either a grab hook or a chain and slip hook. The log chain is hitched to a grab hook or jews-harp ring which is fastened with swivel and clevis to the whiffle-tree. Ground skidding is the most common practice. Sometimes a light sled is used to elevate the front end of the turn.

Such sleds are colloquially termed bobs, bogans, mud-boats, or go-devils. Double sleds used on snow to carry the wood are variously termed drays, scoots, or travois. Skidding chain-wrapped bundles of shortwood sidewise down steep slopes on snow is termed "chaining."

A set of well-fitting harness and the proper type of shoes is essential to obtain good performance from a skidding animal. Shoes with short flat caulks are used on earth and longer sharp caulks on snow or ice. A well-trained skidding horse requires little guidance from the reins and will often perform much of the skidding cycle by voice commands alone. A disadvantage of animal skidding in seasonal or large company operations is the necessity of giving the animals feed and care every day of the year. In some localities men who can handle animals or are willing to work with them are scarce. The farmer-producer can use his animal on the farm when not engaged in pulpwood production. The low capital investment required for animal skidding, as compared with tractor skidding, makes the method attractive to the small producer. Animal skidding is advantageous in farm woodlots and other small tracts of timber and is acknowledged to be the method least damaging to the residual stand where light partial cuts are made.

FORWARDING OR PREHAULING

The operation of moving shortwood from stump site to the landing is termed forwarding in the northern regions and prehauling in the South. A great variety of equipment is used to perform this operation, which may be classified under two general types. In the older type of operation the wood is carried on a trailer, known as a cart or buggy, in a pallet, or on a sled. The carrier is towed behind a tractor. The trailer may be wheel- or crawler-mounted, and every type of tractor is used as the prime mover. The carrier may be loaded by hand or by a mechanical loader mounted on the tractor (Fig. 12–6). In the newer type of operation the wood is carried on the tractor, which is self-loading. The wood may be placed in a rack behind the tractor cab, carried in a forklift in front, or in a sling-wrapped bundle hoisted to the fairlead of a skidder. Wheeled vehicles designed especially for forwarding, as well as conventional tractors, are used. In the southern states prehauling is sometimes

Fig. 12–6. Loading a self-loading pulpwood buggy hitched to a crawler tractor. (Courtesy of the Hiwassee Land Company, Calhoun, Tennessee.)

done with all-wheel drive surplus military trucks or with tilt-bed pallet trucks. In areas where the haul truck is customarily driven to stump site for loading, if ground conditions do not permit a full load, "short-hauling" is done. A half-load is hauled to the truck road, unloaded, and reloaded when the truck returns with another half-load.

Conventional tractor skidding equipment is adapted to forwarding by wrapping a bundle of pulpwood with a sling or chain and hoisting it with the winch line to the fairlead of a wheel sulky, crawler arch, or the integral arch of a wheel skidder. Among the special tractor attachments available for forwarding are the front-mounted forklift, a rear-mounted or crawler trailer-mounted hydraulic grab, and a rear-mounted pulpwood box or bucket. The forklift scoops up a load, lifts it for forwarding, and transfers it to the haul truck. The Continental Emsco pulpwood box is mounted on two lift arms on the rear of a wheel skidder (Fig. 12–7). The box is lowered to ground level for manual loading and raised for forwarding. The one-half cord capacity box can be raised to a height of ten feet to discharge onto a truck or into a pallet. The steel skidding pan has been converted to a pallet pan by a Mississippi operator by welding two lengths of angle iron on the pan to form a channel for the pallet and two one-inch keels on the bottom. Both the pan and the pallet are attached to the tractor drawbar by chains.

Fig. 12–7. Continental Emsco pulpwood box mounted on a Timberjack wheel skidder. (Courtesy of the Hiwassee Land Company, Calhoun, Tennessee.)

Pallets

The most widely used accessory for prehauling is the **pallet.** The Dixie pallet is shaped like the letter U flattened at the base. The members which enclose the wood are two lengths of steel tubing joined by transverse bars. The five standard Dixie pallets range in capacity from 1.72 to 2.5 cords of 6-foot wood. The Pulp-Pal pallet is similarly shaped, but is fabricated from steel I-beams and plates.

The Pulp-Pal is offered in six sizes with capacities of from 1.5 to 3.1 cords of 6-foot wood. Pallets are also home-made of hardwood lumber. The pallet is usually filled at the stump by hand. It may be skidded to the landing behind a tractor, or winched onto a tilt-body pallet cart or truck and towed. At the landing the bed tilts to slide the pallet off (Fig. 12–8). The loading pallet is winched onto the

Fig. 12–8. Tilt-bed truck for prehauling a pulpwood pallet. (Courtesy of the Hiwassee Land Company, Calhoun, Tennessee.)

haul truck or trailer, and the wood remains in the pallet until it reaches the concentration yard. Self-loading pallet carts with a loading boom mounted on the cart and powered from the tractor winch are a labor-saving improvement over hand loading (Fig. 12–9).

Several machines have been specially designed to prehaul pallets. The Harrison Pulpwood Harvester couples a two-wheel cart to a farm tractor from which the front wheels are removed. Steering is accomplished by opposing hydraulic cylinders attached to tractor and cart. The pallet is loaded by a hydraulic grapple mounted at the rear of the tractor and discharged by tilting the cart bed. The

Fig. 12–9. Loading hardwood pulpwood on a self-loading pallet buggy towed by a John Deere crawler tractor. (Courtesy of *Pulpwood Production and Saw Mill Logging Magazine.*)

capacity of the machine is one 6-foot pallet holding 1.6 to 1.8 cords. The Terrain Master is a four-wheel-drive truck with a capacity of two 1.6 cord pallets. The frame tilts to load the pallets with a winch cable and to discharge them at the landing.

A typical landing scene in a southern pine pulpwood operation using the pallet system is shown in Fig. 12–10. At the left a pallet is being winched onto the haul truck. Pallets which have been pre-hauled to the landing are seen in the right background. In the fore-ground are empty pallets awaiting return to the woods.

The popularity of the wheel skidder has led to the development of removable attachments to convert the skidder to a forwarder. When low volumes per acre or tree selection cutting make tree length skidding unfeasible, the change to the shortwood system is made. One example of such an attachment is the "Paypacker" de-signed for mounting on the Hough "Paylogger" wheel skidder. The skidding arch and butt plates are removed, and a cradle bolted on. The cradle holds $1\frac{1}{2}$ cords of 4-foot wood. To load the cradle, it is

Fig. 12–10. A landing in a southern pine pulpwood operation using the pallet system. (Courtesy of *Pulpwood Production and Saw Mill Logging Magazine.*)

tipped down and a pile of wood pulled onto the cradle by the winch cable. The cradle is tipped up for forwarding, and the cable tightened to hold the load secure. At the landing the cable is released and the cradle tipped to discharge the load.

For the "piggy-back" method of forwarding, a frame with two vertical curved arms, one on each side of the arch fairlead, is attached to the rear of the wheel skidder. A sling is placed around a bundle of wood and hooked to the winch line. The bundle is yarded to the skidder and hoisted up against the arms. The winch brake is set to hold the bundle firmly in contact with the arms for forwarding.

Forwarders

The Koehring Forwarder, formerly known as the Dowty Forwarder and the Koehring-Waterous Forwarder, was developed in Canada to transport eight-foot pulpwood. The four-wheel drive, rubber-tired, articulated-steering tractor has a carrier or cradle over the rear axle and a hydraulic knuckle-boom grapple loader mounted between the cab and the cradle (Fig. 12–11). The cradle capacity

Fig. 12–11. The Dowty Forwarder, now marketed as the Koehring Forwarder, with a load of three and one-half cunits of 8-foot pulpwood in Canada. (Courtesy of Dowty Equipment of Canada, Limited.)

is three cunits; and an additional one-half cunit is transported in the grapple, which rests on two arms projecting from the front of the vehicle. Powered with a 100-horsepower diesel engine, the machine has a maximum road speed of 15 miles per hour. The cutters bunch the pulpwood at the stump in piles of one-half cunit or more. The Forwarder is driven alongside the piles, and the wood loaded into the carrier with the grapple which can reach out 13 feet 8 inches at ground level, and rotate 200 degrees to grasp bolts lying at any angle. The boom rotates full circle. The forwarder then transports the $3\frac{1}{2}$-cunit load to the truck road and off-loads onto a waiting truck or trailer, or into pallets. The production potential of the Forwarder is said to range from four cunits per hour under poor operating conditions to eight cunits per hour under excellent conditions, forwarding an average distance of 800 feet.

LOADING PULPWOOD

Most of the mobile machines and methods used for loading logs described in Chapter 9 are also used for loading shortwood. Both the hydraulic **knuckle-boom** with **grapple** and the **forklift** type of tractor loader are widely used in all the pulpwood producing regions. The pulpwood grapple has more tines and greater width of bite than the log grapple, in order to grasp a bundle of loose sticks, and is usually power rotating. It is sometimes called a "clam" for its resemblance to the clam-shell earthmoving bucket (Fig. 12–12).

Fig. 12–12. Loading a pulpwood truck with a Hendrix Special loader. (Courtesy of the Hiwassee Land Company, Calhoun, Tennessee.)

Special pulpwood handling attachments are available for forklift loaders. For example, attachments offered with the Pettibone Cary-Lift include a hydraulic swivel clam, hydraulically closed and side-shifted forks, a hydraulically tripped sling, and an orange peel clam for shortwood, as well as a tree-length baler. Forklifts are mounted on crawler tractors, industrial wheel tractors, and articulated wheel loaders. Hydraulic grapple booms are mounted on self-propelled rubber-tired carriers, on all types of tractors, and on self-loading trucks. Cable-operated straight booms are used as well as hydraulic knuckle-booms. The loader on some of the self-loading prehaulers or forwarders can also be used to reload the wood on trucks. The self-loading truck is more popular with the pulpwood producer than with the logger, since lighter loaders can be used with consequently less reduction in payload. Among recent developments in pulpwood loaders are the Hendrix Special hydraulic knuckle-boom grapple loader mounted on a farm tractor (Fig. 12–12) and the pulpwood bucket mounted on a Timberjack wheel skidder. Manual loading of pulpwood trucks is still practiced by small producers in the southern states. Loading tree length pulpwood with a forklift loader is illustrated in Fig. 12–13.

Fig. 12–13. Loading tree lengths on a pre-loaded trailer with a Drott fork-lift on an International tractor. Ground skidding in the foreground. (Courtesy of the American Pulpwood Association.)

The Melroe "Bobcat," a pulpwood loader which is unique in its small size and maneuverability, is shown in Fig. 12–14. With a wheel base of only 35 inches, width of 60 inches, and weight of 4,200 pounds, the 24-horsepower rubber-tired loader can lift 1,000 pounds or about one-fifth cord to a height of $10\frac{1}{2}$ feet. A hy-

Fig. 12–14. Loading hardwood pulpwood on a truck with a Melroe "Bob-cat" loader. (Courtesy of the Hiwassee Land Company, Calhoun, Tennessee.)

draulic grapple is fixed vertically at the ends of two hydraulic lift arms. It is designed to load wood into pallets or onto trucks at stump site. It will load a 1.7 cord pallet in 6 to 10 minutes if sufficient wood is conveniently available or in 20 minutes from scattered piles. A bumping plate on the front of the grapple is used to align the ends of the loaded sticks. (Cline 1967)

The Universal "Woodtick" is a similar type of pulpwood loader recently introduced in the Southeast. The machine is only 59 inches wide, and the five-time grapple fork raises to a height of 109 inches. Optional engines of 26, 30, or 36 horsepower are offered. Production of 8 cords per hour from stump to truck is reported.

Pallet Loading

The truck used for hauling pallets is equipped with a special bed with L-shaped rails on which the pallet slides for loading, and with two removable or folding skid ramps at the rear, which extend from the truck bed to the ground (Fig. 12–15). The pallets are

Fig. 12–15. Loading a truck with pulpwood pallets by pulling them up skid ramps with a winch line. (Courtesy of the Hiwassee Land Company, Calhoun, Tennessee.)

winched up the ramp and forward on the rails by a line on a power-take-off winch mounted behind the cab and below the bed. A fairlead spool for the winch line is mounted near the rear of the truck frame. If a trailer is used, it is equipped with rails and skid ramps, and the pallets loaded with the truck winch. The Evans pallet trailer has a large roller on the end of the rear ramp. Trailers may be preloaded with a tractor winch to save truck-tractor turn-around time. In a Tennessee Valley study, it was found that truck turn-around time averaged 6 minutes per cord when the truck winch was used for loading pallets. The 1.6-cord pallet capacity of a truck, or truck and trailer, is as follows: single drive axle truck, 2 pallets; tandem axle truck, 3 pallets; single axle combination, 4 pallets; tandem axle combination, 5 pallets.

At the woodyard the wood may be unloaded without removing the pallets, or the pallets may be off-loaded to hold the wood in temporary storage. The Brantford pallet truck bed, made in Canada, has a tipping frame which is raised by twin hydraulic cylinders to slide the pallets off at the haul destination. The pallet system has been adapted to longwood with pallets 32 feet in length holding 8 to 10 cords.

Big Stick Loader

The Big Stick loader is illustrated in Fig. 12–16. A frame 14 feet long is bolted to a bobtail truck with a capacity of 2.6 to 3.5 cords. The loader frame carries a center post, end bulkheads, and a winch powered from the truck engine. A 4-foot horizontal boom swings full circle on the center post. The $\frac{5}{16}$-inch loading line runs through sheaves in the center post and at the boom tip. Models with either manual swing or hydraulic power swing are made. Tongs are used for loading the relatively heavy hardwood sticks. To load pine pulpwood, the end of the loading line is looped around a bundle

Fig. 12–16. Loading hardwood pulpwood on a "bobtail" truck equipped with a Big Stick loader, using tongs. (Courtesy of the Hiwassee Land Company, Calhoun, Tennessee.)

of from 3 to 10 sticks and hooked. A man on the ground pulls out the loading line and chokes the bundle. The truck driver operates the winch controls to hoist the bundle, and the power boom controls. The third member of the loading crew stands on the load, swings the manual boom to position the bundle, and unhooks the line. Bundles can be yarded from a distance of 250 feet, but usually the wood is piled within 50 feet of the truck road. The Big Stick is popular with small producers in the Tennessee Valley area, where it was developed, because it requires less capital investment than other types of mechanical loaders.

Big Stick Highlead. To extend the economical yarding distance and widen the truck road spacing the Big Stick Highlead, shown in action in Fig. 12–17, was developed in Tennessee in 1961. To provide

Fig. 12–17. Yarding a bundle of pulpwood sticks along a cleared path with the Big Stick Highlead. (Courtesy of the Hiwassee Land Company, Calhoun, Tennessee.)

a haulback, a winch with a capacity of 800 feet of ¼-inch line and a haulback mast and fairlead were added to the Big Stick loader equipment. The mainline drum capacity was increased to 400 feet. The pulpwood cutters bunch the wood in piles of about one-tenth of a cord along brushed-out yarding roads 50 to 100 feet apart. The haulback line is threaded through a corner block and a tail block hung at the back end of the yarding road. The mainline is threaded through a light fall block which is attached to the end of the haul-back line. The fall block and mainline is pulled out to a pile of wood by the haulback line. Enough slack is pulled by hand in the mainline to enable it to be looped around a bundle of wood and hooked. The bundle is yarded to the truck and loaded in a continuous operation. The Big Stick Highlead has proved to be an effective system for the small producer operating in mountainous terrain.

Rail Car Loading

A specialized model of the industrial lift truck is used in many pulpwood concentration yards for unloading haul trucks and loading rail cars. A number of manufacturers of lift trucks offer pulpwood attachments. As an example, the Hyster Pulpwood Loader is shown in Fig. 12–18. The upright frame mounted on the 4-wheeled industrial truck tilts forward 14° and backward 6°. A pulpwood

Fig. 12–18. Lowering a sling load of pulpwood, hoisted from the truck at the right, onto a rail car with a Hyster Pulpwood Loader. (Courtesy of the Hyster Company.)

boom rides up and down the upright frame, and is inclined at an angle which centers the lift over the truck. Two wire rope slings hang from the front of the boom. To unload a truck the slings are looped around $1\frac{1}{2}$ to 2 cords and boom raised. The loader moves to the rail car and transfers the sling load. When loading two rows of wood, the upright is tilted forward to load the far side of the car first. When a train of cars is loaded, the loads are trimmed. The upright is tilted back so the boom is vertical, and in a series of short runs the bumping plate is rammed against the ends of the sticks (Fig. 12–19).

Fig. 12–19. Trimming the loaded rail car to align the ends and bind the sticks with the bumping plate of the Hyster Pulpwood Loader. (Courtesy of the Hyster Company.)

This binds the sticks securely in place and aligns the ends. The rail car for 63-inch pulpwood is built with a 6-inch pitch toward the center so the sticks slope inward. The loader travels the length of the train trimming one side, then returns trimming the other side. At a North Carolina yard with a 22-car side track, the Hyster Pulpwood Loader handles 125 cords a day.

For unloading trucks carrying pulpwood loaded lengthwise on the truck, which is termed "mountain style" in the South, a 90° swivel boom is offered by Hyster. The boom is swiveled for unloading and then returned to the normal position for carrying and loading the rail car.

Where longwood is delivered to the concentration yard it must be bucked before it is loaded on shortwood rail cars. The Curry "Cost Cutter," which bucks the wood on the truck, is shown in Fig. 12–20. The cutting is done by a chain saw which travels around a

Fig. 12–20. Curry Cost Cutter bucking 21-foot longwood into 63-inch pulpwood lengths. (Courtesy of the Hiwassee Land Company.)

powered sprocket and three idler sheaves. The machine moves on steel rails to precisely position the saw. In the South, pulpwood to be bucked by this machine is trucked to the yard in 21-foot lengths. The longwood is bucked into four 63-inch shortwood lengths by sawing three vertical cuts through the load. The shortwood is then unloaded from the truck and reloaded on the rail car by the pulp-wood loader.

Load Aligning. When shortwood is loaded crosswise mechanically the ends of the sticks do not line up evenly. It is necessary to push the protruding ends into alignment in order not to exceed the legal width limit on the highway, or to meet the width requirement of the railroad, and for safety. This operation is variously termed aligning, bumping, or trimming. Trimming rail cars with the industrial lift truck has been described. One method of bumping with a hydraulic grapple loader is to hold a short section of a large-diameter log in the grapple and ram it against the ends of protruding sticks. The log section is carried on the rear of the loader when not in use. The rotary load aligner has two large steel drums mounted vertically on axles so that they will rotate freely. They are spaced at a distance equal to the length of the pulpwood bolt. When the loaded truck is driven slowly between the drums, any protruding bolts are squeezed into alignment. The entire load is also realigned symmetrically with the center line of the vehicle.

MAJOR TRANSPORTATION

Every mode of major transportation covered in Chapter 10, "Log Transportation," is also used for transporting pulpwood. The most widely used method of hauling from the forest to the wood yard is by motor truck or truck and trailer. The common carrier railroad is the most commonly used way of transporting pulpwood from the con-centration yard to the mill. Operations tributary to rivers or lakes suitable for towing barges or rafts find water transportation the most economical method. Driving pulpwood down streams or rivers, the oldest method of major transportation, is now obsolete except in eastern Canada and in a few operations in the Northeast.

Motor Trucks

The vehicles used for truck transportation range from the one-ton flatbed truck to trailer trains of up to 160,000 pounds gross combina-tion weight. Load capacities range from 1.44 cords on the one-ton truck to 27 cords on a nine-axle truck and trailer train combination permitted pulpwood haulers on the highway in Michigan. In most states and provinces the legal limit on the highway for the tandem-axle truck and semi-trailer, which is the most economical vehicle for the longer pulpwood hauls, is 74,000 pounds gross combination weight. Payload varies with species density and moisture content, and is usually about 9 or 10 cords of softwood. Other vehicles used for hauling pulpwood are single- and tandem-drive-axle trucks, single- and tandem-axle semi-trailers, and full trailers with two, three, or four axles. The legal width permitted on state highways is 8 feet and wider loads require a special permit. Examples of popular tractor-trailer combinations used for pulpwood transportation are shown in Figs. 12–21, 12–22 and 12–24. Bobtail trucks are shown in the pulpwood loading illustrations.

Fig. 12–21. Single drive axle tractor and Fruehauf tandem-axle pallet trailer. (Courtesy of the Hiwassee Land Company, Calhoun, Tennessee.)

The selection of a vehicle for pulpwood hauling is determined by the length of haul, the road conditions, the terrain and consequent road gradients and curvature, and the public highway weight restrictions. In the case of the small operation, the capital available to the producer to invest in a truck is often a determining factor. The vehicle which is the most popular with such producers in the South is the 1½- or 2-ton bobtail truck carrying 2.8 to 3 cords. In the larger operations the vehicle which will carry the maximum legal load on the highway is favored. The wage rate for truck drivers is an important consideration in selecting the vehicle size, as the larger the payload the lower the labor cost per unit.

The use of preloaded trailers to reduce truck turn-around time appears to be increasing. This is known as **shuttle-hauling** in the southern states. Several methods of supporting the front end of the semi-trailer when it is detached from the truck-trailer are offered by trailer manufacturers. The Evans-Busch pulpwood crib frame trailer has a mobile axle or dolly which is removed when the empty trailer is spotted for loading. The trailer stands on two sets of stiff legs or feet. The truck-tractor backs the mobile axle under the rear of a loaded trailer, engages the front end of the trailer with the fifth wheel and the vehicle is ready to travel. A tandem axle dolly with a crib capacity of 8 cords is also manufactured. A dolly with crawler tracks is also available for off-the-road towing by a tractor. At the truck road the crawler is replaced with a rubber-tired dolly for the truck haul. The change can be made by one man in 3 minutes.

A different type of preloading system is the Nelson Batson Bunk. Longwood is loaded on false bunks in a steel frame (Fig. 12–23). The bunks are lowered to near the ground for ease in loading and placing the wrappers and binders, and raised by hydraulic pistons. The tractor-trailer backs under the load which is then lowered onto the vehicle bunks. (Fig. 12–24). The false bunks are swung to the side, and the haul vehicle driven away. In a thinning operation, where 50 to 70 pieces constituted a legal highway truck load, unbucked tree lengths were skidded and loaded into the Batson. They were then bucked to 50-foot lengths. Built-in bunk scales weighed the load. The average turn-around time of the haul vehicle was 5 minutes.

Railroads

The common carrier railroad is of major importance in the transportation of pulpwood beyond economical trucking distance. Many pulpmills procure wood from areas several hundred miles away and much of it is shipped by rail. One mill in Tennessee obtains wood from six adjacent states as well as from Tennessee sources. Pulpwood has been shipped 800 to 1,000 miles from the Rocky Mountains to Minnesota and Wisconsin mills.

Fig. 12–22. Tandem axle truck and full trailer in the Lake States Region. Combination is self-loading with the Prentice hydraulic boom loader mounted on the rear of the truck. (Courtesy of Prentice Hydraulics, Incorporated.)

Common carrier railroads which haul a large volume of pulpwood provide special freight cars for this purpose. Flat cars with stakes along the sides are used for wood loaded lengthwise. Flat cars with a bulkhead at each end are provided to load wood crosswise. The outer edges of pulpwood cars used in the southern states for 5-foot or 63-inch wood are 6 inches higher than the bed so the two rows of sticks will slope downward to the center line of the car. Standard open-top gondola cars are used with either permanent or temporary stakes at the ends to hold wood piled crosswise above the height of the car sides. The Canadian National Railway has introduced a pulpwood car in which 8-foot bolts are loaded vertically. They are unloaded by an overhead crane with a grab. All of the special

Fig. 12–23. Nelson Batson Bunk lowered ready for loading. The loader is seen in the background at the left, and the departing haul vehicle in the center. (Courtesy of the Nelson Equipment Company.)

pulpwood cars are designed so that they can be loaded and unloaded mechanically.

The Southern Railway pulpwood car in Fig. 12–25 is 38 feet 7 inches inside length, with a load capacity of 110,000 pounds. The average load is about 30 cords of pulpwood. Other Southern Railway pulpwood cars are 36 feet 6 inches inside length carrying 20 cords and 45 feet 9 inches carrying 35 cords. The admonition stenciled on the side of the bulkhead states: "Overhang at side sill must not exceed 10 inches."

Barges

Barging is the most economical method of long-haul transportation of pulpwood on the Mississippi and tributary rivers and on coastal rivers and waterways from Virginia to Louisiana. The many hydro-electric dams built in the Tennessee Valley have created new

Fig. 12–24. Lowering a load onto the tractor-trailer bunks by the pre-loaded Nelson Batson Bunk. (Courtesy of the Nelson Equipment Company.)

routes for barge traffic. The expansion of the paper industry in this area, in localities where there was said to be insufficient supplies of wood to support new mills, was undoubtedly influenced by the availability of barge waterways to bring wood long distances to the mill. The size of barge used is governed by the characteristics of the river or by the size of the lock which raises or lowers the barge between reservoir elevations at dam sites. The capacity of the barges used on Tennessee Valley waterways ranges from 200 to 240 cords with standard hopper to 300 to 350 cords with jumbo hopper (Fig. 12–26). Barges may be either towed or pushed by tow boats depending upon the current and the configuration of the waterway. Both conventional open-top and decked or flat-top barges are used in the southern states. The type used depends upon the method used to load the barge and the conditions at the barge landing where it is moored for loading. Use of the decked type permits the haul truck to drive onto the barge for unloading. Open-top barges are loaded by crane or lift truck of the type used in concentration yards.

Fig. 12–25. Southern railway car with a load of approximately thirty cords of pulpwood. (Courtesy of the Hiwassee Land Company, Calhoun, Tennessee.)

Fig. 12–26. Pulpwood barges on the Tennessee River. Loaded barges at left and unloaded barges at right. (Courtesy of the Hiwassee Land Company, Calhoun, Tennessee.)

Large barges of some 2,000 cords capacity and cargo ships carrying 3,000 cords or more are used for pulpwood transportation on the
Great Lakes and the St. Lawrence waterway. Most of the wood
imported from Canada by American mills is delivered by water
transportation, including booms or rafts as well as barges and ships.
Pulpmills on Puget Sound and the Strait of Juan De Fuca in the
state of Washington import chips in barges and pulp logs in flat
rafts from British Columbia.

Booms

Pulpwood is transported in **booms** or rafts on the Great Lakes,
the St. Lawrence waterway, and on the lakes and rivers of eastern
Canada. For towing by tug boats on Lake Superior the round boom
holding from 4,000 to 12,000 cords is used. The raft of loosely floating
bolts is enclosed by two or three strings of boom sticks. The boom
sticks are Sitka spruce or Douglas-fir logs from British Columbia,
40 to 48 inches in diameter by 20 to 24 feet long, joined at the ends
by 1½ inch chains. Towing round booms is a seasonal operation
limited to the period from mid-June to mid-September.

Several types of fabricated booms are used on Canadian lakes
and rivers. Three or four small logs are bolted together to make boom
sticks of larger bulk and vertical dimension to prevent loose bolts
from bobbing under or being tossed over the boom by waves. The
capacity of towing booms ranges from 700 to 10,000 cords. The bag
boom is used for towing through channels too narrow for a round
boom. The boom is "brailed" or pinched in by cables or chains to
form three "bags." In plan view, the front bag is shaped like a
spherical triangle and the other two bags are barrel-shaped. The bag
boom usually holds about 4,000 cords of wood.

Flumes are V-shaped troughs into which streams are diverted to
carry wood by gravity, used in many mountain regions before the
advent of the motor truck. They are now obsolete except in the province of Quebec.

Chip Pipelines

Research in transporting pulpwood chips from the forest in pipelines is being conducted by the Pulp and Paper Research Institute of
Canada and the United States Forest Service. In initial pilot plant

experiments in Canada, chips were mixed with water and pumped through 8-inch diameter aluminum pipe. One chip pipeline system envisioned in eastern Canada is as follows: Pulpwood would be debarked and chipped at the landing. Chips and water would be metered into a mixing tank, mixed into a slurry and pumped into a branch pipeline laid on the surface of the ground. The branch line would connect to a pumping station for a buried feeder line. The feeder line would connect with a pumping station at a buried main line, which would deliver the chips to the pulpmill. It is estimated that to transport 100,000 cords of chips a year would require 487 acre-feet of water for a 40 percent by volume chip-water mixture, and 647 acre-feet for a 30 percent mixture.

Among the anticipated advantages of the chip pipeline system are lower unit transportation costs, utilization of smaller wood, and reduction of wood inventories. Other possible benefits are the use of the pipelines in forest fire suppression and reuse of the carrier water at the mill.

MOBILE CHIPPING UNITS

One of the problems in handling small diameter logs is the time required to load the many pieces onto a truck and trailer. For economical hauling the vehicle which will carry the legal load limit on the highway is preferred. In loading second-growth thinnings, such a truck may be tied up for an hour or more, even when modern loading machines are used. The bark is a waste product on which freight is paid when logs are hauled. In 1961 the Crown-Zellerbach Corporation initiated the Utilizer program to debark and chip in the forest with a mobile machine and haul chips in large vans to the pulpmill. The material to be chipped was second-growth thinnings, 5 to 12 inches in diameter, and small understory trees in old-growth stands. Three successive models of the Utilizer were designed and built by the Nicholson Manufacturing Company of Seattle for Crown-Zellerbach and placed in service in the paper company's tree farms in Oregon and Washington. By 1969 Nicholson Utilizers were also in use in Canada, New England and the South. The three models manufactured are designated by the maximum diameter of log which could be processed, namely the 14-inch, 18-inch, and 22-inch.

The Nicholson Utilizer. The standard-model 18-inch Nicholson Uti-lizer is a self-propelled mobile debarker and chipper, 40 feet long and 10 feet wide, and mounted on rubber tires (Fig. 12–27). The

Fig. 12–27. Nicholson 18-inch Utilizer operating in Washington. A tree stem is being loaded into the conveyor. Debarker and chipper are behind the operator's cab and a chip van at the rear. (Courtesy of the Nicholson Manufacturing Company.)

maximum travel speed is 10 miles per hour and gradeability is 30 percent. Tree lengths are loaded into a log feed conveyor by a hydraulic knuckle boom and grapple loader with a reach of 30 feet. The conveyor has a 15-foot-long removable extension. The conveyor feeds the tree through a Nicholson Accumat ring-type debarker which also delimbs. Bark and limbs are ground up by a bark hog and blown by a fan to discharge on either side for a distance of 50 to 100 feet. The outlet may be adjusted to discharge to a pile or to scatter the bark as a mulch on the forest floor. From the debarker the tree is fed into a "V" drum chipper. A fan blows the chips into a van which is coupled to the Utilizer. The chipper power unit is 400-horsepower gas turbine. The secondary power unit is a 200-kilowatt diesel electric generator which supplies electric motors driving the barker, hydraulic pump, and rear axles. The conveyor is powered from the chipper shaft to synchronize log feed speed with chipper drum rotation. The debarker will take logs from 2 to 18 inches in diameter.

The Utilizer is operated by one man with controls in an elevated cab on the 360-degree swing loader turntable. The second man on the crew drives a truck-tractor to remove the filled chip van and bring an empty van to the Utilizer. He also adjusts the bark hog outlet to discharge where desired. The filled van is parked along the road where it can be picked up by the truck which hauls it to the pulpmill. The usual method of operation in second-growth stands is to drive the Utilizer along a truck road and feed it with tree-length logs previously decked alongside the road by tractors or wheel skidders. In old-growth stands the Utilizer sits at the landing to chip prelogged or relogged understory trees which have been previously cold-decked by a cable yarding system.

The production of the Utilizer depends upon the diameter of the logs and the number and distance of moves to reach log decks. At a feed speed of 100 feet per minute the chip yield will range from 14.7 units per hour with 6-inch average-diameter logs to 92 units per hour with 15-inch average-diameter logs. The unit is 80 cubic feet of solid wood. At one operation where the Utilizer was chipping thinnings, a van holding 10 cunits of chips was filled in 45 to 60 minutes and production was 6 or 7 van loads per shift.

Pulp industry interest in chipping remote from the pulpmill has led to the development of chipping units by several manufacturers of logging equipment or of stationary debarkers and chippers. The mobile "Bark-Chip-mobile," the portable "Roadside Chipper," and the semi-portable "Chip Mill" all incorporate the Cambio Debarker and the Soderham chipper. The "Chipharvestor" is a portable unit using a Morbark Debarker and a Morbark "Golden-Harvest" chipper. The "Cargator" is similar in type to the Utilizer, with the addition of a bucking saw located $8\frac{1}{2}$ feet ahead of the debarker to buck flared butts or crooks, and to salvage any portion of the tree stem which would yield a product of higher value than chips.

SUGGESTED SUPPLEMENTARY READING

1. BROMLEY, W. S. 1969. *Pulpwood production*. A textbook for use in courses in pulpwood harvesting at the high school and post-high-school levels. Second Edition. 259 pp. The Interstate Printers & Publishers, Inc., Danville, Illinois.

Intended primarily for vocational agriculture courses in pulpwood production in high schools. This well-illustrated book covers the subject in detail.

2. GARDNER, R. B. 1966. Designing efficient logging systems for northern hardwoods using equipment production capabilities and costs. 16 pp. N. Cent. Forest Exp. Sta., St. Paul, Minn., illus. Research Paper NC-7. Describes a typical logging system used in the Lake and Northeastern States, discusses each step in the operation, and presents a simple method for designing an efficient logging system for such an operation.

3. MYERS, J. WALTER JR. 1967. *Forest farmer.* Fifteenth Manual Edition. 120 pp. Forest Farmers' Association, Atlanta, Georgia. The Forest Farmers' Association is an organization of 1900 timberland owners in the South. This manual contains articles on all phases of pulpwood production by fifteen experts. It also includes a directory of "Markets for Your Timber," and of forestry agencies.

4. *Proceedings.* Forest Engineering Seminar II. American Pulpwood Assoc., Lake States Technical Division. March 1968. 206 pp. New York. The purpose of the seminar which was held in Wausau, Wisconsin, January 22–26, 1968, was stated as follows: "To acquaint pulp and paper company woods department field personnel with the knowledge and skills required to manage a modern mechanized logging operation." A faculty of sixteen experts participated in the seminar.

13

Logging Cost Analysis

INTRODUCTION

The purpose of this chapter is to provide a brief survey of some of the analytical techniques used in evaluating logging operations. In recent years considerable effort has been devoted to the development of analytical techniques for guiding production decisions. In general, these developments are associated with the field of operations research. Closely related fields, or substitute titles for similar activities, include management science, managerial economics, decision theory, and systems analysis.

Specialists in all these activities have a common interest in decision-making. Decisions may range from relatively unimportant choices that must be made by an operator of a small logging operation, to issues of national policy or highly complex military operations.

Theories of decision-making, and specialists who apply them, have proven quite useful in a number of industries. In recent years several firms in the forest products industry have established departments of operations research to analyze problems, and to provide guidance in making a wide range of policy and operational decisions. Logging specialists clearly have an interest in knowing something of these techniques, and the extent to which they may be useful in planning logging operations, controlling logging costs, and where necessary, coordinating logging with other activities.

Two Levels of Logging Planning

For the purpose of the present discussion, logging planning is considered at two levels: (1) responses to changing prices of labor, equipment, other inputs used in the logging operation, or log products, and (2) maximizing profits (or minimizing costs) subject to existing market prices and other constraints. The first level of planning requires some knowledge of the economics of the logging industry, and how it is affected by other sectors of the economy. The second level of planning would include much of what might be called **operations research** applications in logging.

Response to Changing Market Prices. Cost items over which a logging operator has little control include the wages which must be paid labor, interest rates that must be charged against **capital,** and the prices paid for other resources used in logging. Prices of these kinds are determined largely by competition in the markets where the logger purchases the inputs used in his operation. Moreover, the price the logger receives for logs (wood), in whatever form he chooses to sell his output, are established in a market. Log prices represent an upper limit, below which unit production costs must be kept if the logger is to make a profit.

The extent to which log prices and price levels fluctuate are important factors in determining the time and effort a logging operator will spend in planning his operation and attempting to control logging costs. If log prices are relatively high and stable, he may be somewhat more complacent about controlling costs than if log prices are low or fluctuating. A negative entry in an income statement usually encourages an operator to watch his costs more carefully.

Rising wage rates, interest rates, stumpage prices, or prices of other factors used in production can cause an increase in the cost structure of the logging operation. If so, this will reduce the margin between log price and unit production cost. Increases in any, or all, of these factors will prompt an operator to determine if less expensive factors can be used in production, or if he can substitute among factors such that costs can be reduced. Quite possibly, his best alternative will be to close down his operation.

Maximizing Subject to Existing Conditions. The response of the logger to changing market prices is a different problem than attempting to maximize profits, or minimize costs for given market conditions or other constraints. The former problem affects all firms in the logging industry. The latter is a problem for the individual company operation, or phase of operation such as felling, skidding, or road location. The latter problems are considered here as problems in operations research.

An important decision logging operators must make is to decide among various methods by which to produce logs. Relative prices of capital and labor, timber type, topography, and restrictions placed on the logging operation will have significant effects on the amounts and kinds of labor a logger will employ, the type of equipment he will use, the location and spacing of roads, and the attention given to site conditions following the logging operation.

The range in wages on logging operations, throughout the United States in particular, is highly varied. For example, logging labor in the Pacific Northwest is paid relatively high wages; logging wages in the South are relatively low. A result is that southern logging operations tend to use considerably more labor per unit output than is used in the Pacific Northwest.

If wages rise relative to capital costs, this generally causes some substitution of capital for labor. Recent examples include portable towers for high climbers, log grapples and powered tongs for loading crews, radios for signal men, and tree shears for felling crews. Further substitution of capital for labor can be expected as wages continue to rise, or rising fringe benefits raise labor costs relative to capital costs.

Timber type, topography, and restrictions on logging operations create further decision problems for the individual operator. Different types of equipment will generally be used to log large timber, or to log steep topography, than will be used to log small timber, or to log on level ground. It should be noted, however, there are exceptions to this generality if wage rates are sufficiently low relative to the **cost of capital.** Logging operations in under-developed countries, where plentiful supplies of cheap labor are available and capital costs are very high, are labor intensive, even where large timber on difficult terrain is being logged.

There is no assurance that a given operation is always being conducted efficiently. Roads may be located in the wrong place, logging show boundaries may be badly placed, equipment may be sitting idle when it should be in use, or work may not be scheduled in a manner that keeps labor crews productive. Problems of these kinds are also the subject of logging cost analysis, and logging planning, by the techniques of operations research.

ECONOMICS OF THE LOGGING INDUSTRY

In many cases an understanding of certain characteristics of the logging industry will be a prerequisite to effective logging cost analyses. This is merely another way of saying that cost analyses may not be very useful unless an operator is familiar with the economic environment in which he operates, and how this environment affects his costs and revenues. The purpose of the present section is to provide a brief description of the logging industry and some of its economic characteristics that may be useful in cost analysis.

Description of Logging Industry

Geographically, the logging industry in North America is located throughout most of the United States and much of Canada. Most of the output is produced in a relatively few states in the Pacific Northwest, in the South, in British Columbia, and in eastern Canada. Logging is done in many states to provide timber products for local use. Timber types range from clear cuts of redwood timber stands in northern California and coniferous forests of the Pacific Northwest and British Columbia to selection logging of hardwoods in Appalachia and pulpwood harvests of spruce and true firs in northern New England and eastern Canada. The topography on which logging operations are conducted ranges from level terrain to steep mountain slopes. Logging is conducted in parts of New England and Canada in the winter months when swamps are frozen over. In the Pacific Northwest winter snows are often a cause of logging closures.

Size of Firms. Partially as a result of the variable conditions cited above, sizes of firms in the logging industry range from very small operations which produce small volumes of timber intermittently to very large firms with continuous high volume output. Number of workers per firm and capital investment vary accordingly. Some firms consist of one or two workers for whom logging is a part-time job. Other companies may employ hundreds of men, including workers with a wide variety of highly specialized skills. Equipment used to log on some operations may consist of farm equipment, on others there may be highly specialized equipment, such as portable towers and wheeled skidders, that are virtually useless for any other purpose.

Fluctuations in Output. Logging is also characterized by wide variation in the number of firms within the industry. This is a result of periodic fluctuations in the demand for timber products. The lumber and plywood industries are particularly susceptible to changes in the demand for housing. The demand for housing, in turn, is very sensitive to changes in interest rates. A small change in the interest rate can add thousands of dollars to the cost of a long-term home mortgage. When interest rates rise, therefore, potential home buyers are likely to wait until rates fall if there is a reasonable prospect they will. A fall in the demand for housing is very quickly felt in the lumber and plywood industries, hence in the logging industry.

Because of the extent to which monetary policy, which affects rates of interest, is used in stabilizing the overall economy, output in the logging industry can be expected to fluctuate relatively more than it does for other industries. In effect, the logging industry bears a disproportionate share of the cost of economic stabilization. If seasonal closures, because of weather conditions, are superimposed on cyclical effects of fluctuations in the demand for housing, the logging industry appears to be one of the more unstable industries in our economy. Fluctuations in the demand for timber products and a highly varied cost structure caused by variation in the sizes of logging operations, numbers of employees, and capital investments, create problems for the logging industry that logging cost analysts may be helpless to solve. Indeed, they may be placed in the unenviable position of a doctor who must tell his patient the outlook is not hopeful.

Fixed and Variable Costs in Logging

The concepts of **fixed** and **variable** costs used in economic analysis are somewhat different from their use in accounting. These differences largely reflect the fact that economic decisions involve the future. The past cannot be changed. One of the uses of accounting is to record what has happened in the past. In economic analysis a fixed cost is a cost that cannot be altered; it is irrevocable. In this sense wages paid to a worker can be a fixed cost if he is on the job and the employer has a contractual obligation to pay him while he is there. Variable costs in economic analysis are costs that can be avoided by deciding not to employ resources. Wages would be a variable cost if they could be avoided by removing employees from the payroll as soon as a work stoppage occurs.

The importance of economic fixed and variable costs can be appreciated if the concept is used to evaluate the relationship between fixed costs in the industry and industry instabilities. It has been noted that firms in the logging industry vary widely in the amounts of capital they have invested in logging equipment. Small operations may use trucks and tractors that are used most of the year for farming. Some operators use specialized logging equipment, but it is of such ancient vintage it would sell for very little in the used equipment market. Large firms may have millions of dollars invested in modern, highly specialized logging equipment.

Capital tied up in logging equipment is considered a fixed cost in the economic sense. The operator either owns the equipment, or he has purchased it with a loan, and he is contractually obligated to repay the loan. For the small, irregular operation with old equipment, in various stages of disrepair, fixed costs will be low. For the large operator with a large investment in logging equipment fixed costs will be high. Loan payments, interest charges against the firm's capital, depreciation, insurance payments, and taxes will be major items in the cost structure of the large firm. In addition to having higher absolute fixed costs than the small firm, a higher proportion of the large firm's total costs are likely to be fixed costs.

Effects of Fixed Costs on Production Decisions. The total cost of the firm that is not anticipating changes in capacity consists of variable costs, which vary with the level of output, and fixed costs, which are constant whatever the level of output. Thus, Total Cost = Variable Cost + Fixed Cost. Dividing both sides of the equation by the quantity of output to express the relationship on an average basis: Average Total Cost = Average Variable Cost + Average Fixed Cost.

If a logger is to make a profit the price he obtains for his output must exceed average total cost. Both price and average total cost are expressed in terms of dollars per unit output. When price is just equal to average total cost his operation is just breaking even. When price is less than average total cost, what can the operator be expected to do? According to the definition used, fixed costs will continue whether or not logs are produced. If price is high enough to cover average variable costs and a portion of average fixed costs, the firm will minimize total losses by continuing to operate. In other words, it costs more to shut down than it does to continue. Needless to say, the operator will be expected to continue production, however unhappy he may be.

The Shut-Down Point. If price falls below the level of average variable cost, a firm will minimize losses if it shuts down. Below average variable cost, price does not cover costs of operating, much less any fixed costs. To continue in operation would be to lose more than fixed costs.

The relationship between price, average costs, and the shut-down point provides an interesting insight into how different firms in the logging industry behave in a fluctuating market. A high proportion of the total cost of a small operation will be variable cost. A high proportion of the total cost of a large operation with a large investment in logging equipment will be fixed cost. In a falling market the large firm can be expected to produce longer than the small firm because price must fall lower before it reaches the larger firm's shut-down point. Accordingly, the large firm will suffer larger losses in a falling market. On the other hand, if average total cost of the large firm is less than the small firm's, it will begin to make profits sooner in a rising market. On the basis of an economic analysis of fluctuating prices, and fixed and variable costs, it would appear large firms in the logging industry produce more or less continuously; however, their profits fluctuate widely with price fluctuations. Small firms adjust to fluctuating prices by entering and leaving the industry.

Ease of Entry into Logging Industry

In a relatively unstable industry such as logging, the number of firms, as well as the capacity of existing firms, can be assumed to be more or less constantly changing. When the price of logs rises, new firms will enter. When prices fall, firms will leave. In this respect the logging industry is quite different from industries such as steel, automobiles, petroleum, and other industries that are characterized by a small number of firms. In the latter industries fluctuations in output, in response to fluctuations in price, typically result in each firm merely changing the level of capacity at which it produces. When demand is high they may operate near 100 percent of capacity. When demand is low, the percent of capacity being used is correspondingly low.

Entry into the logging industry is very easy compared to many other industries. A logger who wants to get into business for himself can generally acquire used equipment for a small down payment. Alternatively, equipment can be leased with minimal financial resources if the logger has reasonable assurance of a job on which he can generate some revenues. Sufficient labor for a small operation is also reasonably easy to find, providing an operator pays the going wage.

Land ownership patterns and the opportunity to contract for large firms in the forest products industry when product markets are good also contribute to ease of entry into the logging industry. Small timber sales, which encourage small operations, are available on farm woodlots and other small timber holdings. Public agencies commonly offer timber sales in small volumes and for short periods of time which enable small operations to compete with larger operations.

Large companies commonly use contract loggers to expand their capacity during periods of peak demand. Thinning operations also provide opportunities for small logging companies to contract their services to the large companies. In the Pacific Northwest, thinning projects often keep small firms employed which would otherwise have to shut down because of the seasonal nature of the regional logging industry. Both the large and small firm benefit from opportunities of these kinds, and labor and equipment are not forced to remain idle.

Asymmetry of Entry and Exit. While ease of entry into the logging industry has numerous advantages, it also creates some problems. The most important is that entry of capital and labor into logging appears to be much easier than exit. Investments in logging equipment are largely sunk costs, and workers are frequently employed in areas where job opportunities may not be available when the log market declines. Moreover, workers tend to be reluctant to leave logging as they acquire logging skills, and as they develop roots in the community. Problems of these kinds are especially difficult for older workers, who are much less mobile geographically, or occupationally, than younger workers.

The problem of exit is particularly difficult for the firm with a large capital investment. Once equipment has been purchased, capital costs become fixed. As noted earlier, payments to creditors must be met, and interest charges must continue against capital invested in logging equipment. If the operator didn't have his capital invested in logging equipment, he could have it working for him elsewhere.

An important result of the ease of entry into logging, and the difficulties of exit, is that an asymmetrical relationship exists between the time it takes for the logging industry to expand capacity, when the demand for logs is increasing, and the time required to reduce capacity in a falling market. A rise in demand which is thought to be reasonably permanent may attract new firms, or may cause old firms to expand their capacity, in a relatively short time. A fall in demand, on the other hand, will not result in a reduction in capacity until some of the equipment on which that capacity is based wears out. Equipment will continue to be used, even though full costs are not being covered by log prices, because it would be more costly to let it stand idle. Needless to say, competition during periods when the industry is reducing capacity is especially painful to many firms in the industry. Competition from firms which will eventually go out of business may force losses for a time on large permanent operators. The extent to which small firms, which are in and out of production, are periodically vilified by spokesmen for large companies is evidence of this situation.

Economic Efficiency in an Unstable Industry

The structure of the logging industry and its response to fluctuations in the demand for logs raises numerous questions about relative efficiency of various firms in the industry. Are small firms efficient? Or, do they contribute to the instability of the industry?

In an industry in equilibrium, such that no exit or entry of firms is occurring, **economic efficiency** is defined in terms of minimizing average total cost of production. To minimize average total cost, inputs must be used in a combination which reflects relative prices and relative productivity. The higher the wage rate relative to the cost of capital, the more equipment will be substituted for labor in production. Conversely, the higher the cost of capital relative to the wage rate, the more labor will be substituted for capital.

When prices fluctuate and some firms expand or contract their capacity, others enter and leave the industry. Both of these tendencies are consistent with economic efficiency. To some extent this is independent of whether or not profits are being made. Indeed, profits will be extremely difficult to make if the industry is reducing capacity. This merely reflects the fact that firms are bidding down prices to the point that profits are not possible, and someone must leave. If all firms continue to profit in a competitive industry like logging, capacity will generally continue to expand.

The economic efficiency of mobile resources is relatively easy to see if one is an outsider, and his profits are not affected. Consider, for example, the economic efficiency of migrant farm workers.

Farmers who use these workers are not likely to want to support them the year round. Similarly, a relatively stable logging operator who knows his profits are being affected by small firms in the industry, who are competing with him in a falling market, would like to see them leave. Individual operators tend to view economic efficiency with the profitability of their own firm. Unfortunately, industry profitability does not always reflect economic efficiency.

It should be apparent from the preceding discussion that both large and small logging operations can be economically efficient. Both serve a highly useful purpose. The large operator maintains a more or less continuous flow of logs. The small operations serve the purpose of providing the capacity required to meet peak demands for logs. When the peaks have passed the small operator leaves the industry.

OPERATIONS RESEARCH IN LOGGING

It is appropriate to consider briefly the content of operations research, and the extent to which it provides new analytical tools for analyzing logging costs. Not surprisingly, there are numerous definitions of operations research. Most agree operations research is a mixture of mathematics, management, and economics. More important, perhaps, there is a considerable emphasis on "**model**" building. Theoretical models are constructed which hopefully will provide assistance in making various kinds of decisions. Operations research overlaps managerial economics, where models and economics are also commonly used in analyzing management problems.

Some writers suggest the "team approach" makes operations research distinctive. The talents of engineers and natural, physical, as well as social scientists, are commonly employed in operations research projects. Since most of the best known operations research techniques are mathematical in character, as opposed to subjective and qualitative techniques commonly employed in business management, a head of an operations research team is likely to be mathematically inclined.

A common characteristic of models employed in operations research is their oversimplification of the real world. The usefulness of models to guide decisions is based on this fact, however. Most situations in logging operations, for example, consist of a bewildering complexity of variables that affect productivity, hence unit production costs. If an attempt were made to evaluate the effects of costs of all these variables, ranging from the temperament of the crew, to slope, timber size, timber species, soil conditions, etc., analysis would become hopelessly complicated, if not impossible. By abstracting away, that is ignoring, less important variables, an analysis can be made, which may provide useful information for making production decisions. Frequently this consists of a model of the logging operation, which can be manipulated to determine effects on costs of various production alternatives, or altering various inputs.

A question obviously arises regarding how to determine which variables are "less important," and can be excluded from the analysis. Not surprisingly, the decision is often one of convenience. Most of the results of operations research analyses are numerical; therefore, variables must be quantified if their effects are to be investigated.

Unfortunately, there are no good criteria for quantifying some variables. In many cases it will be expedient to leave them out. In making this decision one attempts to balance the loss in realism against the gain of reducing the complexity of the analysis. There are no rules for this selection process. One can only hope the operations research specialist will heed the rather unhelpful admonition that he exercise caution and judgment, and check the situation carefully, before he constructs his model.

Anyone familiar with logging operations is well aware of the many variables affecting logging costs that would be extremely difficult to quantify. To ignore them, however, is to compromise seriously the usefulness of results obtained.

Objectives of Operations Research in Logging

Before turning to the objectives of operations research in logging cost analysis, a common element in most decision problems confronting the logger is worth noting. The logger, whether he is considering hiring a new man, purchasing a new piece of equipment or a stand of timber, whether to skid long logs or short logs, or whether to accept the obligation of constructing road for a public agency in conjunction with a timber purchase, will usually consider whether the action in question will add sufficiently to total revenue to make it worth the additional cost. In other words, is it worth the cost? In general, an opportunity will be undertaken only if it will contribute to profit and overhead. The suggested rule is almost a prerequisite for survival in the highly competitive logging industry.

Similar logic is likely to be applied to applications of operations research in logging. If the results of operations research are to be useful, they should lead to more profitable decisions being made than those based on alternative methods, possibly crude rules-of-thumb or repetitions of past behavior. Moreover, the cost of obtaining these results should be less than the increased revenue which is obtained from making better decisions.

Public agencies have conducted a number of logging cost studies that can reasonably be classified as operations research. These studies partially reflect their responsibility to conduct research in various phases of timber production. The U.S. Forest Service in particular has this responsibility. In addition, public agencies employ the results of some of their cost studies in the management of forest lands under their jurisdiction. Public forestry activities range from investments in regeneration, and various timber cultural practices, to the preparation of timber sales. Laws require that an appraisal be made for most public timber sales. Most foresters are aware of the monumental task of collecting **timber appraisal** data.

Many of the studies from which timber appraisal data are obtained are characteristic of the operations research approach. Careful measurements are made of log size, skidding distance, slope, soil conditions, and the effects of these variables on productivity. Cost estimates can be constructed accordingly.

In the Scandinavian countries and in eastern Canada operations research analyses of logging operations appear to be more common than in the United States. Quite possibly this reflects the fact that in the Scandinavian countries and in eastern Canada significantly more of the woods labor force is paid by piece rates. If **piece rates** are used in logging, where conditions are highly varied, adjustments in rates are frequently required to permit workers to maintain a reasonably stable wage rate. Formalized logging cost information can be quite helpful in making these adjustments. This same information may also be useful in negotiating piece rates with labor unions.

Based on the preceding discussion, specific objectives of operations research in logging can be classified as follows: (1) compare alternative logging systems to minimize unit logging costs, (2) improve organization and scheduling of existing crew and equipment to reduce unit logging cost, (3) obtain cost estimates for making timber appraisals, and (4) provide a basis for adjusting piece rates.

Development of Operations Research in Logging in the United States

It is not the purpose of the present chapter to review the history of cost analysis in logging in the United States. Nevertheless, any discussion of the topic would hardly be complete without brief mention of some of the early efforts in the field. A number of studies were made in the 1930's by the U.S. Forest Service in which regional logging costs were evaluated. A major purpose of these studies was to provide information to owners of private timberland which would be useful in making forest management decisions. The intensity of forest management at that time was generally low, and the U.S. Forest Service was attempting to publicize information which would raise the intensity of management. One of the more ambitious of these studies was conducted by Axel J. F. Brandstrom, and has become something of a classic in logging cost analysis. (*Analysis of logging costs and operating methods in the Douglas fir region,* Charles Lathrop Pack Forestry Foundation, 1933.)

In 1942 *Cost Control in the Logging Industry* by Donald M.

Matthews was published. Many of the ideas introduced into logging cost analysis by Matthews at that time are still being used. The **break-even chart,** and the formula for spacing roads are cases in point. Both concepts are discussed in more detail in the following section.

Outside the public agencies, and the academic community, interest in logging studies was generally very low in the United States for many years. Part of the reason was, no doubt, that objectives and management procedures of the public agencies were quite different from the objectives and business practices of private industry. Moreover, information requirements for the management of public agencies were considerably different than for private industry. If one believes competition eliminates inefficiencies, it seems reasonable to conclude that logging cost studies might not have contributed much to the improvement of logging practices in the past. Some of the studies reached conclusions contrary to industry practices. It is a moot point to conclude that a firm which applied the results of available logging studies in planning its operation would have reduced its logging costs.

With the establishment of operations research departments in most vertically integrated firms in the forest products industry in recent years, interest in analyzing logging operations appears to be increasing. No doubt this reflects partially the availability of new specialists with new techniques. New problems, and the desire to schedule logging operations more closely with other activities of the firm, are also major factors. Much of this effort has been proprietary research, hence it is difficult to assess how successful it has been.

Survey of Operations Research Techniques Used in Logging

Numerous techniques are employed in analyses that would appropriately be defined as operations research. Some of the techniques of operations research have not been extensively applied to logging operations. Some of the more common techniques will be briefly discussed and related to the logging industry, or to the analysis of logging problems.

Break-Even Chart. The break-even chart is a standard tool used to help solve certain kinds of management problems. Break-even charts are essentially a practical application of cost and revenue functions. The cost function is assumed to be linear in the relevant output range. The revenue function is also assumed to be linear. The latter assumption reflects the fact price is constant for all units of output produced and sold. Both functions are plotted in Fig. 13–1.

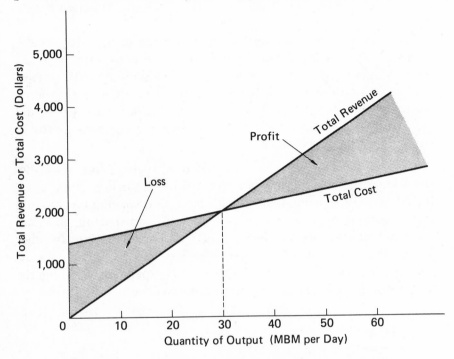

Fig. 13–1. Break-even chart.

The break-even chart in Fig. 13–1 indicates that if production is below 30 thousand board feet per day, the firm would incur a loss. Above this amount profits will be realized. The validity of this conclusion obviously depends upon the validity of the assumptions which underlie its construction.

For most logging operations the assumption of constant prices in the revenue function is realistic. One problem that may be encountered in constructing the function, however, is the necessity of constructing an average price for all output, if logs of different grades and species are being produced. Log prices are typically affected by species and grade. Firms that sell their logs on the basis of "camp run," that is, one price for all logs regardless of species and grade would not have this problem.

In the logging industry, use of the break-even chart is more likely to be difficult because of difficulties in estimating the cost function.

Moreover, if the cost function is not linear, estimates may be subject to considerable error. A curvilinear function can be estimated. Whether or not this would improve cost and profit predictions would depend, of course, on the accuracy with which the function is estimated.

Break-even charts can also be used to make comparative evaluations of different pieces of equipment in the performance of different logging tasks. For example, the cost of skidding with tractors and yarders on different slopes can be compared. On level ground, tractor skidding might be less costly. The break-even chart would indicate the slope at which skidding costs for the two pieces of equipment would be equal.

Optimal Road Spacing. The optimal road spacing formula is derived from the relationship between road construction costs as affected by road spacing, and the effects of road spacing on skidding costs. Skidding costs are a direct function of road spacing, and road construction costs are an inverse function of road spacing. In other words, the wider roads are apart, the less will be spent on road construction, but the more will have to be spent on skidding the longer distances. Formally, the relationship will be

$$TC = aS + b/S \tag{1}$$

where TC = Total Cost
S = Road spacing (100 feet)
a,b = Constants which reflect skidding and road construction costs, respectively, as affected by road spacing distance.

Minimize total cost with respect to S:

$$\frac{d(TC)}{dS} = a - b/S^2 \tag{2}$$

Set Equation (2) equal to zero and solve, for a minimum:

$$aS = b/S \tag{3}$$

The conclusion shown in Equation (3) is that to minimize total cost of skidding and road construction, space roads such that skidding costs and road construction costs are equal (aS = skidding cost; b/S = road construction cost).

Regression Analysis. **Regression analysis** is a statistical technique by which a relationship is developed between a dependent variable and one or more independent variables. Most foresters are exposed to the mechanics of regression analysis, hence it will not be discussed. The purpose of a regression analysis is to develop a means of predicting outcomes of the dependent variable through knowledge about the independent variables. If independent variables can be manipulated, this may permit some control over the dependent variable. Regression analysis may provide some insight into this possibility.

A typical regression relationship in logging cost analysis might be the relationship between skidding time per unit output and skidding distance. More complex relationships might include log size, ground conditions, and other factors that affect skidding time. Once the relationship has been established, predictions can be made on other operations that have a similar range of conditions to those where the original data were collected.

Many regression analyses have been conducted in which logging productivity has been investigated. No doubt this information has been used for some purposes. Follow-up analyses of whether or not this information resulted in better decisions being made than would have been made with other information are unfortunately not generally available. Until they are, the usefulness of regression analyses of logging will remain in doubt.

Equipment Replacement. How much to invest in equipment, the optimal time to replace equipment, and an optimal maintenance policy are all questions a logging operator must face. Moreover, given the amount of equipment used in the logging industry, these are highly important questions. Overcapitalization and equipment down-time have been fatal to numerous logging operations.

Techniques for analyzing decisions for equipment replacement and maintenance have been developed to a high level of mathematical sophistication, but they are beyond the scope of the present discussion. Some of the conclusions of these studies are of interest, however, and are worth noting here.

Equipment investments consist of two kinds: (1) cost reduction, and (2) revenue increasing. Equipment to reduce cost would increase **cash flow** because of a reduction in expenses. Typical examples are pieces of equipment which replace labor but do not increase output. Revenue-increasing investments increase cash flow by expanding capacity so that sales can be increased. Many investments combine elements of both cost reduction and revenue increasing.

Equipment is expected to give rise to a stream of revenues over time, and a stream of operating and maintenance expenses necessary to produce the output. The difference between the revenue and cost streams represents the return on the investment. Discounting procedures are used to compare different investments which generate different life expectancies. As a general rule, it pays to invest in a piece of equipment if the rate of return expected from the investment during its expected life exceeds the market rate of interest.

One of two formulae can be used to evaluate investment decisions.

Present net worth:

$$V = \frac{R_1}{(1 + i)} + \frac{R_2}{(1 + i)^2} + \cdots + \frac{R_n}{(1 + i)^n} + \frac{S}{(1 + i)^n} \qquad (4)$$

in which V = present net worth
                         i = interest rate or cost of capital
$R_1, R_2, \ldots, R_n$ = cash flow after taxes in years $1, 2, \ldots, n$
                         n = life of investment
                         S = salvage value at the end of the nth year

If the present net worth formula is used to evaluate investment decisions, an investment is considered to be profitable if V (present net worth) is greater than the cost of the investment.

Discounted rate of return:

$$C = \frac{R_1}{(1 + r)} + \frac{R_2}{(1 + r)^2} + \cdots + \frac{R_n}{(1 + r)^n} + \frac{S}{(1 + r)^n} \qquad (5)$$

in which C = cost of the investment
$R_1, R_2, \ldots, R_n$ = cash flow after taxes in years $1, 2, \ldots, n$
                         n = life of the investment
                         S = salvage value at the end of the nth year
                         r = rate of return which equates discounted cash
                              flows with the cost of the investment

Using the discounted rate-of-return formula, if the expected rate of return (r) exceeds the cost of capital (i), the investment is considered to be profitable.

In general, one would expect that when $V > C$ it will also be true that $r > i$, and vice versa. Unfortunately, this is not always true. The timing of the expected cash flows can affect the ranking of different investments. It is unlikely this possible difference would affect investments in logging equipment, hence it is not considered further here.

The decision of how long to keep a piece of equipment involves two costs: (1) the interest cost of having funds tied up in the equipment that could otherwise be employed elsewhere, and (2) the loss in market value of the equipment if it is kept in use. Both deterioration and obsolescence can be expected to cause a drop in the equipment's market value. The first cost is the opportunity cost of ownership, the second is depreciation. Depreciation, as used here, is an economic rather than an accounting concept. This distinction shows up as a difference between the market value of equipment, and its book value for accounting purposes.

The principle to be followed in determining the economic life of a piece of equipment is to keep it as long as the cash flow it generates (excess of revenue over operating and maintenance costs) exceeds the sum of the two costs described above (ownership cost plus depreciation).

The principle can be formulated as the inequality:

$$R(t) > iS(t) + S'(t) \tag{6}$$

where $R(t)$ = cash flow after taxes in period t
     $S(t)$ = market value of the equipment in period t
     $S'(t)$ = decline in market value (economic
                   depreciation) in period t
     $t = 1, 2, \ldots, n$
     i = interest rate

Whenever

$$R(n) = iS(n) + S'(n) \tag{7}$$

the contribution to cash flow just equals the cost of ownership, and therefore the equipment contributes nothing to profits. When the nth period is reached, the equipment should be scrapped or sold. If a piece of equipment has no market value, and revenues produced exceed operating and maintenance costs ($R(t) > 0$) the equipment is profitable to keep.

Formula (6) provides insights into two aspects of loggers' behavior that are worth noting. If $R(t)$ is unstable—for example, a small firm in and out of production because of seasonal factors or market conditions—it would be necessary to keep ownership costs and depreciation low if profits are to be made over time. In effect, this means the firm must operate with equipment that has low ownership costs and minimal depreciation. In other words, his equipment is literally ready for the scrap heap. Little cost is involved if it is shut down.

The second point is related to the first. If a new firm is entering the industry, and revenues are expected to be somewhat unstable, or if the cost of capital (i) is high, it may be rational to begin operations with used equipment. The situation is analogous to that of a person who buys a used car as a means of providing transportation.

To summarize this section, equipment replacement models provide valuable insights into whether or not a piece of equipment currently being used should be retained. If a new piece of equipment is being considered, the optimal replacement formula (7) provides an estimate of the length of time that should be used in making an investment analysis.

Linear Programming. Linear programming is a mathematical technique that has been developed primarily since World War II. Applications of linear programming during this period have grown very rapidly. Development of the computer has contributed greatly to the application of linear programming. The technique enables decision-makers to solve maximization or minimization problems where constraints limit what can be done. If multiple objectives exist, or constraints are numerous, computers are necessary to handle the complex calculations required to solve the large linear programming problem that results.

An example of the use of linear programming in the forest products industry is the allocation of logs to various end uses in the vertically integrated, multi-product firm. Typically, the firm would desire to allocate logs to the production of products which would enable them to make the largest gain. Most large firms produce several lumber, plywood, and pulp products. Moreover, these products are produced from a highly varied assortment of logs, by species, size, length, and grade. Needless to say, allocating each log, or possibly portions of each log, to the precise product which enables the firm to make the most profit is a highly complex undertaking.

In order to implement the application of linear programming to the log allocation problem, information is necessary on prices of final products, the cost of converting logs of a specific category (species, grade, and size) to each product, and the quantities of each category of log that are available. Some of this information is relatively difficult and expensive to obtain. Log scaling and grading information, for example, must be highly refined. Product recovery studies must be made in order to determine the quantities of various end products obtainable from a log of a given specification. Finally, studies must be conducted to determine processing costs for converting logs into various kinds of products.

Viewed in the context of linear programming, the firm's production problem is as follows. The firm has certain fixed amounts of inputs (logs) available to process. This limited supply of logs constitutes a constraint on the firm. The firm may also have other constraints, for example, limits on the amount of a particular product that can be sold.

The product output resulting from conversion of a log by a particular process yields the firm a certain amount of profit. The product mix varies from process to process; hence profits will vary among processes. Knowing the profit to be made from each unit of output from each process, and knowing the limited supplies of inputs available, the firm must decide the amounts and kinds of logs to put through each process in order to maximize profits.

Linear programming techniques have not proven to be very useful on small logging operations. Most decisions faced by producers on small operations are relatively straightforward; hence analytical refinements that might be provided by linear programming would not change many of their decisions. In large companies scheduling of trucks to various operations to minimize trucking costs appears to be one area where linear programming techniques may prove useful.

THE FUTURE OF OPERATIONS RESEARCH IN LOGGING

The establishment of operations research departments in most of the major forest products companies, and the involvement of public agencies in operations research in forestry, will ensure the future applications of these techniques in the logging industry. The appropriate question is what kinds of analyses will be undertaken, and how successful they will be in attaining their objectives.

Some of the techniques discussed in this chapter are likely to continue to be the most important techniques for small logging operations. Large companies will undoubtedly be involved in considerably more ambitious operations research projects in which activities of several logging operations are coordinated with other divisions of the large integrated firm.

The use of economic forecasts to schedule production is another possibility for improving logging planning. If the industry is to continue to be faced with a fluctuating demand, forecasts can facilitate budgeting losses during slumps, and preparing for increased production in a rising market. In summary, the techniques of operations research, including economic analysis, are likely to increase rather than diminish in importance. As in most areas of growth, there will be costs as well as gains. A challenge to the logger is to determine which techniques are useful, and which are not.

SUGGESTED SUPPLEMENTARY READING

1. BENNETT, W. D. and H. I. WINER. 1967. Cost analysis in logging. Pulp and Paper Magazine of Canada (Woodlands Rev. Sec.), 68 (3):80–94. Gardenvale 800, Province of Quebec, Canada. Contains a very good review of logging cost analyses and a useful bibliography.
2. International Business Machinery. 1966. *Proceedings: Seminar on operations research in the forest products industry.* IBM, Los Angeles, California. Contains a number of articles, ranging from general to technical, regarding the use of operations research; some apply to logging. 198 pp.
3. LUSSIER, L. J. 1961. Planning and control of logging operations. Laval University For. Res. Foundation, Contr. 8, 135 pp. Forest Research Foundation, Québec City, P. Q., Canada, Université Laval. An introduction to a variety of analytical techniques used in logging cost analysis.

References

ADAMS, T. C. 1967. Production rates in commercial thinning of young growth Douglas fir. U.S. Department of Agriculture, Forest Service, *Pacific Northwest Forest and Range Experiment Station. Research Paper PNW-41.* 35 pp.

ALTMAN, JAMES A. 1965. Cable skidding of hardwoods. *Technical Release No. 65-R-30 American Pulpwood Association,* New York.

American Iron and Steel Institute. 1967. *Handbook of steel drainage and highway construction products.* New York. 368 pp.

American Pulpwood Association. 1968. Plan ahead. *Pulpwood Production and Saw Mill Logging.* Vol. 16, No. 1, p.1.

American Pulpwood Association. 1962. Duplex tires to replace dual tires on trucks. *Technical Release No. 62-R-18. American Pulpwood Association.* New York.

American Pulpwood Association. 1961. Scales for weighing pulpwood trucks. *Technical Release No. 61-R-18. American Pulpwood Association.* New York.

BENNETT, W. D., WINER, H. I., and BARTHOLOMEW, A. 1965. Measurement of environmental factors and their effect on the productivity of tree-length logging with rubber-tired skidders. *Pulp and Paper Institute of Canada, Montreal. Technical Report Series No. 416* (Woodlands Research Index No. 166).

BINKLEY, V. W. 1965. Economics and design of a radio-controlled skyline yarding system. U.S. Department of Agriculture, Forest Service, *Pacific Northwest Forest and Range Experiment Station. Research Paper PNW-25.* 30 pp.

BRANDSTROM, AXEL J. F. 1933. Analysis of logging costs and operating methods in the Douglas fir region. *Pacific Northwest Forest Experimental Station.* U.S. Forest Service, U.S. Dept. of Agriculture, Portland, Oregon. 117 pp.

Bureau of Land Management. 1965. *Roads handbook,* U.S. Department of the Interior, Washington, D.C. Release 9-20.

BURKETT, W. B. 1967. After chain saws—what next? Paper presented to the Farm, Construction and Industrial Machinery Meeting, Milwaukee, Wisconsin. *Society of Automotive Engineers.* September, 1967.

BUSCH, T. N. 1967. The Buschcombine operation. Proceedings, Timber Harvesting and Procurement Short Course, Schools of Forestry and Wildlife Management, *Louisiana State University, Baton Rouge.* May, 1967.

BYRNE, JAMES; ROGER J. NELSON; and PAUL N. GOOGINS. 1960. *Logging road handbook: the effect of road design on hauling costs.* Agriculture Handbook No. 183, U.S. Government Printing Office, Washington, D.C. 66 pp.

CALDER, LESTER E. and DOUGLAS G. CALDER. 1957. *Calders' forest road engineering tables.* Eugene, Oregon. 47 pp.

CAROW, J. 1959. Yarding and loading costs for salvaging in old-growth Douglas fir with a mobile highlead yarder. U.S. Department of Agriculture, Forest Service, *Pacific Northwest Forest and Range Experiment Station. Research Paper 23.* 26 pp.

Caterpillar Tractor Co., Peoria, Illinois. 1966. *Handbook of ripping.* 48 pp.

Caterpillar Tractor Co., Peoria, Illinois, (no date) *Ripping with seismic analysis.* 7 pp.

Caterpillar Tractor Company, Market Division. 1965. *Planned equipment replacement*. Market Division, Caterpillar Tractor Company, Peoria, Illinois. 20 pp.

CLINE, CHARLES E. 1967. The "Bobcat": versatile pulpwood loader. *Pulpwood Production and Saw Mill Logging* Vol. XV, No. 7.

Corps of Engineers, United States Army. 1953. *The unified soil classification system, Technical Memorandum No. 3–357*. 3 volumes. Waterways Experiment Station, Vicksburg, Mississippi.

CRAIG, GEORGE A. 1965. New timber sale contract—construction and concepts. *Forest Industries* 92(6):30–37.

DONNELLY, R. H. 1962. A technique for relating logging costs to logging chances. *The Northeastern Logger*. 2(3):12–13, 34–35, 42.

E. I. DuPont De Nemours & Company (Inc.). 1966. *Blasters' handbook*. 15th ed. Wilmington, Delaware. 524 pp.

DWYER, ROBERT. 1956. Fifth wheel for truck versatility. Proceedings of the 47th Pacific Logging Congress. *Loggers' handbook, 1957*. The Pacific Logging Congress, Portland, Oregon. Pp. 59–61.

ELLIOTT, D. R., and DE MONTMORENCY, W. H. 1963. The transportation of pulpwood chips in pipelines. *Pulp and Paper Institute of Canada, Montreal. Technical Report Series No. 344* (Woodlands Research Index No. 144).

HAZELTON, LESTER. 1966. 1966 Pulpwood annual, *Pulp and Paper Magazine*. April, 1966.

FORBES, R. D. and MEYER, A. B. (eds.) 1955. *Forestry handbook*. The Ronald Press Co., New York. 23 sections.

GULICK, L. H. 1951. *American forest policy*. Institute of Public Administration, Duell, Sloan and Pearce, New York. 292 pp.

HENNES, R. G., and EKSE, M. I. 1955. *Fundamentals of transportation engineering*. McGraw-Hill Book Co., Inc., New York. 520 pp.

JARCK, WALTER. 1967. The case for shortwood. *Pulpwood Production and Saw Mill Logging*. Vol. XV, No. 7. July, 1967.

LYSONS, H. H., and MANN, C. N. 1967. Skyline tension and deflection handbook. U.S. Department of Agriculture, Forest Service, *Pacific Northwest Forest and Range Experiment Station. Research Paper PNW-39*. 41 pp.

LYSONS, H. H., and MANN, C. N. 1967. Single span running skylines. U.S. Department of Agriculture, Forest Service, *Pacific Northwest Forest and Range Experiment Station. Research Note PNW-52*. 7 pp.

McCulloch Corporation. (no date). *How to use a power chain saw*. Los Angeles, California. 16 pp.

MATTHEWS, DONALD M. 1942. *Cost control in the logging industry*. McGraw-Hill Book Company, Inc., New York. 374 pp.

MILLER, ROSWELL K. 1964. Two aids for the design of forest roads. *Journal of Forestry*, 62(6):381–385.

National Corrugated Steel Pipe Assn., Chicago, Illinois. 1965. *Installation manual for corrugated steel structures*. Installation Manual—CMPA 1162; rev. ed. CSPA 1965. 54 pp.

National Forest Products Association. 1965. *Buyer's guide to Forest Service timber sale contracts*. National Forest Products Association, Washington, D.C. 156 pp.

NIXON, R. J. 1958. The asphalt surfacing of logging roads in the Douglas fir region of the Pacific Northwest (Thesis). *Loggers' handbook*, Vol. XVII, 1958 edition: 22 pp.

PEARCE, J. KENNETH. 1960. *Forest engineering handbook*. Bureau of Land Management, State Office, Portland, Oregon.

PEARCE, J. KENNETH. 1954. Better payloads with log truck bunk scales. *New Wood-Use Series, Circular No. 27*. Institute of Forest Products, University of Washington, Seattle, Washington. 4 pp.

Portland Cement Association. 1962. *PCA soil primer*. Portland Cement Association, Chicago, Illinois. 52 pp .

ROLSTON, K. S. 1967. What can pulp company logging operations mean to independent pulpwood producers in the south? *Pulpwood Production and Saw Mill Logging* Vol. XV, No. 8.

RUTH, ROBERT H. 1967. Silvicultural effects of skyline crane and highlead yarding. *Journal of Forestry* 65(4):251–255.

SCHELL, ROBERT L. 1961. Harvesting pine pulpwood in the Tennessee Valley. *Tennessee Valley Authority Report No. 238-61.*

SILEN, R. R. 1955. More efficient road patterns for a Douglas fir drainage. *The Timberman,* Vol. 56, No. 6, p. 82.

SILVERSIDES, C. R. 1964. Development of logging mechanization in Eastern Canada. *H. R. MacMillan Lecture in Forestry, University of British Columbia.*

SIMMONS, FRED C. Revised 1962. *Logging farm wood crops.* Farmers' Bulletin No. 2090. United States Department of Agriculture, Washington, D.C. Pp. 24–25.

SIMMONS, FRED C. 1951. *Northeastern loggers' handbook.* Agriculture Handbook No. 6, U.S. Government Printing Office, Washington, D.C. 160 pp.

STENZEL, G. 1957. Indians operate red alder show. *Western Conservation Journal.* Vol. XIV, No. 5, p. 8.

STENZEL, G. 1953. Uphill felling reduces breakage. *Circular No. 24, New Wood-Use Series.* Institute of Forest Products, University of Washington, Seattle, Washington 98105. 4 pp.

TRACZEWITZ, O. G. 1965. What to expect—1965—1975—from the pulp and paper industry in the forest. The Unit, *Southern Pulpwood Conservation Association,* No. 104. p. 18.

U.S. Department of Agriculture, Forest Service, Forest Resources Report No. 14. 1958. *Timber resources for America's future.* U.S. Government Printing Office, Washington D.C. 713 pp.

U.S. Department of Agriculture. Forest Service, Region 6. 1958. Forest road standards, surveys and plans. *U.S. Forest Service, Pacific Northwest Region,* Portland, Oregon. 22 pp.

U.S. Department of Agriculture, Forest Service, Forest Resources Report No. 17. 1965. *Timber trends in the United States.* U.S. Government Printing Office, Washington, D.C. 235 pp.

U.S. Department of Interior, Bureau of Reclamation. 1956. *Concrete manual.* 6th ed. rev. rep. U.S. Government Printing Office, Washington, D.C., 20402. 491 pp.

U.S. Treasury Department, Internal Revenue Service, Document No. 5050 (11-66) 1966. *Depreciation, investment credit, amortization depletion.* U.S. Government Printing Office, Washington, D.C., 20402. 20 pp.

WALBRIDGE, THOMAS A., JR. 1960. *The design of harvesting systems and machines for use in the pulpwood stands of the Tennessee Valley as dictated by intensive forest management.* Ph. D. dissertation, University of Michigan. (Privately published.)

WALLACE, O. P. 1957. Ratios for determining the economic road and landing or skidding spacing. *Journal of Forestry* 55(5): 378–379.

WEINTRAUB, S. 1959. Price-making in Forest Service timber sales. *American Economic Review* 49(4):628–637.

WOOLRIDGE, DAVID. 1960. Watershed disturbance from tractor and skyline crane logging. *Journal of Forestry.* 58(5):369–372.

Glossary

DEFINITION OF TERMS:

Many terms in this textbook are in common use in logging and timber harvesting parlance. Oftentimes they are defined, described, or illustrated as they appear in the text. However, to aid the forestry student and clarify unfamiliar words for the professional forester a number of terms are defined even though they may tend to be elementary.

SOURCES:

(1) Nichols, Herbert L., Jr. 1962. *Moving the earth*. 2nd ed. North Castle Books, Greenwich, Conn.

(2) Simmons, Fred C. 1951. *Northeastern loggers' handbook*. Agriculture Handbook No. 6 United States Government Printing Office, Washington, D.C.

(3) Society of American Foresters, 1958. *Forestry terminology*. 3rd ed. Monumental Printing Company, Baltimore, Maryland.

(4) U.S. Department of Agriculture, Forest Service, Pacific Northwest Forest and Range Experiment Station. 1969. Glossary of cable yarding terms. United States Government Printing Office, Washington, D.C.

(5) Manufacturers' literature.

A-frame. Two poles mounted in the shape of the letter A to support ₁ead blocks at the upper end.

Abney level. A hand-held clinometer measuring slopes in percent.

Abutment. That part of a bridge which supports the end of the span, and prevents the bank from sliding under it.

Adverse grade. The up gradient in the direction of the loaded log truck travel.

Aggregate. Crushed rock or gravel screened to sizes for use in making concrete or road surfaces.

Angle blade. Blade mounted on a tractor which can be pivoted on a vertical center pin, so as to cast its load to either side.

Arch. Track-mounted framework of heavy steel construction consisting of a yoke, arched axle, reach, and crawler tracks. Used in skidding behind a tractor to carry the front ends of the logs.

Articulated frame. Two-part frame united by a joint.

Back slope. Slope of the cut bank, expressed as a ratio of horizontal distance in feet to one foot of vertical height.

Bank run. Natural mixture of cobbles, gravel, sand, and fines.

Barberchairing. Tree which splits in a vertical direction—generally caused by insufficient undercut.

Bearing strength. The load-carrying ability of a soil.

Bedding ground. The place where a logging balloon is tied down for servicing.

Belt horsepower. Actual horsepower produced by an engine after deducting the drag of accessories.

Bench. The width of a roadbed upon which surfacing material may be placed.

Bent. A timber structure consisting of either sawn timbers or a combination of sawn or round (piling or posts) material arranged to provide a sound foundation for a bridge.

Board foot. A unit of measurement represented by a board 1 foot long, 1 foot wide, and 1 inch thick.

Bolt. A piece of short pulpwood.

Booms. Floating logs or pulpwood enclosed by boom sticks of poles or logs connected end to end by boom chains.

Bos'n's chair. A board on which a man sits to be hoisted up a spar tree.

Brake fade. Condition when the holding ability of brakes is lessened when brakes heat up as a result of friction within the brake mechanism.

Brake horsepower. The horsepower output of an engine measured at the flywheel or belt, usually by some form of mechanical brake.

Break-even chart. A chart which provides a visual portrayal of costs and revenues as affected by level of output or comparative costs of different methods of operation.

Bridge crane. An overhead crane which runs on tracks.

Bridge site survey. Field engineering which provides elevations, locations, and directions by means of instruments as related to an actual or proposed bridge site.

Buck. To saw a felled tree into shorter cuts.

Bull block. The main line lead block in high-lead yarding.

Butt hook. The heavy hook on the butt rigging to which chokers are attached. Hook by which the dragline is attached to tackle on logs.

Butt log. First log above the stump.

Butt rigging. A system of swivels and clevises which connect the haulback and main lines and to which chokers are fastened.

Cable. Wire rope used for lines in yarding systems.

Cable logging. A yarding system employing winches in a fixed position.

Cableway. A transporting system typically consisting of a cable suspended between elevated supports so as to constitute a track along which carriers can be pulled.

Camp run. A designation applied to logs which are sold at a single price per unit volume rather than at differential prices according to grade.

Capillary water. Water held by surface tension and adhesion forces in capillary spaces.

Capital. Plant, equipment, and related facilities used to produce a flow of goods and services.

Cash flow. The difference between cash receipts and cash expenditures over a given time period.

Choked. Condition in which a log is attached to a skidding unit by means of a wire rope or chain choker.

Choker. Short length of flexible wire rope or chain used to attach logs to a winch line or directly to a tractor. A noose of wire rope for hauling a log.

Choker hook. The fastener on the end of a choker that forms the noose.

Chord. The straight line which joins the end points of any arc.

Clear-cutting. All the merchantable trees on a setting to be yarded which are felled.

Clevis. A U-shaped metal fitting with a pin connecting the ends, used for connecting cables and rigging.

Closing line. A line used to close a grapple.

Cold deck. A pile of logs left for later transportation.

Compaction. The act of running heavy weight over new fill or loose material to close loose particles.

Concentration yard. A pulpwood yard providing facilities for unloading trucks, storage, and loading rail cars or barges for shipment to a pulpmill.

Cord. A unit of measure of stacked wood which measures 4 by 4 by 8 feet or 128 cubic feet, of wood, bark, and space.

Cost of capital. A measure of the sacrifice required (alternatives foregone) to create and maintain productive capital.

Crane. A mobile machine mounted on a turntable on wheels or crawler tracks for hoisting material.

Cross-haul. A method of loading logs by rolling them with a cable.

Crotch grabs. Two log dogs, each attached to about 20 inches of $\frac{1}{2}$-inch chain linked together by a $\frac{3}{4}$-inch swivel and ring.

Crotchline. A loading method consisting of two lengths of wire rope suspended from the end of the loading line and terminating in end hooks.

Cross-sectioning. Determining the position of slope stakes marking the points where the bank slopes intersect the ground surface.

Crown slope. The difference in elevation between a road center and the shoulders.

Culvert. A conduit to convey water by gravity through an embankment.

Cunit. Unit of volume measure containing 100 cubic feet.

Cut slope ratio. The relationship between a combination of horizontal and vertical distances which identifies slopes of a certain steepness. such as 1:1 equals 100 percent, and $1\frac{1}{2}$:1 equals 67 percent.

Dealer, pulpwood. A middleman who buys pulpwood from the producer and sells it to the pulpmill company or acts as a commission broker for the company in procuring pulpwood.

Deck. The top portion of a bridge.

Deflection. The vertical distance between the chord and the skyline, measured at midspan; frequently expressed as a percentage of the horizontal span length.

Degree of curvature. The number of degrees at the center of a circle subtended by an arc of 100 feet at its rim.

Dipper stick. The arm on a power shovel to the end of which a digging bucket is attached.

Donkey. Synonym for a yarder.

Drawbar hook. Hook mounted into the tractor drawbar to facilitate attaching chokers.

Drawbar horsepower. A tractor's flywheel horsepower minus friction and slippage losses in the drive mechanism and the tracks or tires.

Dray. Single sled used in dragging logs. One end of the log rests upon the sled, the other drags on the ground. Synonym: *Drag sled.*

Drum barrel. The spool around which cable is wound.

Dummy. A tree rigged to raise a spar tree to be used for yarding.

Economic efficiency. Optimal combinations of resources (minimum cost) required to produce a given level of output.

Embankment or fill. Refers to positioning soil in order to build up those areas which do not coincide with the designed gradient of a roadway.

End hooks. Pointed hooks which are placed against the end of the log for loading.

Equalizer beam. A cross beam added to the underside of log stringers to equalize their vertical deflection.

Excavation or **cut.** Refers to the removal of soil in order to form a roadway which is in keeping with designed gradient.

External yarding distance. Slope distance from landing to farthest point within the cutting unit boundary.

Eye splice. A loop formed by bending a rope's end back and splicing it into the line.

Fall block. A long narrow block with the sheave at the top, generally balanced so most of the weight is at the bottom.

Favorable grade. The down gradient in the direction of the loaded log truck travel.

Felling. Severing a tree at the stump.

Ferrule. A metal band or socket in which the terminal of a wire or wire rope is secured for firm grip.

Fixed cost. That group of costs for an operation which will remain relatively constant for all levels of relevant output.

Flagging. Plastic ribbon available in several eye-catching colors. Used to draw attention to stakes.

Forklift. A device mounted on the front of a tractor for lifting logs for loading or unloading. The device consists of two horizontal arms or forks which slide under the log to lift it.

Forwarding. Moving pulpwood from stump site to truck loading site by carrying it off the ground.

Frustum of a cone. The remainder of a cone when the top is cut off.

Gas turbine. An engine powered by the action of fuel injected into a compressed air stream. After ignition, hot gases pass through a turbine where energy is converted into usable power.

Gin-pole. A loading spar erected in a leaning position so the lead block is centered over the log truck.

Go-devil. Short sled, without a tongue, used in skidding logs.

Grade contour. The layout on a topographical map of the gradient of a road route.

Grade line. A gradient line marked in the field for the road survey line to follow.

Grapple. A hinged mechanism capable of being opened and closed, used to grip logs during yarding or loading.

Grapple yarding. Cable yarding with grapples instead of chokers.

Ground lead. Cable yarding method in which the mainline lead block is hung on a stump. No lift is afforded the log turn.

Growing stock. The sum (by number or volume) of all the trees in a forest or a specified part of it.

Grubbing. The removal of stumps from the ground by any one of several methods or combination of methods.

Guy. A rope, chain, or rod attached to something to brace, steady, or guide it.

Hand level. A metal tube with a plain glass cover at each end and with a spirit level fastened to the top of the tube; used for approximate leveling.

Hardpan. Hard tight soil, like a clay bed.

Hardshell. The sheet of heavy-weight drafting paper on which the preliminary line traverse is plotted and the road designed.

Haunches. The lower part of structure exterior. below widest part.

Heel boom. Loading boom using tongs which heel or force one end of the log against the underside of the boom.

High climber or **head rigger.** The job title of the man who climbs, tops, and rigs spar trees.

High-lead. A cable logging system in which lead blocks are hung on a spar to provide lift to the front end of the logs.

Interlocking yarder. A device which incorporates a means of coupling the main and haulback drums so as to maintain running-line tension.

Intermediate support spar. A spar tree located between the headspar and tailspar to support a multispan skyline.

International log rule. A formula rule which allows $\frac{1}{2}$-inch taper for each 4 feet of length and $\frac{1}{8}$-inch shrinkage for each 1-inch board. An official rule of the U.S. Forest Service.

Jack. A device for suspending a loading line lead block from a skyline.

Jackpot. An unskilled piece of logging work.

Jammer. A lightweight, two-drum yarder usually mounted on a truck with a spar and boom; may be used for both yarding and loading.

Jumping. Moving a spar tree in an upright position to a better location at the landing.

Kerf. The width of cut made by a saw.

Knuckle-boom. A hydraulically operated loading boom which imitates the action of the human arm.

Landing. An area cleared of all standing and down timber where logs are assembled for subsequent transportation. Synonym: *Yard.*

Lay. Position in which a felled tree is lying.

Lead. Direction of movement of lines.

Linear programming. A mathematical technique used to maximize an objective function which is subject to linear constraints.

Location line (L-line). The staked center line of the road to be constructed. The L-line consists of a series of tangents connected by curves.

Log rule. A table intended to show amounts of lumber which may be sawed from logs of different sizes under assumed conditions.

Logging plan. As used in the western regions: the layout on a topographical map of roads, landings, and setting boundaries.

Longwood. Pulpwood 120 inches or more in length.

Machine rate. The cost per unit of time of owning and operating a logging machine.

McLean boom. Loading boom made of two parallel poles and using two tongs.

Main line. The hauling cable.

Main road. A road which supports a high level of traffic; usually well-engineered and designed.

Model. A theoretical abstraction, usually capable of mathematical manipulation, used to evaluate a decision problem or a subject of interest.

Molle. (a) A circle of twisted strands of wire rope used as temporary line to connect the eye splices of two lines. (b) A ring of wire to replace a cotter key.

Multiple use. The practice of forestry which combines two or more objectives.

Multispan skyline. A skyline having one or more intermediate supports.

O & C Lands. Oregon and California Railroad grant lands revested to the Government as a result of violation of agreement by the O & C R.R Co. These lands are administered by the Bureau of Land Management of the Department of the Interior.

Obsolescence. The economic inferiority of a machine to succeeding models as a result of technological improvements.

Operations research. A scientific approach to decision making that involves the operations of organizational systems.

Optimum moisture content. The moisture content which results in the greatest weight per volume.

Optimum road spacing. The distance between parallel roads which will give the lowest combined cost of skidding plus road construction costs per unit of log volume.

Pallet. A U-shaped rack in which shortwood is piled for transportation.

Pass line. A light cable used in hauling gear or the rigger up the spar tree.

Pavement structure. Usually consists of a "base course" of coarse material laid on the subgrade and a "wearing course" of smaller material laid on the base.

Peavy. Stout wooden lever used for rolling logs, fitted with a strong, sharp spike.

Peeler grade. Logs of the size and quality necessary for the production of rotary cut veneer.

Pegging up. Setting slope stakes with hand level and rod; requires more than one turning point and set-up.

Piece rate. A means of payment of labor whereby income is related to output.

Pier. Structure supporting the spans of a bridge.

Pile foundation. An arrangement of round timbers driven into the ground to support other structures.

Plan and profile. A drawing showing both horizontal (plan) and vertical (profile) delineation of a road survey.

Pond value. The market price of logs delivered to a wet site, log pond, or tidewater.

Powder men. Men who engage in blasting activities.

Power shift. Transmissions that can be shifted while transmitting full engine power to the tracks. Synonym: *Shift-on-the-go units.*

Power shovel. A crane equipped with a boom and dipper stick on the end of which a shovel bucket is mounted for earth-moving.

Prehauling. Moving pulpwood from stump site to truck loading site by carrying it off the ground.

Preliminary line (P-line). A traversed line run along a proposed road route from which the location line is designed.

Priority sequence. The consecutive order in which settings are to be yarded.

Processor. A portable machine which limbs and bucks tree lengths into shortwood at the landing.

Producer. An independent operator who produces and delivers pulpwood to a dealer or to a pulpmill company.

Pulaski. Modified hazel hoe carrying a heavy axe blade instead of a pick.

Pulpmill. A mill which converts pulpwood to wood pulp.

Pulpwood. Round wood used as a source of wood fiber in a pulpmill.

Reconnaissance. Inspecting on the ground a proposed road route or logging plan. Initial field investigation.

Reference. A point on the ground surface which is or may be identified in relation to the identifiable location of another point.

Refraction seismograph. An electronic instrument for measuring depth to bed rock.

Regression. A statistical technique used to evaluate relationships among two or more variables.

Reinforced concrete. Steel embedded in concrete in a manner which assists it in supporting imposed loads.

Reloading. Unloading logs from a truck and loading them individually on a railroad car.

Rigging. The cables, blocks, and other equipment used in yarding logs.

Right-of-way. The strip of land within which a road is to be constructed.

Right-of-way plat. A map drawn to scale, showing the alignment of a road, its position in legal subdivisions of land, and the area of the strip of land within which the road is to be constructed.

Riprap. Rough stone of various sizes placed compactly or irregularly to prevent scour by water or debris.

Road ballast. Bank run material with soil binder used throughout the entire surfacing material.

Road pattern. The characteristic arrangement of spur roads in relation to each other.

Road spacing. The distance between parallel truck roads.

Rod. A surveying instrument made of wood and graduated in feet and tenths of a foot. Used in conjunction with various leveling instruments to determine differences between two points.

Rolling resistance. The retarding force of the ground against the wheels of a vehicle.

Route. A way to go.

Route projection. Laying out a proposed road route on a topographic map or aerial photo.

Running skyline. A system of two or more suspended moving lines, generally referred to as main and haulback, that when properly tensioned will provide lift and travel to the load carrier.

Scaling. The determination of the gross and net volume of logs by the customary commercial units for the product involved.

Scarification. Shallow loosening of soil surfaces.

Scoot. Two-runner sled, without tongue or shafts, used to haul logs or bolts out of the woods.

Setting. The area logged to one yarder position.

Shear. A scissors-like device, hydraulically operated, for crosscutting the stem of a tree. One make of tree shear uses a cutting blade which closes parallel to the anvil.

Sheave. A grooved wheel or pulley.

Sheepsfoot roller. Steel drum on the outside of which are attached short metal rods, sometimes shaped like a sheep's foot; used for compacting soil.

Shortwood. Pulpwood less than 120 inches in length.

Shoulder spillway culvert. Corrugated metal culvert which conveys surface water down the face of an embankment.

Show. Any unit of operation in the woods, usually associated with timber harvesting. Synonym: *Chance.*

Shuttle hauling. Using preloaded trailers to reduce truck turn-around time.

Side. A logging unit: the men and equipment needed to yard and load any one unit of an operation.

Sidecasting. Bulldozer action which results in disposing soil onto areas on the lower side of a road location.

Skew or **skew angle.** The acute angle formed by the intersection of the line normal to the centerline of a road with the centerline of a culvert.

Skidding. The act or process of moving logs from stump site to a landing; the term used in all North American regions except the West Coast.

Skidding chain. Length of chain fastened around the end of a log.

Skidding pan. Wide steel sheet used to support end of log in pan skidding.

Skidding tong. Tong used in skidding to grasp a log.

Skid pole. Log or pole, commonly used in pairs on which logs are rolled.

Skyline. A cableway stretched tautly between two spars and used as a track for log carriers.

Skyline carriage. A wheeled device which rides back and forth on the skyline for yarding or loading.

Skyline crane. A yarding system that is capable of moving logs laterally to a skyline as well as transporting logs either up or down a skyline to a landing.

Slackline system. A live skyline system employing a carriage, main line, and haulback line. The main and haulback lines attach directly to the carriage. The skyline is lowered by a slackening of the line to permit the chokers to be attached to the carriage. Lateral movement is provided by side blocking.

Slash. The woody material or debris left on the ground after an area is logged.

Slasher. A machine which bucks longwood into shortwood at the landing.

Slip hook. Rounded hook that will permit a chain to run freely through it.

Sloop. Two-runner sled equipped with a tongue used to haul logs or bolts out of the woods. Similar to a scoot except for the tongue.

Slope stake. A stake marking the line where a cut (excavation) or fill (embankment) meets the original grade.

Snag. A standing dead tree from which the leaves and most of the branches have fallen.

Snubbing line. A line used for lowering a load.

Span. The horizontal distance between skyline supports.

Spar. The tree or mast on which rigging is hung for one of the many cable hauling systems.

Spur road. A road which supports a low level of traffic, such as a level which would serve one or two settings. Little or no engineering is involved.

Stacker. A mobile machine for unloading and stacking or decking logs using the forklift principle and curved top clamps.

Staggered settings. Clear-cut settings, separated by uncut timber.

Standing line. A fixed cable, not running during logging operations; for example, a skyline anchored at both ends.

Station. A 100-ft., length is a full station and any fractional distance is called the "plus," e.g., a point 580.3 feet from the point of beginning or Sta. 0 + 00 is identified as "Station 5 + 80.3."

Stick. A piece of short pulpwood.

Strap. A short cable with an eye in each end.

Stringer. A long horizontal beam running lengthwise in a bridge.

Sub-chord. A distance on a road curve of less than 100 feet.

Stump pull. Slivers of wood remaining attached to the stump after a tree is felled; the slivers are "pulled" from the butt of the log.

Subgrade. The surface produced by grading native earth, which serves as a base for surfacing material.

Sulky. Logging arch equipped with wheels instead of crawler tracks.

Superelevation. The condition where the outer edge of a road is raised or "superelevated" sufficiently to counteract the centrifugal force generated by a vehicle moving around a curve.

Surface course. The final layer of material placed on a road to protect the base course from traffic.

Sustained yield forest management. Those measures which will maintain the productive capacity of the land.

Swamp. To clear the ground of underbrush and other obstructions.

Tagline. (a) A short piece of line added to anything, usually to a main line. (b) A line used to position a loading grapple.

Tangent. The portion of a road which follows a straight line.

Templet. A cut-out the shape of a finished road cross-section.

Tether line. Any line used to restrain a balloon in flight, such as the line from a logging balloon to the butt rigging.

Tilt blade. Blade mounted on a tractor which can be tilted in respect to a vertical position.

Timber appraisal. An economic appraisal of the monetary value of a timber stand.

Tongs (conventional). A pair of curved arms pivoted like scissors so that a pull on the ring connecting the shorter segments will cause the points on longer segments to bite into the log. The tongs are actuated by the pull on the loading line.

Tongs (power). Loading tongs without sharp points which are closed on the log by air or hydraulic cylinders.

Topping. Severing the top of a standing tree to be used as a spar.

Torque converter. A centrifugal pump driven by an engine and rotating in a case filled with oil.

Tower. A steel mast used at the landing for cable yarding instead of a spar tree.

Turbocharger. An air pump designed to put more air into engine cylinders; pump is driven by the heat of exhaust.

Transferring. Lifting an entire truck load of logs from a truck and placing on a railroad car.

Tree shoe. The device in the shape of a segment of a circle used to support a skyline from a spar tree.

Triple drum. A three-drum yarder.

Turn. The logs yarded in any one trip.

Turnout. An area of sufficient size adjacent to a road which serves as temporary parking place for vehicles so that oncoming traffic may be avoided.

Variable cost. That group of costs which vary in some relationship to the level of output.

Veneer grade. Logs of the size and quality necessary for the production of either sawed or sliced veneer.

Vertical curve. Parabolic transition curve where the gradient of the road changes.

Wasting. The process of disposing of excess soil or rock.

Wheel skidder. A four-wheel drive rubber-tired tractor with articulated steering, usually equipped with a winch drum and an integral arch.

Weir. A notched plank set in the crest of a temporary dam to measure stream flow.

Winch. Steel spool connected with a source of power and used for reeling or unreeling cable.

Windfall. A tree which has been uprooted or broken off by the wind.

Wingwall. Wall that guides a stream into a bridge opening.

Wood pulp. Fiber from wood with varying degrees of purification which is used for the production of paper, paper board, and chemical products.

Yard. Place where logs are accumulated. Synonym: *Landing.*

Yarder. A system of power-operated winches used to haul logs from a stump to a landing.

Yarding. The act or process of conveying logs to a landing.

Yarding road. The path followed by a turn of logs yarded by a cable method.

Index